Date Due

PHYSICAL ACOUSTICS

Principles and Methods

VOLUME XI

CONTRIBUTORS TO VOLUME XI

David J. Bergman

R. Besson

D. G. Crighton

J. de Klerk

J. J. Gagnepain

M. B. Lesser

Arthur E. Lord, Jr.

Emmanuel P. Papadakis

PHYSICAL ACOUSTICS

Principles and Methods

Edited by WARREN P. MASON

SCHOOL OF ENGINEERING AND APPLIED SCIENCE
COLUMBIA UNIVERSITY
NEW YORK, NEW YORK

and

R. N. THURSTON

BELL TELEPHONE LABORATORIES
HOLMDEL, NEW JERSEY

VOLUME XI

1975

ACADEMIC PRESS   New York   San Francisco   London

A Subsidiary of Harcourt Brace Jovanovich, Publishers

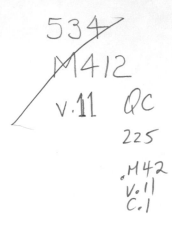

ACADEMIC PRESS, INC.
111 Fifth Avenue, New York, New York 10003

United Kingdom Edition published by
ACADEMIC PRESS, INC. (LONDON) LTD.
24/28 Oval Road, London NW1

Library of Congress Cataloging in Publication Data

Mason, Warren Perry, (date)   ed.
    Physical acoustics.

    Includes bibliographies.
    1.   Sound.   I.   Thurston, Robert N., joint ed.
II.   Title.
QC225.M42   534   63-22327
ISBN 0−12−477911−5 (v.11)

PRINTED IN THE UNITED STATES OF AMERICA

CONTENTS

CONTRIBUTORS   ix

PREFACE   xi

1

Third Sound in Superfluid Helium Films

DAVID J. BERGMAN

I. INTRODUCTION   1

II. THE THEORY OF THIRD SOUND IN FLAT FILMS   3

III. THE PROPERTIES OF THIRD SOUND IN FLAT FILMS—THEORY   23

IV. EXPERIMENTS ON THIRD SOUND   32

V. THE SURFACE ROUGHNESS OF THE SUBSTRATE   38

VI. THIRD SOUND RESONATORS   41

VII. THIRD SOUND IN MIXED He^3–He^4 FILMS   49

VIII. ENERGY IN THIRD SOUND   52

IX. THE NORMAL FLUID MOTION AND ATTENUATION   55

X. MICROSCOPIC THEORIES   57

GLOSSARY OF KEY SYMBOLS AND PHRASES   64

REFERENCES   66

2

Physical Acoustics and the Method of Matched Asymptotic Expansions

M. B. LESSER AND D. G. CRIGHTON

I. INTRODUCTION AND ELEMENTARY ILLUSTRATIONS   70

II. SCATTERING AND DIFFRACTION PROBLEMS   93

III. ACOUSTIC WAVEGUIDES   110

IV. NONLINEAR ACOUSTICS   125
 V. CONCLUSIONS   143
    REFERENCES   147

3

Ultrasonic Diffraction from Single Apertures with Application to Pulse Measurements and Crystal Physics

EMMANUEL P. PAPADAKIS

    I. INTRODUCTION   152
   II. THEORY   153
  III. COMPUTATIONS   160
   IV. EXPERIMENTS   165
    V. DIFFRACTION CORRECTIONS   173
   VI. INPUT AMPLITUDE PROFILE   186
  VII. BROADBAND PULSES   191
 VIII. SPECIMENS OF FINITE WIDTH   205
   IX. SURFACE WAVES   206
    X. SUMMARY   208
       REFERENCES   208
       APPENDIX   211

4

Elastic Surface Wave Devices

J. DE KLERK

    I. INTRODUCTION   213
   II. PHASE CODED SIGNALS   215
  III. 13 BIT BARKER CODE CORRELATOR   223
   IV. PROGRAMMABLE SEQUENCE GENERATOR   231
    V. PULSE COMPRESSION FILTERS   236
       REFERENCES   242

5

Nonlinear Effects in Piezoelectric Quartz Crystals

J. J. GAGNEPAIN AND R. BESSON

I.	INTRODUCTION	245
II.	FUNDAMENTAL EQUATIONS OF QUARTZ	247
III.	CHARACTERISTIC COEFFICIENTS	252
IV.	NONLINEAR EFFECTS IN SHEAR VIBRATING QUARTZ CRYSTAL RESONATORS	266
V.	EQUIVALENT ELECTRICAL CIRCUITS OF A QUARTZ RESONATOR	278
VI.	INFLUENCE OF AN APPLIED DC ELECTRIC FIELD	283
VII.	CONCLUSION	287
	REFERENCES	288

6

Acoustic Emission

ARTHUR E. LORD, JR.

I.	INTRODUCTION	290
II.	HISTORICAL WORK	291
III.	EARLY WORK AND GENERAL BACKGROUND	294
IV.	MATERIALS INVESTIGATED WITH ACOUSTIC EMISSION	301
V.	PROCESSES STUDIED WITH ACOUSTIC EMISSION	320
VI.	STRUCTURAL INTEGRITY	330
VII.	POTPOURRI OF TOPICS (BRIEF DESCRIPTIONS)	333
VIII.	CONCLUSIONS AND SUGGESTIONS FOR FURTHER WORK	336
IX.	APPENDIX	338
	REFERENCES	339
	BIBLIOGRAPHY	345

SUBJECT INDEX	355
CONTENTS OF PREVIOUS VOLUMES	362

CONTRIBUTORS

DAVID J. BERGMAN
Department of Physics and Astronomy,
Tel-Aviv University,
Tel-Aviv, Israel

R. BESSON
Ecole Nationale Superieure de Chronométrie et de
Micromécanique,
Besançon, France

D. G. CRIGHTON
Department of Applied Mathematical Studies,
University of Leeds,
Leeds, England

J. DE KLERK
Westinghouse Research Laboratories,
Pittsburgh, Pennsylvania

J. J. GAGNEPAIN
Ecole Nationale Superieure de Chronométrie et
de Micromécanique,
Besançon, France

M. B. LESSER
Institut CERAC,
Ecublens, Switzerland

ARTHUR E. LORD, JR.
Department of Physics and Atmospheric Science,
Drexel University,
Philadelphia, Pennsylvania

EMMANUEL P. PAPADAKIS
Ford Motor Company,
Manufacturing Development Center,
Detroit, Michigan

PREFACE

As in other recent volumes, the several themes treated here are not directly related to each other, except for their common bond to physical acoustics.

One of the remarkable properties of liquid helium II, the liquid phase existing below $2.172°K$, is that a certain fraction of it can flow without any viscosity. The result is that it can flow from beaker to beaker via the thin film that is absorbed upon the walls. It is possible to excite propagating waves in such films and these waves are called third sound. This first chapter considers the theory of third sound, methods for exciting such waves, third sound resonators, and many other properties. Most of the theories presented are generalizations of hydrodynamic theories but attempts are now being made as described to introduce quantum mechanical properties.

The method of matched asymptotic expansions (MAE) is an established technique in theoretical mechanics but is not yet widely employed in modern acoustics research. The ability of this relatively new method to produce new results in acoustics as well as to provide fresh insight into classical problems is demonstrated in the second chapter. Written as an introduction to the subject, Lesser and Crighton also go far enough to enable the reader to apply the MAE technique to his own problems. The method is of general interest because it is effective not only in resolving particular problems, but also in unifying different mathematical models. For example, the MAE formalism provides derivations of a number of acoustical equations together with estimates of their validity and a definite interpretation of their meaning (e.g., Burgers' equation in relation to the linear wave equation and the Navier–Stokes equations).

One of the principal problems in acoustic measurements at high frequencies is the effect of diffraction in determining the shape of the propagating waves. Diffraction can affect not only the attenuation but also the velocity measurements. The third chapter by Papadakis includes discussions of bulk and surface waves, monochromatic bursts and broadband sources, effects of variation of the displacement of the radiator over its surface, and the effects of the anisotropy of the medium. Calculations are compared with experiments with good agreement.

In recent years, acoustic surface waves have been intensively investigated for a number of signal-processing applications including delay line memories, filters, and correlators. After a discussion of phase coded signals and their generation and detection by interdigital grid structures, the chapter "Elastic Surface Wave Devices" describes a 13 bit Barker code correlator, a programmable sequence generator, and pulse compression filters.

The fifth chapter by Gagnepain and Besson is devoted to an investigation of nonlinear effects in quartz crystals. The extent to which the oscillation level of a quartz crystal unit can be increased (in order to improve the signal to noise ratio and short-term stability) is limited by nonlinear resonator effects. Additional motivation for studying the origin of nonlinearities is provided by the possible use of nonlinear effects in correlators, strain-biased resonators, and other devices. Observed nonlinear behavior is related to nonlinearities in the elastic, piezoelectric, dielectric, and damping properties of the crystal.

Acoustic emission, the subject of the last chapter, deals with the noise produced in materials when they are strained. The first responses seem to have been obtained in rocks, where they were called microseisms and were shown to be connected with strains. These emissions received practical application in providing warnings for rock slides in mines, etc. Starting in 1948, acoustic emissions were observed in metals. Studies indicated that they were of two forms—continuous and burst-type emissions. In all cases, the emissions have been correlated with various types of dislocation motions. Practical applications have been made of these emissions in determining the approach to failure in pressure vessels, bridges, cranes, and other mechanisms. Fatigue in metals and other materials also produces characteristic emissions and these emissions can serve as warnings of this form of degradation. Hence acoustic emissions are becoming a nondestructive test for many types of material degradations.

The Editors owe a debt of gratitude to the contributors who have made this volume possible and to the publishers for their unfailing help and advice.

WARREN P. MASON
ROBERT N. THURSTON

PHYSICAL ACOUSTICS

Principles and Methods

VOLUME XI

—1—

Third Sound in Superfluid Helium Films

DAVID J. BERGMAN

Department of Physics and Astronomy
Tel-Aviv University, Tel-Aviv, Israel

I.	Introduction	1
II.	The Theory of Third Sound in Flat Films	3
	A. Elementary Theory	3
	B. Qualitative Discussion of the Detailed Theory	4
	C. Linearized Equations of Motion for a Superfluid Film	7
	D. Averaging of the Equations Across the Film	9
	E. Equation of Motion for the Substrate	15
	F. Equations of Motion for the Gas	16
	G. The Combined Equations of Third Sound	22
III.	The Properties of Third Sound in Flat Films—Theory	23
	A. General Results	23
	B. Thin Films	27
	C. Thick Films	30
IV.	Experiments on Third Sound	32
V.	The Surface Roughness of the Substrate	38
VI.	Third Sound Resonators	41
VII.	Third Sound in Mixed He^3–He^4 Films	49
VIII.	Energy in Third Sound	52
IX.	The Normal Fluid Motion and Attenuation	55
X.	Microscopic Theories	57
	A. The Average Superfluid Density	57
	B. The Onset of Superfluidity and Attenuation of Third Sound	62
	C. Summary	64
	Glossary of Key Symbols and Phrases	64
	References	66

I. Introduction

One of the remarkable properties of helium II, the liquid phase of helium that exists between $0°$ and $2.172°K$, is that a certain fraction of it can flow without any viscosity. As a result of this, helium II can flow quite freely through very narrow superleaks, or it can flow from beaker to beaker via the

1

thin film that is adsorbed upon the walls. It is also possible to excite propagating waves in such a film, analogous to long waves on the surface of a shallow body of water. These waves are called third sound.

Surface waves on helium films were originally considered as a possible mechanism for the critical velocity of the film. It was speculated by Kuper (1956a,b), Atkins (1957), and Arkhipov (1957) that the critical velocity would be the velocity of flow at which it would become energetically favorable to create a ripplon—one quantum of surface excitation. Careful measurements performed more recently (Pickar and Atkins, 1969) have shown that the velocity of third sound is two or three times larger than the critical velocity of superfluid flow instead of being equal to it, as one would expect from the above mechanism. But in the meantime, third sound has turned out to be a fascinating physical phenomenon in its own right.

Atkins, who did most of the early theoretical work on the properties of third sound (Atkins, 1959), also conducted the first experiment that detected third sound (Everitt *et al.*, 1962). This was done in a helium film formed upon a flat solid substrate which was in equilibrium with helium vapor a few centimeters above a liquid helium bath. This is a relatively thick film, called a saturated film, whose thickness depends mainly on the height above the liquid. Films obtained in this way range in thickness from about 500 Å (140 atomic layers) at a height of 1 cm to about 250 Å (70 atomic layers) at a height of 10 cm. The same group made measurements of the velocity of third sound in these films as a function of temperature, height above the fluid, and frequency, and of the attenuation as a function of frequency (Everitt *et al.*, 1964).

Later, Rudnick and co-workers succeeded in detecting third sound in helium films which are formed on a flat substrate in equilibrium with helium gas whose pressure P is well below the saturated vapor pressure P_v (Rudnick *et al.*, 1968). These are called unsaturated films, and their thickness is determined mainly by the ratio P/P_v. Their thickness ranges from about 180 Å (50 atomic layers) down to 14 Å (4 atomic layers) and even less, depending on the temperature. Groups headed by Rudnick have since made detailed measurements of the velocity (Kagiwada *et al.*, 1969; Rudnick and Fraser, 1970; Fraser, 1969) and of the attenuation (Fraser, 1969; Wang and Rudnick, 1972) of third sound in unsaturated films for various temperatures, film thicknesses, and frequencies. The above mentioned experiments as well as the early theories are described in great detail in an excellent review article (Atkins and Rudnick, 1970). Hence we will not go into them in great detail.

More recently, Ratnam and Mochel (1970a,b, 1974) have developed a new system to investigate the properties of third sound by measuring the response of a third sound resonator. The resonator is made by forming an unsaturated helium film on the inner surface of a hollow cell made of two parallel plates welded together at the edges.

In this article we shall present a detailed discussion of the present state of our understanding of third sound. Section II is devoted to a quite detailed development of the hydrodynamic theory of third sound in flat films. This is done because we feel that any physicist wishing to enter the field should have a good understanding of the continuum theory of third sound. This is especially true since one of the things we look for in third sound is deviation from continuum hydrodynamics. That section can nevertheless still be useful to readers who are not interested in working through the detailed theoretical considerations: They need read only the first two subsections, A and B, omit subsections C–G, and go right on to the results of the theory, which are described in Section III. Section IV describes experimental results to the extent that they can be compared with the theoretical results of Section III. We have not attempted to give an exhaustive account of experimental procedures, and we instead refer the reader to the review article by Atkins and Rudnick (1970). In Section V we discuss the problem of surface roughness of the substrate. Section VI is about third sound resonators—both theory and experiments. Section VII is about mixed He³–He⁴ films. Section VIII is about the energy content of a third sound wave. Section IX discusses the contribution of normal fluid motion to the attenuation of third sound. In Section X we describe briefly some attempts to go beyond hydrodynamics in describing third sound. That section, the last one in the article, is incomplete, since work on a microscopic theory of third sound is still going on in several places at this time.

In order to assist the reader, we have compiled a glossary of some of the more important symbols and standard phrases that are used in this article.

II. The Theory of Third Sound in Flat Films

A. Elementary Theory

An elementary discussion of third sound makes the enormously simplifying assumptions that (a) there are no temperature variations in the helium film, (b) there is no interaction between the film and its surroundings, (c) there is no normal fluid motion in the film, (d) dissipative processes in the film are unimportant and, (e) the properties of the film are completely constant in the direction perpendicular to the plane of the film. As a consequence of these assumptions, the hydrodynamic equations of superfluid motion in the film reduce to an equation for the conservation of mass

$$\rho_f \, \partial h/\partial t = -h\rho_s \, \partial v_{sx}/\partial x \tag{1}$$

and an equation of motion for the superfluid velocity v_{sx} in a direction parallel to the film

$$\partial v_{sx}/\partial t = -\partial \mu/\partial x = -f \, \partial h/\partial x, \tag{2}$$

where

$$f \equiv (\partial \mu/\partial h)_T. \tag{3}$$

In these equations ρ_f is the total mass density of the film, ρ_s is the superfluid mass density, $h(x, t)$ is the instantaneous thickness of the film (see

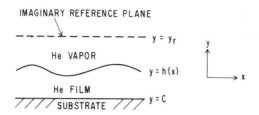

IMAGINARY REFERENCE PLANE

$y = y_r$

He VAPOR

$y = h(x)$

He FILM

SUBSTRATE $\quad y = C$

FIG. 1. Schematic drawing of a third sound wave. The imaginary reference plane at $y = y_r$ is placed outside the range of the substrate–helium forces, but y_r is still much less than the mean free path in the gas. From Bergman (1969), by permission of the American Institute of Physics and *Physical Review*.

Fig. 1), $v_{sx}(x, t)$ is the superfluid velocity, assumed to be entirely parallel to the plane of the film, and $\mu(x, t)$ is the chemical potential per unit mass of the helium.

Equations (1) and (2) can be combined to give a wave equation

$$\ddot{h} = hf\frac{\rho_s}{\rho_f}\frac{\partial^2 h}{\partial x^2} \tag{4}$$

which describes third sound as an unattenuated wave propagating with a velocity u_3 given by

$$u_3{}^2 = hf\frac{\rho_s}{\rho_f}. \tag{5}$$

B. Qualitative Discussion of the Detailed Theory

Even this very simplified theory describes third sound rather well—Eq. (5) is usually in good agreement with experiments. It does not explain, however, the origin of the rather large attenuation that is observed experimentally. In order to get a better theory which includes attenuation, and also to understand why the simplified theory works so well, we will have to reexamine and modify some of the drastic assumptions we have been making.

It is clear, for example, that contrary to assumption (a) of the previous section, there will in fact be temperature variations in the film if assumption (c) about the absence of normal fluid motion has any validity: The peaks of the wave must then be associated with an increase of ρ_s above its average value and with a decrease of the entropy density S/V below its average value. These variations will bring about a variation of the temperature as well. The variations in temperature or entropy satisfy an equation of motion which was ignored in the simplified treatment, but which should be included in a more detailed theory. The temperature variations will also contribute to the gradient of the chemical potential in Eq. (2).

The periodic variations of the temperature, as well as those of the chemical potential μ, will cause energy to flow from the film to its surroundings (the helium gas and the substrate) and helium particles to evaporate into and condense from the gas phase. While these surface phenomena would be of minor importance as far as bulk properties of helium are concerned, it turns out that they are very important in the case we are considering where all of the liquid helium is very near to the two surfaces of the film.

The importance of the temperature variations and of evaporation and condensation phenomena in third sound was first realized by Atkins (1959). The decisive role of heat flow to the surroundings of the film in determining some of the properties of third sound was first realized by Bergman (1969).

The normal fluid motion in the film parallel to the surfaces is damped because of the boundary condition that requires it to vanish at the substrate, and the finite shear viscosity that is encountered when different layers of helium slide past each other with different normal fluid velocities. When the thickness of the film is much less than the viscous penetration depth, $(\eta_f/2\rho_n\omega)^{1/2}$, this damping is very effective and $v_{nx} \ll v_{sx}$. The possibility that it is still important in accounting for the attenuation of third sound was considered by Pollack (1966a,b).

In the next subsection we will give a rather detailed treatment of the theory of third sound based only on hydrodynamics. In this treatment we will not make any of the assumptions made in the previous subsection. We shall find that, although the simple theory given before is usually adequate to give the velocity of third sound, a calculation of the attenuation requires a careful consideration of all the interactions with the surrounding media. The intrinsic dissipative processes of the liquid helium film—viscosity and thermal conductance—still turn out to be unimportant.

Before closing this subsection, we will summarize qualitatively the results of the detailed theory.

When the interactions of the helium film with its surroundings are taken into account, we find that third sound is a phenomenon which is not confined to the film: Along with the wave traveling in the film there are companion waves in the adjoining substrate and gas. In the substrate this is simply a thermal conduction wave in which only the temperature oscillates. In the gas we have a combination of three waves: an ordinary acoustic wave, a viscous wave, and a thermal wave. Each of the companion waves has its own wave vector and propagates in a different direction, though always away from the film. The wavefronts of these waves are drawn qualitatively in Figure 2. The viscous and thermal waves travel nearly perpendicular to the film with a wave vector whose real and imaginary parts are nearly equal, and are usually greater than the third sound wave vector k_3. Their amplitude decreases exponentially as one moves away from the film. Nevertheless, the distance that they penetrate into the gas and substrate is always much greater than the thickness of the film. Whereas typical values for the penetration depth are 0.003 cm for the viscous and thermal lengths in the gas, and 0.1 cm for the thermal length in the substrate, the film thickness is typically 15–500 Å.

The acoustic wave travels nearly parallel to the film and, contrary to what one might expect, its amplitude increases exponentially as one moves away from the film in a perpendicular direction. This happens because the

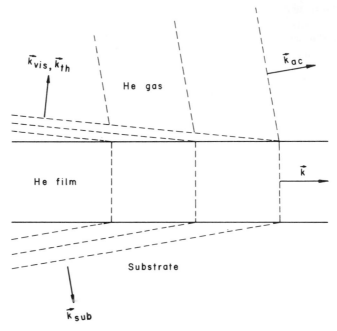

FIG. 2. Schematic drawing of the wavefronts of third sound and its companion waves: a thermal wave in the substrate; a thermal, a viscous, and an acoustic wave in the gas. The thermal and viscous waves travel nearly perpendicular to the film. The acoustic wave travels nearly parallel to the film.

attenuation of third sound is much larger than the intrinsic attenuation of the acoustic wave in the gas. The acoustic wave at a point further away from the film thus reflects the third sound intensity at an earlier point in the film, where it was greater. The characteristic perpendicular distance over which the acoustic mode increases is about equal to the third sound wavelength. But since it travels nearly parallel to the film it must cover a large distance (i.e. until it is significantly attenuated) before it can get that far away from the film.

While third sound is thus seen to penetrate rather far into the surroundings of the film, most of the energy of the wave resides within the film, in the form of kinetic energy of the superfluid flow and potential energy in the force field of the substrate in equal amounts. The importance of the companion waves is that they supply most of the attenuation of third sound. It turns out that for thin films (the precise meaning of thin and thick films in this context will be defined in Section III, A) the attenuation is due to the excitation of thermal waves in the substrate and the gas. Energy is lost by being radiated from the film in the form of thermal waves. This energy is eventually transformed into an increase of the total entropy by the dissipative processes of thermal conduction in the gas and substrate and the Kapitza resistance at the film–substrate interface. For thick films (the precise meaning of this will also be defined in Section III,A) the important dissipative process is evaporation and condensation of helium atoms between the film and the gas.

C. Linearized Equations of Motion for a Superfluid Film

The hydrodynamic theory of third sound in superfluid helium films starts out from the ordinary two-fluid hydrodynamic equations of bulk superfluid helium in linearized form (e.g., see Khalatnikov, 1965, p. 66):

$$\dot{\rho}_f + \text{div } \mathbf{J} = 0 \tag{6}$$

$$\frac{\partial J_i}{\partial t} + \frac{\partial P}{\partial x_i} = \frac{\partial}{\partial x_k}\left\{\eta_f\left[\frac{\partial v_{ni}}{\partial x_k} + \frac{\partial v_{nk}}{\partial x_i} - \frac{2}{3}\delta_{ik}\text{ div }\mathbf{v}_n\right]\right.$$

$$\left. + \delta_{ik}\zeta_1\text{ div}(\mathbf{J} - \rho_f\mathbf{v}_n) + \delta_{ik}\zeta_2\text{ div }\mathbf{v}_n\right\} \tag{7}$$

$$\dot{\mathbf{v}}_s + \boldsymbol{\nabla}\mu = \boldsymbol{\nabla}[\zeta_3\text{ div}(\mathbf{J} - \rho_f\mathbf{v}_n) + \zeta_1\text{ div }\mathbf{v}_n] \tag{8}$$

$$(\partial/\partial t)(\rho_f S_f) + \text{div }(\rho_f S_f\mathbf{v}_n - (\kappa_f/T)\boldsymbol{\nabla}T_f) = 0 \tag{9}$$

$$\text{curl }\mathbf{v}_s = 0, \tag{10}$$

where

$$\rho_f = \rho_s + \rho_n, \tag{11}$$

$$\mathbf{J} = \rho_s\mathbf{v}_s + \rho_n\mathbf{v}_n. \tag{12}$$

We look for a solution of these equations which describes a wave with a frequency ω, traveling in the positive x-direction with a (generally complex) wave vector k. All the dynamic variables of the film thus have the form

$$T_f = T + T_f'e^{-i\omega t + ikx}, \tag{13}$$

where

$$\omega/k \equiv c_3 \tag{14}$$

defines the (complex) velocity of third sound, c_3. As defined up to now, our problem is essentially a two-dimensional one, since nothing happens in the z-direction. We will consequently ignore the z-coordinate and the z-components of all vectors throughout the rest of our discussion.

We first analyze these equations in the zero frequency limit. In that case, third sound reduces to a dc superfluid flow. Hence all velocity components except for v_{sx} must vanish as $\omega \to 0$, and we may write

$$v_{sy} \sim \omega v_{sx} \qquad v_{ny} \sim \omega v_{sx} \qquad v_{nx} \sim \omega v_{sx}. \tag{15}$$

Looking at Eq. (7) (the Navier–Stokes equation) in this limit, we can clearly simplify its x-component to read

$$\rho_s\dot{v}_{sx} + (\partial P/\partial x) = \eta_f(\partial^2 v_{nx}/\partial y^2). \tag{16}$$

We will now write the x-component of Eq. (8) (the equation of motion for the superfluid velocity) in the same limit, using also the approximation of neglecting the temperature gradient which, as we shall see in Sections III,B and C, is at least a good first approximation. We thus get

$$\dot{v}_{sx} = -\frac{\partial \mu}{\partial x} = -\frac{1}{\rho_f} \frac{\partial P}{\partial x}. \tag{17}$$

Substituting this in Eq. (16) we immediately get

$$\rho_n \dot{v}_{sx} = -\eta_f \frac{\partial^2 v_{nx}}{\partial y^2}. \tag{18}$$

Finally, Eq. (10) together with (15) leads us to the result

$$\frac{\partial v_{sx}}{\partial y} = \frac{\partial v_{sy}}{\partial x} \sim \omega^2 v_{sx} \cong 0, \tag{19}$$

which means that v_{sx} is constant across the film.

We now use (13) to calculate $\dot{v}_{sx}$, and integrate Eq. (18) across the thickness of the film to get

$$\left. \frac{\partial v_{nx}}{\partial y} \right|_y - \left. \frac{\partial v_{nx}}{\partial y} \right|_h = \frac{i\omega \rho_n}{\eta_f} (y - h) v_{sx}. \tag{20}$$

But at the free surface $y = h$ we have to satisfy a boundary condition to make the shear force vanish (the shear force exerted by the gas on the other side is negligible, as we will show later in Sections III, B and C)

$$\partial v_{nx} / \partial y |_h = \partial v_{ny} / \partial x |_h = i k v_{ny} \sim \omega^2 v_{sx}. \tag{21}$$

Thus, $\partial v_{nx} / \partial y$ is of higher order in ω at $y = h$ than elsewhere, and it can be neglected there. Another integration then leads to the desired result

$$v_{nx}(y = h) = -\frac{i\omega \rho_n h^2}{\eta_f} v_{sx} = -i \frac{h^2}{l_\eta^2} v_{sx}, \tag{22}$$

where

$$l_\eta \cong \left(\frac{2\eta_f}{\omega \rho_n} \right)^{1/2} \tag{23}$$

is the viscous penetration depth for liquid helium. Its values range from $l_\eta = 2 \times 10^{-4}$ cm at $T = 1.9°$K, $\omega = 10^4$ sec^{-1} to $l_\eta = 7 \times 10^{-3}$ cm at $T = 1.3°$K, $\omega = 10^2$ sec^{-1}. The thickest films in which third sound has been observed are 500 Å thick. This is still 40 times less than the smaller l_η. Hence we will in our further discussions always neglect v_{nx} as compared to v_{sx}.

As for the two perpendicular velocities v_{sy}, v_{ny}, it seems fairly obvious that in the limit $\omega \to 0$ they cannot depend on either η_f, ζ_1, ζ_2, ζ_3, or κ_f.

The only dimensionless constant one has in the theory that is proportional to ω but independent of any of the dissipative coefficients is

$$hk = h\omega/c_3.$$

Remembering that the wave vector k must always be accompanied by i, we may write

$$v_{sy} \cong ihkv_{sx}; \qquad v_{ny} \cong ihkv_{sx}. \tag{24}$$

These equalities are expected to hold as far as the order of magnitude is concerned.

Having thus determined v_{nx}, v_{sy}, v_{ny} in terms of v_{sx} we do not need Eq. (7) (the Navier Stokes equation) any longer and we will base our subsequent considerations on Eqs. (6) and (8)–(10) alone.

We note however that merely to say that, e.g., $v_{nx} \ll v_{sx}$ is not enough in order to justify neglecting v_{nx} altogether. The important question is whether v_{nx} makes any sizable contribution to the dissipative processes which govern the attenuation of third sound. We will return to this question in Section IX.

D. Averaging of the Equations Across the Film

Equations (6) and (8)–(10) and the functions appearing therein still depend on the y-coordinate. But there is no practical way to measure any of these dependences. In practice what is always measured is some quantity that is averaged across the film and propagates in the x-direction. We accordingly try to eliminate any explicit reference to y in these equations by integrating them from $y = 0$ to $y = h$. This type of procedure was first used by Sanikidze *et al.* (1967) in the treatment of fourth sound. We will follow and slightly simplify the discussion given by Bergman (1969) for the third sound equations.

Because the velocity of third sound is much less than the velocity of first sound in helium, and because all the other velocities (i.e. $\mathbf{v}_s$, $\mathbf{v}_n$) are certainly much less than that for small amplitude waves, we may assume that the liquid helium is incompressible. Equation (6) thus becomes

$$\operatorname{div} \mathbf{J} = 0. \tag{25}$$

Integrating this over y we find

$$0 = \int_0^{h(x)} dy \left(\frac{\partial J_x}{\partial x} + \frac{\partial J_y}{\partial y} \right) = \frac{\partial v_{sx}}{\partial x} \int_0^h dy \, \rho_s(y) + J_y(x, h), \tag{26}$$

where we have ignored nonlinear terms in the oscillating amplitudes, as well as the y-dependence of v_{sx} which, according to Eqs. (9) and (24), is given by

$$\frac{\partial v_{sx}}{\partial y} \cong hk^2 v_{sx}. \tag{27}$$

We have also used the fact that

$$J_y(x, y = 0) = 0, \tag{28}$$

as well as the fact that $v_{nx} \ll v_{sx}$. There is, however, no need to assume that ρ_s is independent of y (its dependence on x is a second order term and thus may be consistently ignored): We can define an average superfluid density

$$\bar{\rho}_s \equiv \frac{1}{h} \int_0^h dy \rho_s(y), \tag{29}$$

in terms of which Eq. (26) becomes

$$h\bar{\rho}_s \frac{\partial v_{sx}}{\partial x} + J_y(x, h) = 0. \tag{30}$$

We also note that the rate at which helium evaporates from the film into the gas per unit area of surface, J_M, is connected to $J_y(x, h)$ and the vertical velocity of the liquid–gas interface $\dot{h}$ as follows:

$$J_M = J_y(x, h) - \dot{h}\rho_h, \tag{31}$$

where ρ_h is the liquid density at the interface. Consequently, Eq. (30) now becomes

$$\boxed{\dot{h}\rho_h + h\bar{\rho}_s \frac{\partial v_{sx}}{\partial x} + J_M = 0} \ . \tag{32}$$

We note at this point that mass conservation allows us to write an alternative expression for J_M in terms of variables of the gas:

$$J_M = [\rho_g(v_{gy} - \dot{h})]_{y = y_r}. \tag{33}$$

These variables are taken not at $y = h$ but at an imaginary reference plane $y = y_r$ (see Fig. 1) which is far enough away from the film so that all quantities such as ρ_g have their bulk values and are not influenced by the short range potential exerted by the substrate on atoms of the gas. Since the mean free path in the gas

$$l_g \simeq \frac{3\kappa_g}{\rho_g C_p c}, \tag{34}$$

where c is the velocity of sound in the gas and C_p is the constant pressure heat capacity per unit mass, is at least 10^{-5} cm and thus much greater than the range of these forces, a plane $y = y_r$ can be found where the gas is still in equilibrium with the film. In this connection it has been pointed out by Rudnick in a private communication that there is no other consistent way to describe the equilibrium at the film–gas interface: Because the range of the substrate potential is less than the mean free path, the gas atoms do not reach equilibrium within the potential in the sense that their velocity distribution is not Maxwellian, and their pressure does not satisfy the barometric formula. The true situation is in fact at the other extreme, where one can neglect the effect of the substrate potential on the gas and consider the bulk gas outside the potential to be directly in equilibrium with the liquid film.

A similar treatment is now given to Eq. (9): We integrate it across the thickness of the film, obtaining

$$
0 = \int_0^{h(x)} dy \left(\frac{\partial(\rho_f S_f)}{\partial t} + \frac{\partial(S_f v_{ny})}{\partial y} - \frac{\kappa_f}{T} \frac{\partial^2 T_f}{\partial y^2} - \frac{\kappa_f}{T} \frac{\partial^2 T_f}{\partial x^2} \right)
$$
$$
= \frac{\partial}{\partial t} \int_0^{h(x)} \rho_f S_f \, dy - \rho_h S_f(y=h)\dot{h} + \left[\rho_f S_f v_{ny} - \frac{\kappa_f}{T} \frac{\partial T_f}{\partial y} \right]_0^{h(x)} - \int_0^{h(x)} dy \frac{\kappa_f}{T} \frac{\partial^2 T_f}{\partial x^2}.
$$
$$(35)$$

This is further transformed as follows: We introduce an average film temperature $\bar{T}_f$

$$
\bar{T}_f(x) \equiv \frac{1}{h} \int_0^h dy \, T_f(x, y), \tag{36}
$$

in terms of which the last term in (35) becomes

$$
-\frac{h\kappa_f}{T} \frac{\partial^2 \bar{T}_f}{\partial x^2}. \tag{37}
$$

We express the entropy (as well as other thermodynamic quantities of the film) as functions of h and $\bar{T}_f$. This leads to the following expression for the first term of (35)

$$
\frac{\partial}{\partial t} \int_0^h \rho_f S_f \, dy = \dot{h} \frac{\partial}{\partial h} \int_0^h \rho_f S_f dy + \dot{\bar{T}}_f \frac{\partial}{\partial \bar{T}_f} \int_0^h \rho_f S_f \, dy
$$
$$
= \dot{h}\rho_h \bar{S} + h\bar{\rho}_f C_h \frac{\dot{\bar{T}}_f}{T}, \tag{38}
$$

where

$$
\bar{S} \equiv \frac{1}{\rho_h} \left(\frac{\partial}{\partial h} \int_0^h \rho_f S_f \, dy \right)_T \tag{39}
$$

is the partial (as distinct from average) entropy per unit mass of the film, i.e. it is the rate at which the total entropy changes as more mass is added to the film. As opposed to the average entropy of the film, which is always positive, the partial entropy is sometimes negative, as shown by Fraser (1969) (see also Atkins and Rudnick, 1970) following a suggestion by Bergman. This is an indication of the fact that, as the film is made very thin, some order (perhaps the superfluid order) is destroyed. The other new quantities in (38) are the average heat capacity per unit mass at constant h

$$
C_h \equiv \frac{T}{\bar{\rho}_f} \left(\frac{\partial}{\partial \bar{T}_f} \int_0^h \rho_f S_f \, dy \right)_h, \tag{40}
$$

the average mass density $\bar{\rho}_f$, and the ambient temperature T.

The second term and the term in square brackets in Eq. (35) represent the heat fluxes that flow out of the film and into its surroundings. Energy conservation at the film–substrate interface is expressed by

$$\left[\rho_f S_f v_{ny} - \frac{\kappa_f}{T}\frac{\partial T_f}{\partial y}\right]_{y=0} = -\frac{\kappa_{sub}}{T}\frac{\partial T_{sub}}{\partial y}\bigg|_{y=0} \equiv -J_{sub}, \qquad (41)$$

where J_{sub} is the heat current flowing into the substrate. Energy conservation at the film–gas interface is expressed by

$$\left[\mu(J_y - \dot{h}\rho_f) + T\rho_f S_f(v_{ny} - \dot{h}) - \kappa_f\frac{\partial T_f}{\partial y}\right]_{y=h}$$

$$= \left[(\mu + TS_g)\rho_g(v_{gy} - \dot{h}) - \kappa_g\frac{\partial T_g}{\partial y}\right]_{y=y_r}. \qquad (42)$$

By using mass conservation at this interface, i.e. equating (31) and (33), the terms including μ in (42) are seen to cancel. [In reality, the chemical potential in the gas differs from that in the film, but that would be a second order effect in (42).] We are thus left with an equality of heat flows:

$$\left[T\rho_f S_f(v_{ny} - \dot{h}) - \kappa_f\frac{\partial T_f}{\partial y}\right]_{y=h} = \left[T\rho_g S_g(v_{gy} - \dot{h}) - \kappa_g\frac{\partial T_g}{\partial y}\right]_{y=y_r}$$

$$= TS_g J_M + J_g, \qquad (43)$$

where we have separated the total heat flow in the gas into a sum of a convective flow $TS_g J_M$ and a thermal conduction flow

$$J_g \equiv -\kappa_g\frac{\partial T_g}{\partial y}\bigg|_{y=y_r} \qquad (44)$$

If we now substitute (37), (38), (41), and (43) into (35), we obtain

$$\boxed{\dot{h}\rho_h T\bar{S} + h\bar{\rho}_f C_h \dot{T}_f + TS_g J_M + J_g + J_{sub} - h\kappa_f\frac{\partial^2 \bar{T}_f}{\partial x^2} = 0} \qquad (45)$$

Subtracting from this Eq. (32) multiplied by $T\bar{S}$ we get

$$h\bar{\rho}_f C_h \dot{T}_f - h\bar{\rho}_s T\bar{S}\frac{\partial v_{sx}}{\partial x} + LJ_M + J_g + J_{sub} - h\kappa_f\frac{\partial^2 \bar{T}_f}{\partial x^2} = 0, \qquad (46)$$

where

$$L \equiv T(S_g - \bar{S}) \qquad (47)$$

is the latent heat of evaporation per unit mass from the film to the gas. Dividing (46) by L and subtracting it from Eq. (32), we get another form for this equation

$$\dot{h}\rho_h + h\bar{\rho}_s\left(1 + \frac{T\bar{S}}{L}\right)\frac{\partial v_{sx}}{\partial x} - \frac{h\bar{\rho}_f C_h}{T}\dot{T}_f - \frac{J_g + J_{sub}}{L} - \frac{h\kappa_f}{L}\frac{\partial^2 \bar{T}_f}{\partial x^2} = 0. \qquad (48)$$

In order to develop Eq. (8), we first rewrite its x-component in detail, taking into account the fact that div $\mathbf{J} = 0$:

$$\dot{v}_{sx} = -\frac{\partial \mu}{\partial x} + (\zeta_1 - \rho_f \, \zeta_3)\left(\frac{\partial^2 v_{nx}}{\partial x^2} + \frac{\partial^2 v_{ny}}{\partial x \, \partial y}\right). \tag{49}$$

On the right-hand side, one of the terms is

$$(\zeta_1 - \rho_f \, \zeta_3)\frac{\partial^2 v_{nx}}{\partial x^2} = -k^2(\zeta_1 - \rho_f \, \zeta_3)v_{nx}$$

$$\cong \frac{i\omega\rho_n h^2}{\eta_f} \, k^2(\zeta_1 - \rho_f \, \zeta_3)v_{sx}, \tag{50}$$

where we have used (22). Comparing this with the left-hand side we find:

$$(\zeta_1 - \rho_f \, \zeta_3)(\partial^2 v_{nx}/\partial x^2)/\dot{v}_{sx} \cong -\frac{\rho_n(\zeta_1 - \rho_f \, \zeta_3)}{\eta_f} \, h^2 k^2. \tag{51}$$

The dimensionless number

$$hk = h/\lambda = h\omega/c_3$$

is extremely small, being between 10^{-6} and 10^{-3} in all cases. We do not have any numerical information about ζ_1 or ζ_3, but we do have rather good information about η_f, showing that its values lie between 10 and 20 μP for temperatures in the range 1.1–2.1°K. We also have some information about ζ_2 from attenuation measurements of first sound which indicate that it can be up to 100 times greater than η_f (see Dransfeld *et al.*, 1958). If we assume that both $\rho_n \zeta_1$ and $\rho_n \rho_f \zeta_3$ are not much greater than ζ_2, then the term $\partial^2 v_{nx}/\partial x^2$ can be ignored in (49).

We integrate the remainder of Eq. (49) across the film to get

$$\int_0^{h(x)} \dot{v}_{sx} \, dy = h\dot{v}_{sx} = -\int_0^{h(x)} \frac{\partial \mu}{\partial x} \, dy + (\zeta_1 - \rho_f \, \zeta_3)ikv_{ny}\bigg|_0^h. \tag{52}$$

Comparing the last term with the left-hand side, and using (24), we find

$$(\zeta_1 - \rho_f \, \zeta_3)ikv_{ny}/h\dot{v}_{sx} \cong i\omega(\zeta_1 - \rho_f \, \zeta_3)/c_3{}^2. \tag{53}$$

Again, having no better estimate for ζ_1 and ζ_3, we assume

$$\rho_f(\zeta_1 - \rho_f \, \zeta_3) = \zeta_2 \leqslant 10^{-3} \text{ P}, \tag{54}$$

and find that the ratio calculated in (53) is very small, being no larger than 10^{-3}. We can therefore rewrite Eq. (52) in the form

$$\dot{v}_{sx} = -\partial\mu_f/\partial x, \tag{55}$$

where μ_f is now the chemical potential averaged across the film

$$\mu_f \equiv \frac{1}{h}\int_0^h \mu \, dy. \tag{56}$$

As before, we treat μ_f as a function of h and $\bar{T}_f$. We have already used one of its partial derivatives, which we now redefine as

$$\left(\frac{\partial \mu_f}{\partial h}\right)_T \equiv f. \tag{57}$$

The other one is determined as follows: The increment of internal energy per unit area of the film dE_A is given by

$$dE_A = \bar{T}_f \, dS_A + \mu_f \, dM_A,$$

where S_A is entropy per unit area and M_A is mass per unit area. Taking into account the fact that one can write

$$dM_A = \rho_h \, dh,$$

and making a Legendre transformation we find

$$d(E_A - \bar{T}_f S_A) = -S_A \, d\bar{T}_f + \mu_f \, \rho_h \, dh. \tag{58}$$

Hence, assuming that ρ_h is independent of T and recalling (39), we get

$$\left(\frac{\partial \mu_f}{\partial T}\right)_h = -\frac{1}{\rho_h} \left(\frac{\partial S_A}{\partial h}\right)_T = -\bar{S}. \tag{59}$$

We can summarize (57) and (59) in the form

$$d\mu_f = -\bar{S} \, d\bar{T}_f + f dh, \tag{60}$$

and use this to write (55) in the form

$$\boxed{\dot{v}_{sx} = \bar{S} \frac{\partial \bar{T}_f}{\partial x} - f \frac{\partial h}{\partial x}} \ . \tag{61}$$

     The three equations we have obtained in this section, namely (32), (45) [or one of its alternative forms (46), (48)], and (61) are the basic linearized hydrodynamic equations for a thin superfluid helium film adsorbed upon a flat solid substrate and in contact with helium vapor. They include all of the existing interactions with these two neighboring media. They do not include any intrinsic dissipative mechanisms except for the horizontal thermal conductivity term $-\kappa_f \, \partial^2 \bar{T}_f / \partial x^2$ in (45) and in its alternative equations (46) and (48). All of the other dissipative terms, i.e. the various types of viscous forces, were discarded in the process of setting up these equations because we found that they were small. While this means that one can leave them out when solving the equations to a lowest approximation, we have not ruled out the possibility that they might make an important contribution to more subtle features, such as the attenuation of third sound. This possibility must be considered since, as we have seen in the introduction, the crudest approximation, which neglects both the intrinsic dissipative mechanisms as well as interactions with the surrounding media, leads to a zero attenuation.

We will return to examine the possibility that η_f contributes to the attenuation in Section IX.

At this point we will summarize the conditions which must be satisfied for the analysis of this section to hold:

1. $hk \ll 1$: h is much less than the wavelength of third sound.
2. $h/l_\eta \ll 1$: h is much less than the viscous penetration depth in the film.
3. $\omega(\zeta_1 - \rho_f \zeta_3)/c_3^2 \ll 1$.

Of these, the first two are known to hold very well in all the experiments which have been performed on third sound. The last one is not known for lack of information about ζ_1 and ζ_3, but is estimated to hold from the assumption that $\rho_f \zeta_1 \cong \rho_f^2 \zeta_3 \leqslant \zeta_2$.

The three hydrodynamic equations of motion that we derived in this section for h [Eq. (32)], for $\bar{T}_f$ [Eq. (46)], and for v_{sx} [Eq. (61)], include also some variables of the substrate and the gas: J_M, J_{sub}, J_g. In order to solve these equations, we must first find explicit expressions for these quantities or some more equations of motion. We do this in the next two subsections, where we discuss the hydrodynamic equations for the substrate and the gas and transport processes between them and the film.

E. Equation of Motion for the Substrate

Consideration of the substrate is necessary in order to obtain an explicit expression for J_{sub}. Hydrodynamically speaking, the substrate is a relatively simple system, with only the temperature variations to worry about. These satisfy a diffusion type equation

$$\rho_{sub} C_{sub} \dot{T}_{sub} = \kappa_{sub} \nabla^2 T_{sub}, \tag{62}$$

whose solution will also depend on the boundary conditions at the film–substrate interface, and on the other side of the substrate.

Ordinarily in hydrodynamics or in thermal conduction theory we would take $T_{sub} = \dot{T}_f$ at the interface. But in superfluid helium we know that this is usually not a good approximation, due both to the the very efficient heat conducting processes of the superfluid and to the relatively high value of the thermal boundary resistance between helium and all solids, caused by the large mismatch in phonon velocities. We therefore allow for a discontinuity in temperature at the interface and write

$$J_{sub} \equiv \kappa_{sub} \, \partial T_{sub}/\partial y \big|_{y=0} = B_1(\bar{T}_f - T_{sub}) \big|_{y=0} \tag{63}$$

for the boundary condition. $1/B_1$ is the well known Kapitza resistance. If the other side of the substrate is sufficiently far away (compared to the thermal penetration depth, shortly to be defined), then we may take the other boundary condition to be that the wave travels away from the film without any reflections from the far side. We can therefore write the solution in the form

$$T_{sub} = T + T'_{sub} \exp(-i\omega t + ikx + q_{sub} y), \tag{64}$$

where q_{sub} satisfies the following dispersion equation

$$\frac{q^2_{sub}}{\omega^2} = \frac{1}{c_3{}^2} - \frac{i\rho_{sub}C_{sub}}{\omega\,\kappa_{sub}}\; ; \; \text{Im } q_{sub} < 0. \tag{65}$$

In this equation, the second term on the right-hand side usually dominates and in that case we have

$$q_{sub} \cong e^{-i\pi/4}\left(\frac{\rho_{sub}C_{sub}\,\omega}{\kappa_{sub}}\right)^{1/2} \tag{66}$$

The real (and the imaginary) part of q_{sub} is then equal to the reciprocal thermal penetration depth $1/l_{sub}$, where

$$l_{sub} = \left(\frac{2\kappa_{sub}}{\rho_{sub}C_{sub}\,\omega}\right)^{1/2} \tag{67}$$

As long as $h_{sub} \gg l_{sub}$, where h_{sub} is the thickness of the substrate, there will be no reflections to worry about.

If we now substitute (64) into the other boundary condition (63), we can solve for T_{sub} in terms of $\bar{T}_f$. We can then write

$$J_{sub} = B(\bar{T}_f - T), \tag{68}$$

where

$$\frac{1}{B} \equiv \frac{1}{B_1} + \frac{1}{\kappa_{sub}q_{sub}} \tag{69}$$

is an effective Kapitza resistance.

We have not included in our discussion the possibility of exciting an acoustic wave in the substrate. The coupling between third sound and such a wave is expected to be quite negligible since the pressure oscillations in the film are very small. Though we have not included this acoustic wave in our analysis, we can learn by analogy with the acoustic mode in the gas (see Sections II,B and III) that it will make a negligible contribution to the properties of third sound. This is all the more so for the acoustic mode in the substrate because there is less interaction with it than with the acoustic mode in the gas, and because the velocity mismatch with third sound is even greater.

F. Equations of Motion for the Gas

To calculate J_g and J_M is far more complicated, because the hydrodynamic equations of the gas are more numerous and they have not one but three distinct wave-type solutions similar to (64), which are excited in the gas when a third sound wave travels in the film. The equations, linearized but including all of the dissipative mechanisms, are

$$\dot{\rho}_g + \rho_g \text{ div } \mathbf{v}_g = 0, \tag{70}$$

$$\rho_g \dot{\mathbf{v}}_g + \mathbf{\nabla} P - \eta_g \nabla^2 \mathbf{v}_g - (\zeta_g + \tfrac{1}{3}\eta_g)\mathbf{\nabla} \text{ div } \mathbf{v}_g = 0, \tag{71}$$

$$\rho_g T\dot{S}_g - \kappa_g \nabla^2 T_g = 0, \tag{72}$$

plus the ideal gas equation of state

$$P = \rho_g k_B T_g / m. \tag{73}$$

The boundary condition away from the film, if the gas is thick enough, will again be that the wave travels away from the film. But the boundary condition at the gas–film interface is more complicated than before: Whereas before only energy could be transported across the film–substrate interface (in the form of heat), we now have to consider both energy and mass transport. These will be the result of small discontinuities in both T and μ across the interface. There is also the added complication that the interface is moving with velocity $\dot{h}$ relative to the gas far away. We will now calculate formulas for these transport processes, analogous to Eq. (63), in terms of $T_g - \bar{T}_f$ and $\mu_g - \mu_f$, by using simple kinetic theory for the gas.

We do this by starting from the simple kinetic theory formulas for the mass and energy fluxes $J_M{}^g$ and $J_E{}^g$, respectively, that hit the wall of a vessel containing a classical ideal gas in equilibrium but moving away from the wall with a velocity u_g:

$$J_M{}^g = P_g \left(\frac{m}{2\pi k_B T_g} \right)^{1/2} - \frac{1}{2} \rho_g u_g \tag{74}$$

$$J_E{}^g = 2 P_g \left(\frac{k_B T_g}{2\pi m} \right)^{1/2} - \frac{5}{4} P_g u_g. \tag{75}$$

The first term on the right-hand side of these equations is the usual formula for a stationary gas. The second term takes into account the fact that the gas is assumed to be moving away from the wall.

For the analogous fluxes in the film phase no such simple treatment is possible. Since we assume, however, that local equilibrium holds everywhere in the film, we calculate instead the fluxes in the gas phase that would be in equilibrium (locally) with the film. These are equal to the corresponding fluxes in the film if we assume that there are no reflections of gas particles striking the interface. For a film in equilibrium with gas at temperature $\bar{T}_f$, pressure P_{gf}, and density ρ_{gf}, but moving at a velocity u_f towards the wall, we thus find for the fluxes of mass and energy that come out of the film

$$J_M{}^f = P_{gf} \left(\frac{m}{2\pi k_B \bar{T}_f} \right)^{1/2} + \frac{1}{2} \rho_{gf} u_f, \tag{76}$$

$$J_E{}^f = 2 P_{gf} \left(\frac{k_B \bar{T}_f}{2\pi m} \right)^{1/2} + \frac{5}{4} P_{gf} u_f. \tag{77}$$

The net mass and energy fluxes flowing from the film and into the gas, J_M and J_E respectively, are obtained by subtracting (74) and (75) from (76) and (77), remembering that u_g and u_f are in fact determined by J_M:

$$\rho_g u_g = J_M, \tag{78}$$

$$\rho_h u_f = J_M. \tag{79}$$

plus the ideal gas equation of state

$$P = \rho_{\mathrm{g}} k_{\mathrm{B}} T_{\mathrm{g}}/m. \tag{73}$$

The boundary condition away from the film, if the gas is thick enough, will again be that the wave travels away from the film. But the boundary condition at the gas–film interface is more complicated than before: Whereas before only energy could be transported across the film–substrate interface (in the form of heat), we now have to consider both energy and mass transport. These will be the result of small discontinuities in both T and μ across the interface. There is also the added complication that the interface is moving with velocity $\dot{h}$ relative to the gas far away. We will now calculate formulas for these transport processes, analogous to Eq. (63), in terms of $T_{\mathrm{g}} - \bar{T}_{\mathrm{f}}$ and $\mu_{\mathrm{g}} - \mu_{\mathrm{f}}$, by using simple kinetic theory for the gas.

We do this by starting from the simple kinetic theory formulas for the mass and energy fluxes $J_{\mathrm{M}}{}^{\mathrm{g}}$ and $J_{\mathrm{E}}{}^{\mathrm{g}}$, respectively, that hit the wall of a vessel containing a classical ideal gas in equilibrium but moving away from the wall with a velocity u_{g}:

$$J_{\mathrm{M}}{}^{\mathrm{g}} = P_{\mathrm{g}}\left(\frac{m}{2\pi k_{\mathrm{B}} T_{\mathrm{g}}}\right)^{1/2} - \frac{1}{2}\rho_{\mathrm{g}} u_{\mathrm{g}} \tag{74}$$

$$J_{\mathrm{E}}{}^{\mathrm{g}} = 2P_{\mathrm{g}}\left(\frac{k_{\mathrm{B}} T_{\mathrm{g}}}{2\pi m}\right)^{1/2} - \frac{5}{4}P_{\mathrm{g}} u_{\mathrm{g}}. \tag{75}$$

The first term on the right-hand side of these equations is the usual formula for a stationary gas. The second term takes into account the fact that the gas is assumed to be moving away from the wall.

For the analogous fluxes in the film phase no such simple treatment is possible. Since we assume, however, that local equilibrium holds everywhere in the film, we calculate instead the fluxes in the gas phase that would be in equilibrium (locally) with the film. These are equal to the corresponding fluxes in the film if we assume that there are no reflections of gas particles striking the interface. For a film in equilibrium with gas at temperature $\bar{T}_{\mathrm{f}}$, pressure P_{gf}, and density ρ_{gf}, but moving at a velocity u_{f} towards the wall, we thus find for the fluxes of mass and energy that come out of the film

$$J_{\mathrm{M}}{}^{\mathrm{f}} = P_{\mathrm{gf}}\left(\frac{m}{2\pi k_{\mathrm{B}} \bar{T}_{\mathrm{f}}}\right)^{1/2} + \frac{1}{2}\rho_{\mathrm{gf}} u_{\mathrm{f}}, \tag{76}$$

$$J_{\mathrm{E}}{}^{\mathrm{f}} = 2P_{\mathrm{gf}}\left(\frac{k_{\mathrm{B}} \bar{T}_{\mathrm{f}}}{2\pi m}\right)^{1/2} + \frac{5}{4}P_{\mathrm{gf}} u_{\mathrm{f}}. \tag{77}$$

The net mass and energy fluxes flowing from the film and into the gas, J_{M} and J_{E} respectively, are obtained by subtracting (74) and (75) from (76) and (77), remembering that u_{g} and u_{f} are in fact determined by J_{M}:

$$\rho_{\mathrm{g}} u_{\mathrm{g}} = J_{\mathrm{M}}, \tag{78}$$

$$\rho_{\mathrm{h}} u_{\mathrm{f}} = J_{\mathrm{M}}. \tag{79}$$

We thus obtain

$$\left(1 - \frac{\rho_{\mathrm{g}}}{\rho_{\mathrm{h}}}\right) J_{\mathrm{M}} = 2\left(\frac{m}{2\pi k_{\mathrm{B}} T}\right)^{1/2} \left[P_{\mathrm{gf}} - P_{\mathrm{g}} - \frac{P}{2T}(\bar{T}_{\mathrm{f}} - T_{\mathrm{g}})\right] \tag{80}$$

$$J_{\mathrm{E}} = \frac{5}{2}\frac{k_{\mathrm{B}} T}{m} J_{\mathrm{M}} - \frac{1}{2}\left(\frac{k_{\mathrm{B}} T}{2\pi m}\right)^{1/2} \left[P_{\mathrm{gf}} - P_{\mathrm{g}} + \frac{9}{2}\frac{P}{T}(\bar{T}_{\mathrm{f}} - T_{\mathrm{g}})\right], \tag{81}$$

where we have only kept terms that are linear in $\bar{T}_{\mathrm{f}} - T_{\mathrm{g}}$ or $P_{\mathrm{gf}} - P_{\mathrm{g}}$, and written T, P for the average temperature and pressure. Since P_{gf} is not a direct property of the film, a more convenient form for these equations is in terms of the discontinuity in the chemical potential $\mu_{\mathrm{f}} - \mu_{\mathrm{g}}$:

$$\left(1 - \frac{\rho_{\mathrm{g}}}{\rho_{\mathrm{h}}}\right) J_{\mathrm{M}} = 2\rho_{\mathrm{g}}\left(\frac{m}{2\pi k_{\mathrm{B}} T}\right)^{1/2} \left[\mu_{\mathrm{f}} - \mu_{\mathrm{g}} + \left(S_{\mathrm{g}} - \frac{k_{\mathrm{B}}}{2m}\right)(\bar{T}_{\mathrm{f}} - T_{\mathrm{g}})\right], \tag{82}$$

$$J_{\mathrm{g}} \equiv J_{\mathrm{E}} - \frac{5}{2}\frac{k_{\mathrm{B}} T}{m} J_{\mathrm{M}}$$

$$= -\frac{1}{2}\rho_{\mathrm{g}}\left(\frac{k_{\mathrm{B}} T}{2\pi m}\right)^{1/2} \left[\mu_{\mathrm{f}} - \mu_{\mathrm{g}} + \left(S_{\mathrm{g}} - \frac{9}{2}\frac{k_{\mathrm{B}}}{m}\right)(\bar{T}_{\mathrm{f}} - T_{\mathrm{g}})\right], \tag{83}$$

where J_{g} is the heat flux into the gas. We can now use the thermodynamic equations

$$d\mu_{\mathrm{f}} = -\bar{S}\, d\bar{T}_{\mathrm{f}} + f\, dh \tag{84}$$

$$d\mu_{\mathrm{g}} = -S_{\mathrm{g}}\, dT_{\mathrm{g}} + \frac{1}{\rho_{\mathrm{g}}}\, dP_{\mathrm{g}}$$

$$= \left(-S_{\mathrm{g}} + \frac{k_{\mathrm{B}}}{m}\right) dT_{\mathrm{g}} + \frac{k_{\mathrm{B}} T}{m}\frac{d\rho_{\mathrm{g}}}{\rho_{\mathrm{g}}} \tag{85}$$

to expand both μ_{f} and μ_{g} in (82) and (83) around the average μ. We thus get

$$\left(1 - \frac{\rho_{\mathrm{g}}}{\rho_{\mathrm{h}}}\right) J_{\mathrm{M}} = \frac{4mA}{k_{\mathrm{B}} T}\left[\left(\frac{L}{T} - \frac{k_{\mathrm{B}}}{2m}\right) T_{\mathrm{f}}' - \frac{k_{\mathrm{B}}}{2m} T_{\mathrm{g}}' - \frac{k_{\mathrm{B}} T}{m}\frac{\rho_{\mathrm{g}}'}{\rho_{\mathrm{g}}} + fh'\right] \tag{86}$$

$$J_{\mathrm{g}} = -A\left[\left(\frac{L}{T} - \frac{9}{2}\frac{k_{\mathrm{B}}}{m}\right) T_{\mathrm{f}}' + \frac{7}{2}\frac{k_{\mathrm{B}}}{m} T_{\mathrm{g}}' - \frac{k_{\mathrm{B}} T}{m}\frac{\rho_{\mathrm{g}}'}{\rho_{\mathrm{g}}} + fh'\right], \tag{87}$$

where

$$A \equiv \tfrac{1}{2}\rho_{\mathrm{g}}(k_{\mathrm{B}} T/2\pi m)^{1/2}. \tag{88}$$

We would like to point out that our treatment is by no means faultless. One glaring example is found if one writes J_{E} and $-J_{\mathrm{M}}$ in terms of

$$\Delta(1/T) \equiv (1/T_{\mathrm{g}}) - (1/\bar{T}_{\mathrm{f}}), \tag{89}$$

$$\Delta(\mu/T) \equiv (\mu_{\mathrm{g}}/T_{\mathrm{g}}) - (\mu_{\mathrm{f}}/\bar{T}_{\mathrm{f}}). \tag{90}$$

Neglecting ρ_g/ρ_h compared to 1, these become

$$-J_M = 2\rho_g T \left(\frac{m}{2\pi k_B T}\right)^{1/2} \left[\Delta\frac{\mu}{T} - 2\frac{k_B T}{m}\Delta\frac{1}{T}\right] \qquad (91)$$

$$J_E = 2\rho_g T \left(\frac{m}{2\pi k_B T}\right)^{1/2} \left[-\frac{9}{4}\frac{kT}{m}\Delta\frac{\mu}{T} + \frac{11}{2}\frac{k_B T}{m}\Delta\frac{1}{T}\right]. \qquad (92)$$

We then note that the rate of increase of entropy per unit area due to the transport of mass and energy through the interface is given by

$$\dot{S}/A = J_E \cdot \Delta(1/T) - J_M \cdot \Delta(\mu/T). \qquad (93)$$

Onsager's relations then require that the coefficient of $\Delta(\mu/T)$ in J_E be equal to the coefficient of $\Delta(1/T)$ in $-J_M$. This is clearly not the case in (91) and (92). Nevertheless, the quadratic expression for $\dot{S}$ in terms of $\Delta(1/T)$ and $\Delta(\mu/T)$ that is obtained when (91) and (92) are substituted in (93) is positive definite. The violation of Onsager's relations means that we have erred in some of the assumptions made in calculating J_M and J_E. There is clearly room here for work to investigate the properties of a liquid gas interface under conditions where simple continuity of temperature, pressure, and chemical potential does not hold.

If we just substituted (68), (86), and (87) into the three equations for the film, (32), (46), and (61), our troubles would not be over: instead of the unknown quantities J_M, J_g, and J_{sub} we would now have the unknown quantities T_g' and ρ_g' in addition to the film variables. What we must do is to enlarge our system of equations so as to include also those equations that are implied by the film–gas boundary conditions, i.e.

$$J_M = \rho_g(v_{gy} - \dot{h})|_{y=y_r}, \qquad (94)$$

$$J_g = -\kappa_g \,\partial T_g/\partial y|_{y=y_r}, \qquad (95)$$

$$v_{nx} = v_{gx}|_{y=y_r}. \qquad (96)$$

The last condition is the one usually assumed to hold between velocities at any boundary of superfluid helium, including a liquid–gas interface. It has been verified experimentally for a gas–liquid helium interface by Osborne (1962) in a direct experiment, and by Henkel *et al.* (1968) indirectly, by the observation of persistent currents in a helium film in contact with helium gas. We have not included a boundary condition on the pressure or on the tangential forces acting on the interface, and this omission requires some discussion.

In conventional treatments of an interface between a liquid and its own vapor, it is always assumed that the temperature, the pressure, and the shearing stresses are continuous across the interface. The mass and energy currents are then determined by the bulk properties of the two media. This is not always good enough for our situation, even though there are cases where the discontinuities are very small. In order to investigate what happens in all cases we had to calculate J_M and J_E microscopically without assuming local equilibrium across the interface. These direct calculations replace the assumptions of continuous T and μ. The pressure and the shear stress exerted on the interface by the helium film are not exactly calculable within the framework of our rather primitive theory for the film. However, since we have already assumed that they have no influence on the values of J_M and J_E, we may also consistently assume that they simply adjust themselves so as to equal the corresponding stresses exerted by the gas, and then stop concerning ourselves with them.

The new equations, (94)–(96), also include two new gas variables: v_{gy} and v_{gx}. What must now be done is to express all four gas variables T_g', ρ_g', v_{gx}, and v_{gy} in terms of independent amplitudes representing the independent wave modes of the gas. These are the solutions of (70)–(73) that have the form

$$e^{-i\omega t + ikx - qy}; \ \text{Im} \ q < 0. \tag{97}$$

There are three such modes: an acoustic mode M_3 (corresponding to ordinary sound waves), a viscous mode M_1, and a thermal mode M_2. The appropriate dispersion equations for q are

$$M_1: \ \frac{1}{c_{01}} \equiv \frac{q_{\text{vis}}}{\omega} = \left(\frac{1}{c_3{}^2} - \frac{i\rho_g}{\eta_g \omega}\right)^{1/2}, \tag{98}$$

$$M_2: \ \frac{1}{c_{02}} \equiv \frac{q_{\text{th}}}{\omega} = \left(\frac{1}{c_3{}^2} - \frac{i\rho_g C_p}{\kappa_g \omega}\right)^{1/2}, \tag{99}$$

$$M_3: \ \frac{1}{c_{03}} \equiv \frac{q_{\text{ac}}}{\omega} = \left(\frac{1}{c_3{}^2} - \frac{1}{c^2}\right)^{1/2}, \tag{100}$$

$$\text{Im} \ 1/c_{0i} < 0 \quad \text{for} \quad i = 1, 2, 3, \tag{101}$$

where c is the velocity of sound in the gas. In (99) and (100) we have neglected terms of the order

$$\kappa_g \omega / \rho_g C_p c^2 \quad \text{or} \quad \eta_g \omega / \rho_g c^2 \tag{102}$$

compared to 1, and we shall continue to do this throughout.

Equation (101), which selects one of the two square roots in each of Eqs. (98)–(100), expresses the requirement that all the modes in the gas have wave vectors whose real parts point away from the film. Actually, it is the group velocity which must always point away from the film, but this too is satisfied by the solutions we have selected. In all the interesting cases, we find that

$$\text{Im} \ 1/c_3{}^2 \ll \rho_g/\omega\eta_g \quad \text{and} \quad \rho_g C_p/\omega\kappa_g,$$

$\text{Re} \ 1/c_3{}^2$ is at most not much greater than either $\rho_g/\omega\eta_g$ or $\rho_g C_p/\omega\kappa_g$,

$$\text{Im} \ 1/c_3{}^2 \gg \text{Im} \ 1/c^2,$$

and

$$\text{Re} \ 1/c_3{}^2 > \text{Re} \ 1/c^2.$$

It follows that

$$\operatorname{Re} q_{\text{vis}} \quad \text{and} \quad \operatorname{Re} q_{\text{th}} > 0,$$

but

$$\operatorname{Re} q_{\text{ac}} < 0.$$

This means that in M_1 and M_2 the amplitude decreases exponentially as one moves away from the film in the y-direction, the characteristic distances being of the order of the viscous penetration depth of the gas

$$l_{\text{vis}} \equiv \left(\frac{\eta_{\text{g}}}{2\rho_{\text{g}}\,\omega}\right)^{1/2} \tag{102a}$$

and the thermal penetration depth of the gas

$$l_{\text{th}} \equiv \left(\frac{\kappa_{\text{g}}}{2\rho_{\text{g}} C_{\text{p}}\,\omega}\right)^{1/2}, \tag{102b}$$

respectively. The real part of the wave vector in these modes is likewise of the same order as $1/l_{\text{vis}}$ and $1/l_{\text{th}}$, so that these waves have a very short wavelength, and decay after traveling at most only a few wavelengths.

In the acoustic mode M_3, the amplitude increases exponentially as one moves away from the film. The characteristic distance is of the order of $1/k$, i.e. the wavelength of third sound. This increase does not continue indefinitely, however. It stops as soon as one gets to a point where the beginning of the acoustic wave train has just arrived. Moreover, it can easily be shown from (100) that at any point moving with the wavefronts of the acoustic mode the amplitude is constant: Its increase in the y-direction, determined by $\operatorname{Re} 1/c_{03}$, is exactly canceled by its decrease in the x-direction, determined by $\operatorname{Im} 1/c_3$. This happens because the attenuation of the acoustic mode is due to small terms such as (102), which we have neglected.

Choosing for the three independent amplitudes of the gas $T_{\text{g}2}$, $T_{\text{g}3}$ (the temperature amplitudes of the thermal and the acoustic mode, respectively), and $v_{\text{g}x}$, we find the following expressions for the various gas quantities that appear in our equations, all of them at $y = y_{\text{r}}$:

$$\frac{\rho_{\text{g}}{}'}{\rho_{\text{g}}} = -\frac{T_{\text{g}2}}{T} + \frac{1}{\gamma - 1}\frac{T_{\text{g}3}}{T}, \tag{103}$$

$$\frac{v_{\text{g}x}}{c_3} = \frac{ic_{01}}{c_3}\frac{v_{\text{g}x}}{c_3} - \frac{\kappa_{\text{g}}\,\omega}{\rho_{\text{g}} C_{\text{p}} c_3{}^2}\left(\frac{c_3}{c_{02}} - \frac{c_{01}}{c_3}\right)\frac{T_{\text{g}2}}{T}$$

$$+ \frac{iTC_{\text{p}}}{c_3{}^2}\left(\frac{c_3}{c_{03}} - \frac{c_{01}}{c_3}\right)\frac{T_{\text{g}3}}{T}, \tag{104}$$

$$-\kappa_{\text{g}}\frac{\partial T_{\text{g}}}{\partial y} = \kappa_{\text{g}}\,\omega\left(\frac{T_{\text{g}2}}{c_{02}} + \frac{T_{\text{g}3}}{c_{03}}\right), \tag{105}$$

where

$$\gamma \equiv C_{\text{p}}/C_{\text{v}}. \tag{106}$$

In the following discussion we will follow Bergman (1971) in assuming that

$$v_{\mathrm{g}x}\big|_{y=y_{\mathrm{r}}} = 0 \tag{107}$$

instead of (96) and (22). This has the advantage of basing our discussion on a published calculation. This is in any case expected to be a good approximation since, according to (22), $v_{\mathrm{n}x} \ll v_{\mathrm{s}x}$. We have in fact also solved the equations of the film using (96), and found that, indeed, there are no significant changes in the result.

G. THE COMBINED EQUATIONS OF THIRD SOUND

If we use (103)–(105), (94), (95), (107), and (13) and its analogs to substitute in (32), (48), (61), (86), and (87), we obtain the following system of equations in the corresponding order:

$$\left(1 - \frac{\rho_{\mathrm{g}}}{\rho_{\mathrm{h}}}\right)\frac{h'}{h} - v - \frac{i\kappa_{\mathrm{g}}}{h\rho_{\mathrm{h}}C_{\mathrm{p}}}\left(\frac{1}{c_{02}} - \frac{c_{01}}{c_3{}^2}\right)\frac{T_{\mathrm{g}2}}{T}$$
$$- \frac{\rho_{\mathrm{g}}}{\rho_{\mathrm{h}}}\frac{TC_{\mathrm{p}}}{h\omega}\left(\frac{1}{c_{03}} - \frac{c_{01}}{c_3{}^2}\right)\frac{T_{\mathrm{g}3}}{T} = 0, \tag{108}$$

$$\frac{h'}{h} - \left(1 + \frac{T\bar{S}}{L}\right)v - \frac{T}{h\rho_{\mathrm{h}}L}\left(\frac{iB}{\omega} + h\bar{\rho}_{\mathrm{f}}C_{\mathrm{h}} + \frac{i\kappa_{\mathrm{f}}h\omega}{c_3{}^2}\right)\frac{T_{\mathrm{f}}'}{T}$$
$$- \frac{i\kappa_{\mathrm{g}}T}{h\rho_{\mathrm{h}}L}\left(\frac{1}{c_{02}}\frac{T_{\mathrm{g}2}}{T} + \frac{1}{c_{03}}\frac{T_{\mathrm{g}3}}{T}\right) = 0, \tag{109}$$

$$\frac{hf\bar{\rho}_{\mathrm{s}}}{c_3{}^2\rho_{\mathrm{h}}}\frac{h'}{h} - v - \frac{T\bar{S}}{c_3{}^2}\frac{\bar{\rho}_{\mathrm{s}}}{\rho_{\mathrm{h}}}\frac{T_{\mathrm{f}}'}{T} = 0, \tag{110}$$

$$\left[2hf\frac{m}{k_{\mathrm{B}}T} - \frac{ih\omega}{a}\left(\frac{\rho_{\mathrm{h}}}{\rho_{\mathrm{g}}} - 1\right)\right]\frac{h'}{h} + \frac{ih\omega}{a}\left(\frac{\rho_{\mathrm{h}}}{\rho_{\mathrm{g}}} - 1\right)v$$
$$+ \left(\frac{2L}{k_{\mathrm{B}}T/m} - 1\right)\frac{T_{\mathrm{f}}'}{T} + \frac{T_{\mathrm{g}2}}{T} - \frac{\gamma+1}{\gamma-1}\frac{T_{\mathrm{g}3}}{T} = 0, \tag{111}$$

$$hf\frac{m}{k_{\mathrm{B}}T}\frac{h'}{h} + \left(\frac{L}{k_{\mathrm{B}}T/m} - \frac{9}{2}\right)\frac{T_{\mathrm{f}}'}{T} + \left(\frac{9}{2} + \frac{2\kappa_{\mathrm{g}}\omega T}{aPc_{02}}\right)\frac{T_{\mathrm{g}2}}{T}$$
$$+ \left(\frac{7}{2} - \frac{1}{\gamma-1} + \frac{2\kappa_{\mathrm{g}}\omega T}{aPc_{03}}\right)\frac{T_{\mathrm{g}3}}{T} = 0, \tag{112}$$

where

$$v \equiv (\bar{\rho}_{\mathrm{s}}/\rho_{\mathrm{h}})(v_{\mathrm{s}x}/c_3); \qquad a \equiv (k_{\mathrm{B}}T/2\pi m)^{1/2}. \tag{113}$$

These equations are identical to the system (23) of Bergman (1971). Equations (108)–(112) are a system of five linear, homogeneous equations for the five amplitudes h', v, T_f', T_{g2}, and T_{g3}. The determinant of this system, when set equal to zero, gives a very complicated dispersion equation for c_3 in terms of ω. The exact dispersion equation has been calculated by Bergman (1971). There one can also find a discussion of the possible approximations that can be made to simplify it.

III. The Properties of Third Sound in Flat Films—Theory

A. GENERAL RESULTS

A simple approximate form for the dispersion equation, which seems to be adequate over the entire range of experiments that have been performed, is

$$
\frac{u_{30}^2}{c_3{}^2} = \left\{ 1 + \left[1 - \frac{9}{32} \frac{k_B T}{m} \frac{i\omega\rho_h}{Af} \frac{T\bar{S}}{L} \left(1 - \frac{\rho_g}{\rho_h} \right) \left(1 + \frac{T\bar{S}}{L} \right)^{-1} \right] \right.
$$
$$
\times \frac{Tf}{\rho_h L^2} \left[\frac{iB}{\omega} + e^{i\pi/4} \left(\frac{\rho_g C_p \kappa_g}{\omega} \right)^{1/2} \right] \right\}
$$
$$
\times \left[1 - \frac{9}{32} \frac{k_B T}{m} \frac{i\omega\rho_h}{Af} \left(\frac{T\bar{S}}{L} \right)^2 \left(1 - \frac{\rho_g}{\rho_f} \right) \left(1 + \frac{T\bar{S}}{L} \right)^{-2} \right]^{-1}, \qquad (114)
$$

where

$$
u_{30}^2 \equiv hf(\bar{\rho}_s/\rho_h)[1 + (T\bar{S}/L)]^2. \qquad (115)
$$

For this approximation to hold it is only necessary that

$$
\kappa_g \omega/\rho_g C_p cc_3 \ll 1, \qquad (116)
$$

and that

$$
\kappa_g \omega/\rho_g C_p c_3{}^2 \quad \text{and} \quad \eta_g \omega/\rho_g c_3{}^2 \qquad (117)
$$

be at most not much greater than 1.

Even simpler approximate dispersion equations are obtained in two special cases: (a) the limit of very thin films or low frequencies, which is characterized by

$$
\frac{\kappa_g \omega}{\rho_g C_p c_3{}^2} \ll 1 \quad \text{and} \quad \frac{\eta_g \omega}{\rho_g c_3{}^2} \ll 1. \qquad (118)
$$

(b) The region where the thickness or the frequency are large enough so that

$$
\frac{\kappa_g \omega}{\rho_g C_p c_3{}^2} \gtrsim 1 \quad \text{and} \quad \frac{\eta_g \omega}{\rho_g c_3{}^2} \gtrsim 1, \qquad (119)
$$

but still small enough for our whole treatment to be valid (i.e. (102), (116), and (117) must be satisfied). For these two cases, which we will call "thin films" and "thick films," respectively, the dispersion equation becomes

$$\frac{u_{30}^2}{c_3{}^2} = 1 + \frac{Tf}{L^2\rho_{\rm h}}\, e^{i\pi/4} \left\{ \left(\frac{\rho_{\rm g} C_{\rm p} \kappa_{\rm g}}{\omega}\right)^{1/2} \right.$$
$$\left. + \left[\left(\frac{\rho_{\rm sub} C_{\rm sub} \kappa_{\rm sub}}{\omega}\right)^{-1/2} + e^{-i\pi/4}\,\frac{\omega}{B_1}\right]^{-1}\right\} \tag{120}$$

and

$$\frac{u_{30}^2}{c_3{}^2} = \left[1 - \frac{9}{32}\frac{k_{\rm B} T}{m}\frac{i\omega\rho_{\rm h}}{Af}\left(\frac{T\bar{S}}{L}\right)^2\left(1 - \frac{\rho_{\rm g}}{\rho_{\rm h}}\right)\left(1 + \frac{T\bar{S}}{L}\right)^{-2}\right]^{-1}. \tag{121}$$

These equations all lead to a complex value for c_3. Hence we can derive not only the velocity of third sound u_3, but the coefficient of attenuation α as well:

$$u_3 = 1/{\rm Re}(1/c_3) = u_{30}/{\rm Re}(u_{30}/c_3), \tag{122}$$

$$\alpha = 2\omega\, {\rm Im}(1/c_3) = (2\omega/u_{30})\, {\rm Im}(u_{30}/c_3). \tag{123}$$

In many cases, (122) leads to results which are not very different from those of Eq. (5) of the elementary theory, because both $T\bar{S}/L$ and the second term on the right-hand side of (120) or (121) are small compared to 1. For thick films, where $\bar{S}$ is equal to the bulk entropy per unit mass of liquid helium $S_{\rm bulk}$, and L is equal to the latent heat of bulk helium $L_{\rm bulk}$, $TS_{\rm bulk}/L_{\rm bulk}$ is an increasing function of T. Typical values are

$$TS_{\rm bulk}/L_{\rm bulk} = 0.081 \quad \text{for} \quad T = 2°{\rm K},$$
$$TS_{\rm bulk}/L_{\rm bulk} = 0.013 \quad \text{for} \quad T = 1.5°{\rm K}.$$

The second term on the right-hand side of (120) becomes large when either h or ω are very small. In that case it leads not only to changes in u_3 but also to some dispersion, with u_3 an increasing function of ω. In Fig. 3 we have plotted the ratio u_3/u_{30} versus h for thin films at two different frequencies. We should point out that u_3 is the phase velocity of third sound, whereas some experiments (i.e., the time of flight experiments of Rudnick's group—Rudnick *et al.*, 1968; Kagiwada *et al.*, 1969; Rudnick and Fraser, 1970; Fraser, 1969) measure the group velocity. The region of thicknesses where u_3/u_{30} deviates seriously from 1 is also where the dispersion becomes appreciable, and it works in such a way that it cancels much of the deviation. Other techniques of measurement, such as the third sound resonators used by Ratman and Mochel (1970a,b, 1972), measure the phase velocity and they should be able to detect this dispersion.

For sufficiently thick films or high frequencies u_3/u_{30} will again deviate from 1 as well as exhibit dispersion. This will occur when the second term on the right-hand side of (121) is large. Here again u_3/u_{30} will be an increasing function of both ω and h, but the dispersion will now serve to increase the deviation in the expression for the group velocity.

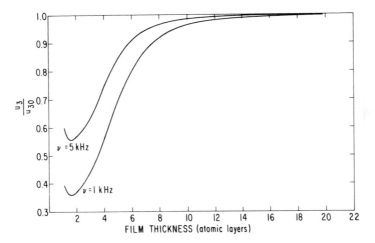

FIG. 3. Plot of u_3/u_{30} versus the film thickness for thin films at $T = 1.5°$K and frequencies of 1 kHz and 5 kHz. From Bergman, (1969), by permission of the American Institute of Physics and *Physical Review*. [*Note*: Since this article was written, Rudnick and his co-workers decided to revise the value of α (the coefficient of the Van der Waals potential) that they use to calculate the thickness of their films (Scholtz *et al.*, 1974). This leads to a reduction of all their film thicknesses by a factor .677. The new value of α was obtained from ultrasonic interferometric measurements of the thickness of helium films (Sabisky and Anderson, 1973a,b). Since all the thicknesses displayed in Figs. 3, 4, 8, 9, 11 were calculated using the old value of α, they should now all be corrected by multiplying by this factor .677.]

Another point to remember is that u_3 can deviate from the value given by Eq. (5) also because of $\bar{\rho}_s$ being different from the bulk value $\rho_{s,bulk}$. This effect does not appear in the ratio u_3/u_{30}, which is independent of $\bar{\rho}_s$. We will discuss this effect later in Section X,A.

Equation (123) always leads to interesting results, since the simple theory of Section II,A gave no attenuation at all. In Fig. 4 we plot the results of a typical calculation of the attenuation coefficient α versus h, as derived from Eqs. (114), (120), and (121) and from a numerical solution of the exact dispersion equation derived by Bergman (1971). In this calculation we have taken $\rho_s = \rho_{s,bulk}$, which will turn out to be not so for the small thicknesses. (In practice it would be preferable to plot αu_{30} which is independent of $\bar{\rho}_s$, but from Fig. 4 one can get a clearer picture of the order of magnitude of α.) From Fig. 4 it is clear that for most values of h either (120) or (121) (i.e. the equation that gives the larger value for α) gives a sufficiently precise value for α, (114) being required only in a small transition region. Even there, a simple sum of the values obtained from (120) and (121) is fairly accurate. As for (114), it gives results which are indistinguishable (in the figure) from the numerical results, the largest difference being 2% and occuring near the minimum in the transition region.

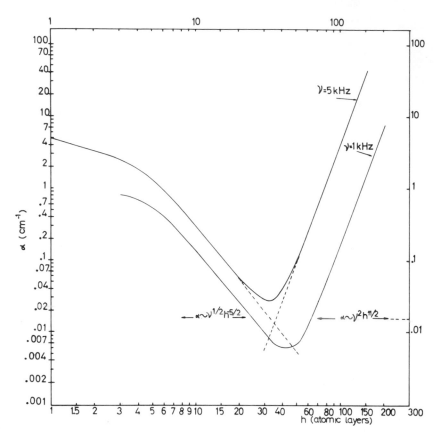

FIG. 4. Attenuation coefficient α versus film thickness h for $T = 1.3°$K. The two solid lines are the results of a numerical solution of the third sound equations at frequencies of 1 kHz and 5 kHz. They also represent, to within the accuracy of the drawing, the results calculated from Eq. (114). The two dashed lines are the result of the approximate equations (120) and (121). For thick films (not too thick though) α is proportional to $\omega^2 h^{11/2}$, while for thin films (again, not too thin) α is proportional to $\omega^{1/2}$ and also, approximately, to $h^{-5/2}$. From Bergman (1971), by permission of the American Institute of Physics and *Physical Review*. See note in caption to Fig. 3.

In the region of thick films shown in Fig. 4 and in most of the region of thin films (as long as the attenuation over one wavelength is sufficiently small), α has a simple dependence on both the thickness and the frequency:

$$\alpha \sim \omega^{1/2} h^{-5/2} \quad \text{for thin films,} \tag{124}$$

$$\alpha \sim \omega^2 h^{11/2} \quad \text{for thick films.} \tag{125}$$

The strong dependence on h results mostly from the presence of the substrate-helium force f both explicitly and implicitly, i.e. in u_{30}.

Not all of the physical processes we took into account in setting up the equations of motion in fact contribute to these results. In order to see which of them are important we will write down the solution for the various oscillating quantities in the two limits (118) and (119). We will also write down simplified equations of motion for the two limits that will reproduce the dispersion equations (120) and (121).

B. Thin Films

For the thin film case the physical quantities which take part in the third sound wave are related to each other as follows

$$\frac{v_{sx}}{c_3} = \frac{hf}{c_3{}^2}\left(1 + \frac{T\bar{S}}{L}\right)\frac{h'}{h} \tag{126}$$

$$\frac{T_f'}{T} = -\frac{hf}{L}\frac{h'}{h} \tag{127}$$

$$\frac{T_g' - T_f'}{T} = -\frac{ih'\omega\rho_h}{16A}\frac{1 - \rho_g/\rho_h}{1 + T\bar{S}/L}\left\{\frac{T\bar{S}}{L} - \frac{Tf}{\rho_h L^2}\left[\frac{iB}{\omega} + \bar{\rho}_f h C_h\right.\right.$$
$$\left.\left. + e^{i\pi/4}\left(\frac{\rho_g C_p \kappa_g}{\omega}\right)^{1/2}\left(1 - \frac{4L}{k_B T/m}\frac{1 + T\bar{S}/L}{1 - \rho_g/\rho_h}\right)\right]\right\}, \tag{128}$$

$$\frac{T_{g3}}{T} = -\frac{h'\omega c_{03}}{TC_p}\frac{\rho_h}{\rho_g}\frac{1}{1 + T\bar{S}/L}\left\{-\frac{T\bar{S}}{L} + \frac{\rho_g}{\rho_h}\left(1 + \frac{T\bar{S}}{L}\right)\right.$$
$$\left. + \frac{Tf}{\rho_h L^2}\left[\frac{iB}{\omega} + \bar{\rho}_f h C_h + e^{i\pi/4}\left(\frac{\rho_g C_p \kappa_g}{\omega}\right)^{1/2}\left(1 - \frac{L}{TC_p}\left(1 + \frac{T\bar{S}}{L}\right)\right)\right]\right\}, \tag{129}$$

$$J_M = ih'\omega\rho_h\left\{\frac{T\bar{S}}{L}\left(1 - \frac{\rho_g}{\rho_h}\right) - \frac{Tf}{\rho_h L^2}\left[\frac{iB}{\omega} + \bar{\rho}_f h C_h + e^{i\pi/4}\left(\frac{\rho_g C_p \kappa_g)}{\omega}\right)^{1/2}\right]\right\}, \tag{130}$$

$$J_g = e^{-i\pi/4}(\omega\kappa_g \rho_g C_p)^{1/2}T_f', \tag{131}$$

$$\mu_f - \mu_g = (S_g - \tfrac{9}{2}k_B/m)(T_g' - T_f') + (ih'\omega f T/AL)e^{i\pi/4}(\rho_g C_p \kappa_g/\omega)^{1/2}, \tag{132}$$

$$\mu_f' \equiv \mu_f - \mu_{eq} = h'f[1 + (T\bar{S}/L)], \tag{133}$$

$$T'_{sub} - T_f' = -J_{sub}/B_1 = -(B/B_1)T_f'. \tag{134}$$

Some consequences follow immediately from these equations:

$$T_g' - T_f' \ll T_f', \tag{135}$$

$$\mu_g' - \mu_f' \ll \mu_f', \tag{136}$$

so that in fact the temperature and the chemical potential of the gas and the film follow each other closely at the interface. On the other hand, $T_{\text{sub}} - T_{\text{f}}'$ is not much smaller than T_{f}'. It is also evident that

$$T_{\text{g3}} \ll T_{\text{g}}', \tag{137}$$

so that most of the temperature variations in the gas are due to the thermal mode. However, $v_{\text{g}y}$ has comparable contributions from both the thermal and the acoustic mode. Similarly, $v_{\text{g}x}$, although it vanishes by assumption, is made up of mutually cancelling contributions from both the acoustic and the viscous modes. We have summarized the main contributors to various gas quantities in Table I. This table shows that all three modes of the gas

TABLE I

PRINCIPAL CONTRIBUTORS TO VARIOUS GAS
VARIABLES[a]

Gas variable	Principal contributing modes
$T, \rho, \partial T/\partial y$	M_2
v_y	M_2, M_3
v_x	M_1, M_3

[a] From Bergman (1969), by permission of the American Institute of Physics and *Physical Review*.

must be excited in order to satisfy the boundary conditions. Nevertheless, not all of these modes make a significant contribution to the dispersion equation for third sound. From Eq. (120) it is clear that the right-hand side depends only on the two heat fluxes J_{g} and J_{sub}, and these are connected mainly with the thermal wave modes of the gas (see Table I) and the substrate.

We also note that, from (130) and (33), we get that

$$v_{\text{g}y} = (J_{\text{M}}/\rho_{\text{g}}) + \dot{h} \ll -(\rho_{\text{h}}/\rho_{\text{g}})\dot{h}. \tag{138}$$

We also note that in the mode M_3 (the acoustic mode)

$$v_{\text{g}x3} \cong v_{\text{g}y3} \cong v_{\text{g}y}, \tag{139}$$

that in the mode M_1 (the viscous mode)

$$v_{\text{g}x1} = -v_{\text{g}x3}, \\
q_{\text{vis}} \gg q_{\text{ac}} \cong k, \tag{140}$$

and that therefore we can estimate the tangential shear force exerted by the gas on the surface of the film by

$$\eta_{\mathrm{g}} \left(\frac{\partial v_{\mathrm{g}x}}{\partial y} - \frac{\partial v_{\mathrm{g}y}}{\partial x} \right) \bigg|_{y=y_{\mathrm{r}}} \simeq \eta_{\mathrm{g}}(q_{\mathrm{vis}} - ik)v_{\mathrm{g}y} \ll \eta_{\mathrm{g}} q_{\mathrm{vis}} \frac{\rho_{\mathrm{h}}}{\rho_{\mathrm{g}}} h$$

$$\simeq \left(\frac{\omega \eta_{\mathrm{g}}}{\rho_{\mathrm{g}} c_3{}^2} \right)^{1/2} \rho_{\mathrm{h}} h \omega v_{\mathrm{s}x} . \tag{141}$$

This must be equal to the shear force exerted on the interface fom the other side

$$\eta_{\mathrm{f}}(\partial v_{\mathrm{n}x}/\partial y - \partial v_{\mathrm{n}y}/\partial x)|_{y=\mathrm{h}} . \tag{142}$$

If we rewrite Eq. (20) in the form

$$\eta_{\mathrm{f}}(\partial v_{\mathrm{n}x}/\partial y|_{y} - \partial v_{\mathrm{n}x}/\partial y|_{\mathrm{h}}) = i\rho_{\mathrm{n}} \omega(y - h)v_{\mathrm{s}x} , \tag{143}$$

we can easily see by comparing the right-hand side of (141) and (143) that, except when $y = h$, (142) is much less than the right-hand side of (143). We can therefore add (142) to the left-hand side of (143) without changing anything. This is equivalent to replacing $\partial v_{\mathrm{n}x}/\partial y$ by $\partial v_{\mathrm{n}y}/\partial x$ at $y = h$, as we did in (21) without the proper justification.

With the knowledge we now have, we can set up a simplified system of equations which will have (120) as their exact dispersion equation: These are Eqs. (109) and (110) and $\mu_{\mathrm{f}} = \mu_{\mathrm{g}}$. In these equations we set

$$C_{\mathrm{h}} = \kappa_{\mathrm{f}} = J_{\mathrm{M}} = J_{\mathrm{g}} = T_{\mathrm{g}3} = 0, \tag{144}$$

$$T_{\mathrm{g}2} = T_{\mathrm{g}}' = T_{\mathrm{f}}' . \tag{145}$$

We thus get the following system

$$\frac{h'}{h} - \left(1 + \frac{T\bar{S}}{L} \right) v - \frac{T}{h\rho_{\mathrm{h}} L} \left[\frac{iB}{\omega} + e^{i\pi/4} \left(\frac{\rho_{\mathrm{g}} C_{\mathrm{p}} \kappa_{\mathrm{g}}}{\omega} \right)^{1/2} \right] \frac{T_{\mathrm{f}}'}{T} = 0, \tag{146}$$

$$\frac{hf\bar{\rho}_{\mathrm{s}}}{c_3{}^2\rho_{\mathrm{h}}} \frac{h'}{h} - v - \frac{T\bar{S}\,\bar{\rho}_{\mathrm{s}}}{c_3{}^2\,\rho_{\mathrm{h}}} \frac{T_{\mathrm{f}}'}{T} = 0, \tag{147}$$

$$fh' + L\frac{T_{\mathrm{f}}'}{T} = 0. \tag{148}$$

The first of these equations, (146), depends on the nonreflecting boundary conditions we have assumed to hold at the far ends of both the substrate and the gas. Since we will later apply this theory to situations where the gas and substrate are not infinite and where there are reflections, we rewrite it in a more general form that follows directly from Eq. (48) if we make the assumptions (144) and (145):

$$\frac{h'}{h} - \left(1 + \frac{T\bar{S}}{L} \right) v + \frac{J_{\mathrm{g}} + J_{\mathrm{sub}}}{ih\omega\rho_{\mathrm{h}} L} = 0, \tag{149}$$

where

$$J_g \equiv -\kappa_g \, \partial T_g/\partial y\big|_{y=y_r} \tag{150}$$

and

$$J_{sub} \equiv \kappa_{sub} \, \partial T_{sub}/\partial y\big|_{y=0} \tag{151}$$

have to be calculated taking into account only the thermal modes of the gas and substrate which must however satisfy the correct boundary conditions.

Note added in proof: It has been pointed out by Scholtz *et al.* (1974) that for the thinnest films used in experiments (i.e., films approaching a total thickness of two atomic layers) the third-sound velocity is close enough to the first-sound velocity in liquid helium so that compressibility corrections become non-negligible. Under these conditions the value we obtained for $1/c_3^2$ should be multiplied by a factor

$$1 + hf \frac{\bar{\rho}_f}{\rho_h}\left(\frac{\partial \bar{\rho}_f}{\partial P}\right)_T \simeq 1 + \frac{\bar{\rho}_f}{\bar{\rho}_s}\frac{u_{30}^2}{c_1^2}$$

This lowers the calculated velocity u_3 by as much as 8% for the thinnest films and considerably improves the agreement with experiment.

C. THICK FILMS

For the thick film case, the amplitudes of oscillating physical quantities are related to each other as follows:

$$\frac{v_{sz}}{c_3} = \frac{\rho_h}{\bar{\rho}_s}\left(1+\frac{T\bar{S}}{L}\right)^{-1}\frac{h'}{h}, \tag{152}$$

$$\frac{T_f'}{T} = -\frac{hf}{L}\frac{1}{J_{10}}\frac{h'}{h} \tag{153}$$

$$\frac{T_g' - T_f'}{T} = \frac{i\omega h'\rho_h}{16A}\frac{T\bar{S}}{L}\frac{1-\rho_g/\rho_h}{1+T\bar{S}/L}, \tag{154}$$

$$\frac{T_{g3}}{T} = \frac{h'\omega c_3}{TC_pJ_1}\frac{\rho_h}{\rho_g}\frac{1}{1+T\bar{S}/L}\left[\frac{T\bar{S}}{L}-\frac{\rho_g}{\rho_h}\left(1+\frac{T\bar{S}}{L}\right)\right], \tag{155}$$

$$J_M = ih'\omega\rho_h(T\bar{S}/L)/(1+T\bar{S}/L), \tag{156}$$

$$v_{gy} = ih'\omega\frac{\rho_h}{\rho_g}\left(\frac{T\bar{S}/L}{1+T\bar{S}/L}-\frac{\rho_g}{\rho_h}\right) \ll -\frac{\rho_h}{\rho_g}h, \tag{157}$$

$$J_g = \frac{\kappa_g\omega T_f'}{c_{02}}\left[1+\frac{i\omega\rho_h}{16Af}\frac{T\bar{S}}{}J_{10}\left(1-\frac{\rho_g}{\rho_h}\right)\left(1+\frac{T\bar{S}}{L}\right)^{-1}\right], \tag{158}$$

$$\mu_f - \mu_g = (S_g - \tfrac{9}{2}k_B/m)(T_g' - T_f'), \tag{159}$$

$$\mu_f' = h'f\left(1+\frac{T\bar{S}}{L}J_{10}\right), \tag{160}$$

$$T'_{sub} - T_f' = -(B/B_1)T_f', \tag{161}$$

where

$$J_{10} \equiv \left[1 - \frac{9}{32}\frac{k_B T}{m}\frac{i\omega\rho_h}{Af}\frac{T\bar{S}}{L}\left(1 - \frac{\rho_g}{\rho_h}\right)\left(1 + \frac{T\bar{S}}{L}\right)^{-1}\right]^{-1}, \qquad (162)$$

$$J_1 \equiv 1 - \frac{c_{01}c_{03}}{c_3^2} - \frac{i\kappa_g\,\omega}{\rho_g C_p^2 T}\left(\frac{c_{03}}{c_{02}} - \frac{c_{01}c_{03}}{c_3^2}\right). \qquad (163)$$

Some consequences of these equations are:

(a) $T_g' - T_f'$ is not small compared to T_f'. All three quantities T_f', T_g', and $T_g' - T_f'$ are comparable.

(b) $T_{g3} \ll T_g'$ so that again most of the temperature variations in the gas are due to the thermal mode.

(c) $\mu_g' - \mu_f'$ is comparable to μ_f'.

(d) In the expression (82) for J_M the contributions of $\Delta\mu \equiv \mu_g - \mu_f$ and of $\Delta T \equiv T_g - \bar{T}_f$ are comparable to the final result.

(e) In the expression (83) for J_g the contributions of $\Delta\mu$ and ΔT separately are much greater than the final result. These two terms in J_g thus nearly cancel each other.

(f) Because of (157), we can repeat the argument following Eq. (138) and again justify the assumption that was made in Eq. (21).

We see that in many respects the situation is quite different from the one prevailing in the thin film regime, where μ_g, T_g followed μ_f, $\bar{T}_f$ very closely.

From the dispersion equation (121) it is clear that none of the modes excited in the gas makes a significant contribution to the properties of third sound in thick films. The fact that the parameter A appears in (121) indicates that one or both of the transport processes through the film–gas interface must be important.

A simplified system of equations which has (121) as their exact dispersion equation is obtained from (32), (46) [or (48)], (61), (86), and (87) by setting

$$C_h = \kappa_f = J_g = J_{sub} = T_{g3} = 0. \qquad (164)$$

We thus get the following equations:

$$\rho_h \dot{h} + h\bar{\rho}_s(\partial v_{sx}/\partial x) + J_M = 0, \qquad (165)$$

$$-h\bar{\rho}_s T\bar{S}(\partial v_{sx}/\partial x) + LJ_M = 0, \qquad (166a)$$

or

$$\rho_h \dot{h} + h\bar{\rho}_s[1 + (T\bar{S}/L)]\,\partial v_{sx}/\partial x = 0, \qquad (166b)$$

$$\dot{v}_{sx} - \bar{S}(\partial T_f/\partial x) + f(\partial h/\partial x) = 0, \qquad (167)$$

$$J_M = \frac{4mA}{k_B T}\left(1 - \frac{\rho_g}{\rho_h}\right)^{-1}\left(\frac{L}{T}T_f' + \frac{k_B}{2m}\frac{T_g' - T_f'}{T} + fh'\right), \qquad (168)$$

$$0 = J_g = -A[(L/T)T_f' + \tfrac{9}{2}(k_B/m)(T_g' - T_f') + fh']. \qquad (169)$$

From these equations it is clear that J_M, i.e. the evaporation and condensation of helium, is the main cause of third sound attenuation in thick films. This was first realized by Atkins (1959).

To conclude this section, we would like to point out that for the thin film case, the detailed expressions (82) and (83) that we derived for J_M and J_g were unimportant, as they only served in practice to determine that $\bar{T}_g = \bar{T}_f$ and $\mu_g = \mu_f$. But in the thick film case the detailed expressions are important, since the fact that $J_g = 0$ determines a certain ratio between $\Delta\mu$ and ΔT. This, in turn, determines a definite expression for J_M, on which the dispersion equation depends.

That J_g must vanish in all cases can also be seen from the following consideration: if we try to calculate the contribution to J_g of only one of the four terms, say the third one, appearing in (83), we find, in order of magnitude,

$$\rho_g c k_B T_f'/m. \tag{170}$$

Calculating J_g from (44), assuming that only the thermal mode contributes, we get, in order of magnitude,

$$(\omega \kappa_g \rho_g C_p)^{1/2} T_g'. \tag{171}$$

The ratio of the total flux in (171) to the partial flux in (170) is of the order

$$(\kappa_g \omega / \rho_g C_p c^2)^{1/2},$$

which is always much less than 1 [see (102)]. Hence the conclusion is that either $\Delta T = \Delta\mu = 0$, as in the thin films, or else ΔT and $\Delta\mu$ cancel each other in the expression for J_g, as in the thick films.

IV. Experiments on Third Sound

Experiments on third sound in flat films have been performed using both saturated and unsaturated films. Both types of film are formed on a flat solid substrate in equilibrium with helium gas.

For saturated films the experiment is performed a small distance H above the surface of a liquid helium bath. Therefore the pressure of the gas is very close to saturated vapor pressure, the difference being due only to gravitational effects. Equilibrium is usually maintained by having the lower part of the substrate immersed in the liquid so that there is direct contact between the film and the bulk liquid. The thickness of the film is determined by the substrate–helium atom potential at the film–gas interface $\phi(h)$. This potential is usually assumed to have the form

$$\phi(h) = -\alpha/h^n, \tag{172}$$

where n is either 3 or 4, and α is a positive constant. The potential is due in part to a sum of all the Van-der-Waals interactions of atoms of the substrate with an atom of helium at a distance h, minus the sum of the interactions that the helium atom would have with a bulk liquid helium bath replacing the substrate. If this were the only effect, it would lead to a $1/h^3$ potential. In practice, $\phi(h)$ is determined from experimental measurements of adsorption isotherms of helium (e.g., see McCormick *et al.*, 1968; Anderson and Sabisky, 1970a,b). The sum of this potential and of the gravitational potential must vanish, leading to

$$\alpha/h^n = gH, \tag{173}$$

where g is the acceleration of gravity. In practice, the film thicknesses obtained in this way range from 500 Å (140 atomic layers) at $H = 1$ cm to 250 Å (70 atomic layers) at $H = 10$ cm.

Unsaturated films are formed when the substrate is in equilibrium with helium gas that is well below its saturation pressure. There is no bulk liquid in the neighborhood of the film. In this case the potential of Eq. (172) must be cancelled by the additional chemical potential of the gas over and above that of the saturated gas. This leads to the following equation to determine the thickness

$$\alpha/h^n = (k_B T/m) \log[P_V(T)/P], \tag{174}$$

where the right-hand side was calculated assuming that the helium gas is a classical ideal gas. In practice the thicknesses obtained in this way range from about 180 Å (50 atomic layers) down to about 14 Å (4 atomic layers) and even less, depending on the temperature. For very thin films or very low temperatures it eventually becomes impossible to control the film thickness by means of the pressure because even the saturation pressure becomes too small. One then has to resort to a method where one bleeds a known amount of helium into the system whose total surface area is known.

We would like to point out that the classification of films into either saturated or unsaturated films is not the same as their classification into either thick or thin films. The latter classification differentiates between regimes where different physical processes determine the properties of third sound. The former classification differentiates between regimes where different experimental methods have to be used to create the films and to make measurements. In fact, as we can see from Fig. 4, the regime of unsaturated films includes part of the regime of thick films in addition to the entire regime of thin films.

The saturated film experiments have mostly been done by Atkins and coworkers (Everitt *et al.*, 1962, 1964). They used a periodically interrupted infrared beam shining on a small segment of the film to excite the third sound waves. Detection of the waves was achieved by an optical method that could sense the small periodic variations in the film thickness. The velocity u_3 was measured as a function of frequency, temperature, and H and was found to be independent of frequency, in agreement with theory. The dependence on H and T can be obtained by noting that

$$hf = h(\partial/\partial h)(-\alpha/h^n) = n\alpha/h^n = ngH. \tag{175}$$

Substituting this into the approximate formula

$$u_3{}^2 \cong u_{30}^2 \equiv hf(\bar{\rho}_s/\rho_h)[1 + (T\bar{S}/L)]^2, \tag{176}$$

and noting that for saturated films $\bar{\rho}_s = \rho_{s,\text{bulk}}$ and $\bar{S} = S_{\text{bulk}}$, we get

$$u_3{}^2 = ngH(\rho_{s,\text{bulk}}/\rho_h)[1 + (TS_{\text{bulk}}/L)]^2. \tag{177}$$

The results indicate that $u_3{}^2 \sim H$ except for $H < 0.3$ cm, and that n is between 3 and 4 (see Fig. 5).

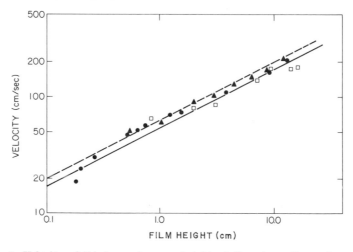

Fig. 5. Velocity of third sound versus height of film above the surface of bulk helium; ▲, highly polished stainless steel; ●, roughly polished stainless steel; □, nickel; ——theoretical curve with $n = 3$; – – – theoretical curve with $n = 4$. From Everitt *et al.* (1964), by permission of the American Institute of Physics and *Physical Review*.

The temperature dependence of (177) is mostly due to $\rho_{s,\text{bulk}}$. This dependence is in approximate though not exact agreement with experiment (see Fig. 6). The reason for the discrepancy is not clear. One possibility is that due to the experimental method of exciting the waves, the film is not at its equilibrium thickness. Another possibility is that the exponent n depends on the temperature. More experiments need to be done to clarify this.

The same group also measured the attenuation coefficient of third sound as a function of frequency at $T = 1.2°\text{K}$ and $H = 9$ cm (see Fig. 7). The measurements, though not accurate nor reproducible from day to day, seem to show an attenuation that is two or three orders of magnitude greater than the theoretical predictions. Despite the poor quality of the experimental results, it seems quite clear that the theory is inadequate here.

Experiments on unsaturated films have been done mostly by Rudnick and co-workers (Rudnick *et al.*, 1968; Kagiwada *et al.*, 1969; Rudnick and Fraser, 1970; Fraser, 1969). They utilized thin strips of superconducting aluminum deposited upon the flat substrate as both transmitters and detectors

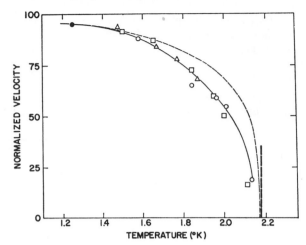

FIG. 6. Normalized third sound velocity versus temperature for different film heights: ○, 0.44 cm; □, 12.3 cm; △, 13 cm; ●, normalization point; – – – proportional to $(\rho_{s,\,bulk}/\rho)^{1/2}\ (1 + TS_{bulk}/L)^{1/2}$. From Everitt *et al.* (1964), by permission of the American Institute of Physics and *Physical Review*.

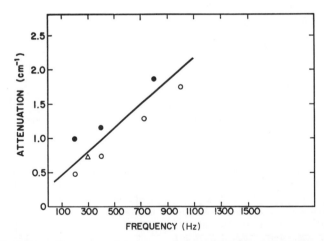

FIG. 7. Attenuation versus frequency at a constant film height $H = 9$ cm. ○, △, and ● represent three different experiments performed on three different days. From Everitt *et al.* (1964), by permisssion of the American Institute of Physics and *Physical Review*.

for third sound pulses. One strip was operated as a fast response electric heater in order to emit a third sound pulse. The other strip was operated near its superconducting transition as a very sensitive resistance thermometer in order to pick up the temperature oscillations of third sound. The velocity was measured by recording the time of flight of a third sound pulse.

Since the film thickness was controlled by varying the pressure of the helium gas, results were obtained as a function of P, T, and to a certain extent also frequency. Using (174) and (175) to substitute into (176), we get

$$u_3{}^2 \cong n(\bar{\rho}_s/\rho_h)[1 + (T\bar{S}/L)]^2(k_B T/m) \log[P_V(T)/P]. \qquad (178)$$

In Fig. 8 we show experimental results and a theoretical curve for u_3 versus P at a fixed temperature. The theoretical curve was calculated from (178)

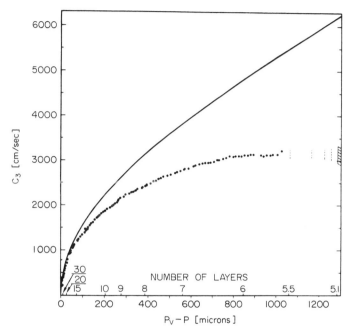

FIG. 8. The velocity of third sound u_3 plotted versus the pressure P_g of the helium gas in contact with the adsorbed film. P_V is the saturated vapor pressure at this temperature. The broken vertical lines indicate that third sound signals were visible but were too small to be measured accurately. The hatched area at $P_V - P_g = 1290\ \mu$ indicates where all signals disappeared. The full curve is based on Eq. (178) with $\bar{\rho}_s = \rho_{s,\,\text{bulk}}$, $\rho_h = \rho_{\text{bulk}}$, $T\bar{S}/L = 0$. From Rudnick *et al.* (1968), by permission of the American Institute of Physics and *Physical Review Letters*. See note in caption to Fig. 3.

assuming $\bar{\rho}_s = \rho_{s,\text{bulk}}$. Three things are at once evident: (a) For films thicker than about 15 layers the agreement of theory with experiment is very good. (b) For thinner films, a large discrepancy quickly develops—the theoretical values are too high. (c) For sufficiently thin films, third sound vanishes altogether. The large discrepancy was interpreted as being due to a size effect that makes $\bar{\rho}_s$ depend on h (and therefore on P) for sufficiently thin films. This is to be expected since, if h is too small, we know that superfluidity disappears completely, i.e. $\bar{\rho}_s = 0$. This suppression of $\bar{\rho}_s$ as well as the

onset of superfluidity cannot be understood merely on the basis of hydrodynamics. We will return to a discussion of these phenomena in Section X.

The attenuation was also measured for this regime by using the same
basic arrangement of a transmitting strip and a detecting strip deposited on a
cylinder rather than on a flat substrate (Wang and Rudnick, 1972). In this
way they could observe and measure the decaying amplitude of the third
sound pulse as it made its way several times around the cylinder. In Fig. 9 we

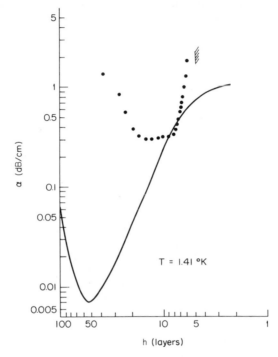

FIG. 9. The coefficient of attenuation α versus the film thickness h for $T = 1.41°$K.
The points are the measured attenuation of third sound pulses. The line is calculated from
Eq. (114). The vertical hatching marks the onset thickness of third sound propagation.
The frequency of maximum spectral intensity for the pulses is 230 Hz, and this was used
in calculating the line. From Wang and Rudnick (1972), by permission of Plenum
Publ. Co. See note in caption to Fig. 3.

see a comparison of the experimental results and the theoretical prediction for
a typical case. There is a rather small intermediate range of thicknesses
where the two agree, while everywhere else, both above and below, the experimental attenuation is greater, sometimes by as much as two or three orders
of magnitude. The fact that it is never significantly less than the theoretical
result is encouraging. We interpret it to mean that the hydrodynamic sources
of attenuation are always there. But in many of the regions there are apparently other sources of attenuation that have to be taken into account. These

must, in our opinion, be looked for outside the realm of continuum hydrodynamics, in the framework of which we have tried to take into account all
possible processes. In particular, it would not surprise us if for sufficiently
thin films, continuum theory broke down and one needed to develop a microscopic theory. It is, however, surprising that there is a large (in fact the
largest) discrepancy in the thick film region. We will return to a discussion of
this in Section X.

As for the frequency dependence, that has been checked close to the
experimental attenuation minimum, where we expect the hydrodynamic
attenuation to dominate, and found to obey a $\omega^{1/2}$ law (Wang and Rudnick,
1972) as predicted by the theory for thin films.

Goodstein and Saffman (1971) claim to have found a new mechanism for
attenuation of third sound which gives values for α that have the right
order of magnitude for thick films. Closer examination proves, however, that
their mechanism is not new but had been included in the detailed theory
developed by Bergman (1969, 1971) and described in Section II. The way they
obtained their result was by arbitrarily selecting one of the terms that
comprise the expression for J_M [see Eq. (86)], and discarding all the rest. In
fact, however, some of the other terms are just as important and tend to
cancel the term that was singled out. Their results are therefore, unfortunately, incorrect, as has been pointed out by Bergman (1973).

V. The Surface Roughness of the Substrate

When one wishes to apply the theory to experimental situations, one should
be concerned about one of the idealizations which has been made from the
beginning: We have assumed all along that the substrate, as well as the film,
are ideally flat whereas, in fact, we know that even the best polished surfaces
have irregularities when viewed on a microscopic scale. In most physical
situations one can ignore such irregularities if their scale of size is much
smaller than the size of the relevant physical phenomena. In our case, the
size of the irregularities is certainly small compared to the wavelength of
third sound, but it is not small compared to the film thickness. Electron
micrographs of polished surfaces of optical glass show a fairly jagged structure, with peaks of a few hundred angstroms jutting up occasionally.

These peaks and irregularities in the substrate will have a complicated
influence on the helium film: To a certain extent the film will simply follow
the contour of the more gradual variations. The steeper peaks, however, will
have less helium coating them, whereas the steeper crevices will have more
than the average thickness of helium coating them. This will have two main
effects: (a) Some of the third sound wave will be scattered and this will
appear as additional attenuation of the wave. (b) The effective optical path
between the emitter and the detector will be increased, reducing somewhat
the measured velocity.

The first of these effects can be estimated by noting that a single circular
obstruction of radius a exhibits the following cross section for the scattering

of long wavelength sound in two dimensions (e.g., see Morse and Feshbach, 1953, pp. 1377 and 1382)

$$\sigma = \tfrac{3}{4}\pi^2 a(ka)^3; \qquad ka \ll 1. \tag{179}$$

The contribution of this cross section to the attenuation coefficient is at most equal to $N_A \sigma$, where N_A is the total number of scatterers per unit area (this occurs when they scatter independently). N_A is at most equal to $(\pi a^2)^{-1}$; hence an upper bound to this contribution is given by

$$(3\pi/4)(ka)^3/a \simeq 0.5 \times 10^{-3} \text{ cm}^{-1}$$

for $a = 100$ Å, $k = (2\pi/10^{-2})\text{cm}^{-1}$, which is an extremely bad case. This is much less than the experimentally observed attenuation (see Fig. 9).

The second effect can be analyzed by looking at a vertical section of the film (see Fig. 10) and noting that a portion that has a slope of θ contributes an

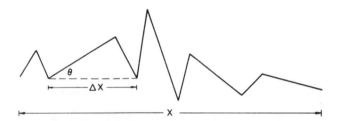

FIG. 10. Vertical section of the film showing in a schematic way the jagged peaks and crevices that the film must climb over. The effective optical path between the end-points is increased above x because each segment Δx making an angle θ with the plane of the film contributes an amount $\Delta x/\cos\theta$ to the path.

amount $\Delta x/\cos\theta$ to the optical path between the two endpoints. A path x which contains many sections will thus be effectively increased by the average of $1/\cos\theta$. This average, while always greater than 1, can be very close to it if the large angle segments are rare. In that case, we may write

$$\langle 1/\cos\theta \rangle \simeq 1 + \tfrac{1}{2}\langle\theta\rangle^2. \tag{180}$$

Such an effect would cause the measured velocity u_3 to appear smaller than it is in a flat film. One has to consider the possibility that this is the cause of the discrepancy in Fig. 8.

In order to do this, we reproduce a graph taken out of Atkins and Rudnick (1970) where u_3 is plotted versus h for two different temperatures (see Fig. 11). If the deviation of u_3^2 from the "bulk" value $u_3^2{}_{,\text{bulk}}$, given by

$$u_3^2{}_{,\text{bulk}} \equiv hf \frac{\rho_{s,\text{bulk}}}{\rho_h}\left(1 + \frac{T\bar{S}}{L}\right)^2, \tag{181}$$

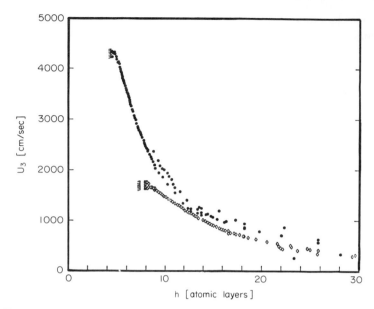

F<small>IG</small>. 11. Measurements of u_3 plotted versus h at two temperatures $T_1 = 1.18°$K($\bullet$) and $T_2 = 1.77°$K($\Diamond$). The graph exhibits the fact that the ratio $u_3(T_1, h)/u_3(T_2, h)$ begins to vary as a function of h for h smaller than 12 atomic layers. This is about the same thickness where u_3 begins to deviate from $u_{3,\text{bulk}}$ (see Fig. 8). From Atkins and Rudnick (1970), by permission of North-Holland Publ. Co. See note in caption to Fig. 3.

were entirely due to an increase in the effective optical path, i.e. a purely geometrical effect, we could take it into account by means of an appropriate h-dependent factor $G(h)$

$$u_3{}^2 = hf \frac{\rho_{\text{s,bulk}}}{\rho_{\text{h}}} \left(1 + \frac{T\bar{S}}{L} \right)^2 G(h). \tag{182}$$

From this it would follow that the ratio of velocities at two different temperatures for the same thickness h should be independent of h

$$\frac{u_3(T_1, h)}{u_3(T_2, h)} = \frac{\rho_{\text{s,bulk}}(T_1)}{\rho_{\text{s,bulk}}(T_2)} = \text{independent of } h. \tag{183}$$

From Fig. 11 it is clear that while this ratio is constant for large h, it changes quite considerably when the thickness begins to approach the onset thickness. Moreover, the ratio begins to vary at about the same thickness where u_3 begins to deviate from $u_{3,\text{bulk}}$ (see Fig. 8). Hence we may conclude that the deviations are not primarily due to optical path effects.

To summarize this section, we conclude that surface roughness of the substrate does not appear to have any observable effects on the velocity or attenuation of third sound. The idealization of a geometrically flat film is justified. One must look elsewhere for an explanation of the discrepancies shown in Figs. 8 and 9.

VI. Third Sound Resonators

The theory described in detail in Section II for third sound in a flat film can be extended to other configurations. We now do this for third sound resonators operating in the thin film regime. Third sound resonators have been developed by Ratnam and Mochel (1970a,b).

Such a resonator is usually made up of two thin quartz or Pyrex plates that are spaced a certain distance apart from each other. The volume between them is then sealed off hermetically by fusing the plates together at the edges. All this is performed in an atmosphere of argon, some of which is thus sealed within the cell. A certain amount of helium is then admitted into the cell by placing it in a room temperature helium atmosphere at a filling pressure, P_{fill}, and allowing helium to diffuse through the walls. Subsequent experiments are performed at low temperatures, where diffusion through the walls is negligible, so that the total amount of helium in the cell is fixed. When the system is cooled down from room temperature, the argon is completely adsorbed on the walls in the form of a solid film long before any helium begins to be deposited. The solid argon coating thus forms the substrate for the liquid helium film. Third sound is generated by periodic heating of the helium film by means of a carbon strip resistor placed at one end of the cell, and is detected by means of another carbon strip resistor operated as a thermometer (see Fig. 12). It is thus possible to measure the response of the resonator at

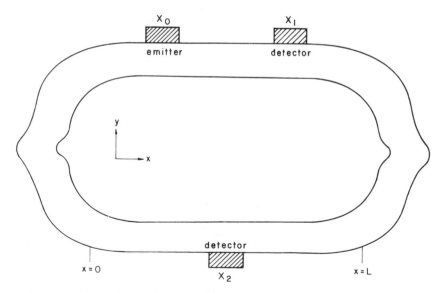

Fig. 12. Schematic drawing of a vertical section of a third sound resonator of length L obtained by sealing the edges of a double film. Also shown are the emitter and detector of third sound, placed at the points x_0 and x_1, respectively, on the wall of the cell. It is also possible to place the detector on the other wall of the cell, at x_2.

different frequencies. Various resonances were found and their properties investigated (see Ratnam and Mochel, 1970a,b, 1974).

As a prelude to considering the properties of such a resonator, we will discuss the properties of third sound in a pair of thin, infinite, parallel liquid helium films, deposited on a pair of identical parallel substrates of thickness h_{sub}, and separated by helium gas of thickness h_g (see Fig. 13).

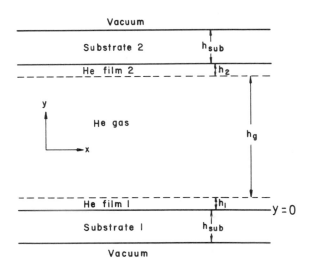

Fig. 13. Schematic drawing of a vertical section of a helium double film: Two flat, parallel substrates, separated by helium gas and coated on the inside by an adsorbed liquid helium film.

Both h_g and h_{sub} are not assumed to be large compared to the appropriate (i.e. the thermal) penetration lengths, so that reflections will have to be taken into account. Outside the two substrates we assume that there is a vacuum. This immediately fixes one boundary condition, namely, that no heat flows from the substrate into the vacuum (radiation can be shown to be negligible), i.e.

$$\partial T_{sub}/\partial y = 0 \text{ at substrate vacuum interfaces.} \tag{184}$$

In order to write down the equations of motion for this system, we take advantage of the insight gained from our detailed treatment of the single, infinite, thin film. We will thus, from the beginning, use the simplified system of equations (147)–(151). In order to calculate J_{sub} we note that T_{sub} is now a combination of the two modes

$$\exp(-i\omega t + ikx + q_{sub} y), \tag{185}$$

$$\exp(-i\omega t + ikx - q_{sub} y). \tag{186}$$

From Eq. (184) we find that the right combination for the lower film in Fig. 13 is

$$T_{\text{sub}} = T + T'_{\text{sub}} e^{-i\omega t + ikx} 2 \cosh[q_{\text{sub}}(y + h_{\text{sub}})].$$ (187)

Using (63), we can now express J_{sub} in terms of T'_{f} alone

$$J_{\text{sub}} = BT'_{\text{f}},$$ (188)

where

$$1/B \equiv (1/B_1) + (\kappa_{\text{sub}} q_{\text{sub}} \tanh q_{\text{sub}} h_{\text{sub}})^{-1}.$$ (189)

The calculation of J_{g} is, as usual, more complicated. T_{g} is also a combination of two waves similar to (185) and (186). But there is no boundary condition this time to determine the combination. We must therefore leave it in the general form

$$T_{\text{g}} = T + T'_{\text{g1}} \exp(-i\omega t + ikx - q_{\text{th}} y) + T'_{\text{g2}} \exp[-i\omega t + ikx + q_{\text{th}}(y - h_{\text{g}})].$$ (190)

The actual combination will only be known when we have solved the equations. There are now eight variables

$$T'_{\text{f1}}, \; T'_{\text{f2}}, \; h_1', \; h_2', \; v_1, \; v_2, \; T'_{\text{g1}}, \; T'_{\text{g2}},$$ (191)

where the subscripts 1 and 2 added to T_{f}', h', and v, refer to the two films (see Fig. 13). The eight equations are

$$fh_i' + L(T'_{\text{f}i}/T) = 0,$$ (192)

$$\frac{hf\bar{\rho}_{\text{s}}}{c_3^2 \rho_{\text{h}}} \frac{h_i'}{h} - v_i - \frac{T\bar{S}}{c_3^2} \frac{\bar{\rho}_{\text{s}}}{\rho_{\text{h}}} \frac{T'_{\text{f}i}}{T} = 0,$$ (193)

where $i = 1, 2$, and

$$\frac{h_1'}{h} - \left(1 + \frac{T\bar{S}}{L}\right)v_1 - \frac{i\kappa_{\text{g}} q_{\text{th}}}{h\omega\rho_{\text{h}} L}[T'_{\text{g1}} - T'_{\text{g2}} \exp(-q_{\text{th}} h_{\text{g}}) - \frac{iBT'_{\text{f1}}}{h\omega\rho_{\text{h}} L} = 0,$$ (194)

$$\frac{h_2'}{h} - \left(1 + \frac{T\bar{S}}{L}\right)v_2 - \frac{i\kappa_{\text{g}} q_{\text{th}}}{h\omega\rho_{\text{h}} L}[T'_{\text{g2}} - T'_{\text{g1}} \exp(-q_{\text{th}} h_{\text{g}})] - \frac{iBT'_{\text{f2}}}{h\omega\rho_{\text{h}} L} = 0,$$ (195)

$$T'_{\text{f1}} = T'_{\text{g1}} + T'_{\text{g2}} \exp(-q_{\text{th}} h_{\text{g}}),$$ (196)

$$T'_{\text{f2}} = T'_{\text{g2}} + T'_{\text{g1}} \exp(-q_{\text{th}} h_{\text{g}}).$$ (197)

These coupled equations can be separated into two uncoupled sets of four equations each for the two sets of variables

$$h_1' + h_2', \; T'_{\text{f1}} + T'_{\text{f2}}, \; v_1 + v_2, \; T'_{\text{g1}} + T'_{\text{g2}},$$ (198)

$$h_1' - h_2', \; T'_{\text{f1}} - T'_{\text{f2}}, \; v_1 - v_2, \; T'_{\text{g1}} - T'_{\text{g2}}.$$ (199)

The dispersion equations for the two sets are

$$\frac{u_{30}^2}{c_3{}^2} = 1 + \frac{Tf}{\rho_{\mathrm{h}} L^2} \left[\frac{iB}{\omega} + \frac{i\kappa_{\mathrm{g}} q_{\mathrm{th}}}{\omega} \tanh \frac{q_{\mathrm{th}} h_{\mathrm{g}}}{2} \right] \tag{200}$$

and

$$\frac{u_{30}^2}{c_3{}^2} = 1 + \frac{Tf}{\rho_{\mathrm{h}} L^2} \left[\frac{iB}{\omega} + \frac{i\kappa_{\mathrm{g}} q_{\mathrm{th}}}{\omega} \coth \frac{q_{\mathrm{th}} h_{\mathrm{g}}}{2} \right], \tag{201}$$

respectively.

For $q_{\mathrm{th}} h_{\mathrm{g}} \gg 1$ and $q_{\mathrm{sub}} h_{\mathrm{sub}} \gg 1$, both of these revert to the equation we got previously for a single thin film with infinitely thick gas and substrate [see Eq. (120)]. For $q_{\mathrm{th}} h_{\mathrm{g}} \ll 1$, the mode described by Eq. (201) is very strongly damped. The other mode, by contrast, will have the attenuation due to the gas greatly diminished, because

$$\frac{i\kappa_{\mathrm{g}} q_{\mathrm{th}}}{\omega} \tanh \frac{q_{\mathrm{th}} h_{\mathrm{g}}}{2} \simeq \frac{i\kappa_{\mathrm{g}} q_{\mathrm{th}}{}^2 h_{\mathrm{g}}}{2\omega} \simeq \text{a real number}, \tag{202}$$

i.e. the lowest order in h_{g} gives almost no attenuation. This will then be the mode that is observed experimentally. At intermediate values of $q_{\mathrm{th}} h_{\mathrm{g}}$, both will be present with nearly the same velocities but different attenuations, and interference effects may occur.

In a resonator these modes will first propagate, then get reflected at the edges. A complicated interference pattern will usually appear, depending on the precise geometry of the resonator. We will assume a very long resonator, and that only longitudinal modes are excited in it. This is effectively a one-dimensional resonator. In it, each of the modes (200), (201) appears twice—as a wave moving either in the positive (e.g. $v_1{}^+$) or in the negative (e.g. $v_1{}^-$) x-direction. The boundary condition at the edges of the resonator is obtained by noting that when the superfluid flow v_{s} in one film reaches the end of the resonator, it simply turns around the edge (which is, of course, also covered by the helium film) and reappears as a superfluid flow in the opposite direction in the other film. We thus obtain the following conditions at the edges of the resonator

$$v_{s1}^+ = -v_{s2}^- \quad \text{and} \quad v_{s2}^+ = -v_{s1}^- . \tag{203a}$$

Using Eqs. (148), (61) [(we cannot use (147) instead of (61) because (147) is only correct for a wave traveling in the positive x-direction], and (145) we can translate these into boundary conditions for either T_{f}', h', or T_{g}'. All of these quantities have the same behavior at the edge, exemplified by

$$T_{\mathrm{f}1}^+ = T_{\mathrm{f}2}^- \quad \text{and} \quad T_{\mathrm{f}2}^+ = T_{\mathrm{f}1}^- . \tag{203b}$$

From Eqs. (203a) and (203b) we can easily derive the manner in which each of the two propagating modes that we have found gets reflected at the edges.

We can use these boundary conditions to calculate the response of the resonator, i.e. the steady state temperature amplitude at a point x_1 that results when a periodic signal of unit amplitude in the temperature, $e^{-i\omega t}$, is inserted at a point x_0, as shown in Fig. 12. Such a calculation gives the result (details of this calculation will be published elsewhere)

$$\frac{i}{2} \frac{\cos kx_0 \cos k(L - x_1)}{\sin kL} \mp \frac{i}{2} \frac{\sin k'x_0 \sin k'(L - x_1)}{\sin k' L}, \tag{204}$$

where k is the complex wave vector resulting from (200), while k' is the complex wave vector resulting from (201). The upper sign in (204) refers to the case where both the emitter and the detector are on the same plate of the resonator, while the lower sign refers to the case where they are on opposite plates (see Fig. 12: in the second case we have denoted the position of the detector by x_2 in the figure). The first term of (204) clearly comes from the symmetric mode of (198), while the second term comes from the antisymmetric mode of (199). As we noted above, there will in general be interference effects between the two terms. The second term will, however, be suppressed if we place either the emitter or the detector at the edge of the cell, so that either $x_0 = 0$ or $x_1 = L$. The second term will also be less important when $q_{th} h_g \ll 1$, since the antisymmetric mode is then strongly damped, while the symmetric mode has its damping reduced.

We therefore focus our attention on the first term in (204). This term will have resonances when the denominator is very small. If we consider only the denominator (the other parts are slowly varying functions of k) we can write

$$1/\sin kL = Ae^{-i\phi}, \tag{204a}$$

where

$$A \equiv |1/\sin kL| = [\sin^2(\text{Re } kL) + \sinh^2(\text{Im } kL)]^{-1/2}, \tag{205}$$

and

$$\tan \phi = \tanh(\text{Im } kL)/\tan(\text{Re } kL). \tag{206}$$

Assuming that the attenuation is small, i.e. that

$$\text{Im } kL \equiv \omega L \, \text{Im}(1/c_3) \ll 1, \tag{207}$$

we can write

$$A = \{\sin^2[\omega L \, \text{Re}(1/c_3)] + [\omega L \, \text{Im}(1/c_3)]^2\}^{-1/2} \tag{208}$$

$$\tan \phi = \omega L \, \text{Im}(1/c_3)/\tan[\omega L \, \text{Re}(1/c_3)]. \tag{209}$$

The response function obviously has resonances whenever $\omega = \omega_n$, where ω_n is given by

$$\omega_n L \, \text{Re} \, 1/c_3 = n\pi; \qquad n = 1, 2, 3, \ldots. \tag{210}$$

This condition just means that the length of the cell L must as usual be equal to an integral number of half-wavelengths of third sound. The half-maximum of these resonances will occur at $\omega = \omega_{1/2}$, given by

$$|\sin[\omega_{1/2} L \, \text{Re}(1/c_3)]| = \omega_{1/2} L \, \text{Im}(1/c_3). \tag{211}$$

Because of (207) and (210), this can be written as

$$|\omega_{1/2} - \omega_n| \, L \, \mathrm{Re}(1/c_3) = \omega_n \, L \, \mathrm{Im}(1/c_3). \tag{212}$$

The Q value of each resonance is thus given by

$$Q^{-1} \equiv 2 \, |\omega_{1/2} - \omega_n| / \omega_n = 2 \, \mathrm{Im}(1/c_3)/\mathrm{Re}(1/c_3). \tag{213}$$

Experiments on third sound resonators have been made by Ratnam and Mochel (1970a,b, 1974). By using a heater at one end to excite the resonator and a carbon thermometer at the other end to detect the temperature oscillations, they could observe a series of resonances in the response of the cell. Measurements of both the resonance frequency (see Fig. 14) and the Q values (see Fig. 15) have been made from the onset of superfluidity down to 0.3°K.

These measurements also include mixed He^3–He^4 films, which we shall discuss in Section VII. Looking for the moment only at the results for pure He^4, we can deduce from Fig. 14 that below about 0.8°K, the film thickness is constant. This also agrees with a direct calculation, which shows that only about 1% of the He^4 is in the gas phase at that temperature. It is therefore a little surprising to find in Fig. 15 that the Q-value is still strongly increasing

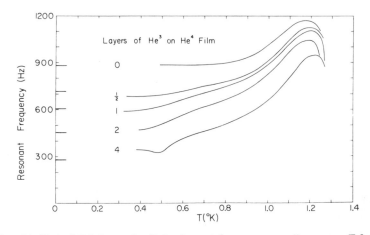

Fig. 14. Plot of third sound cell fundamental resonance $\omega_1/2\pi$ versus T for a cell containing a fixed amount of He^4 and varying amounts of He^3. The He^4 content is equivalent to a film 4.87 atomic layers thick at 0°K. Curves are plotted for amounts of He^3 equivalent to about 0, $\frac{1}{2}$, 1, 2, and 4 atomic layers at 0°K. The filling pressures at 300°K for the He^4 and He^3 were 552; 551, 50; 554, 100; 559, 199; and 557, 400 mm Hg, respectively. The length of the cell is $L = 1.79$ cm, and the third sound velocity is given by $u_3 = 3.58 \, \omega_1/2\pi$. The curves terminate at the high temperature side at the points above which no third sound is observed. (After Ratnam and Mochel, 1974, by permission of Colorado University Press.) On the ordinate axis we have marked by elongated straight lines the theoretical predictions for $\omega_1/2\pi$ at 0°K for the various curves, normalized to the observed resonance frequency of the pure He^4 film.

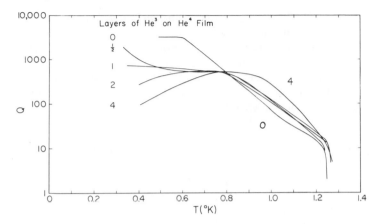

Fig. 15. Plot of third sound cell $Q \equiv \omega_1/\Delta\omega_{1/2}$, where ω_1 is the resonance frequency and $\Delta\omega_{1/2}$ the resonance width, versus T for the same experimental setup as in Fig. 14. From Ratnam and Mochel (1974), by permission of Colorado University Press.

as one goes from 0.8°K to 0.6°K, leveling off only below 0.6°K. The only physical quantity that changes appreciably over that region is the density of the gas ρ_g, which decreases by about a factor 4 from 0.8°K to 0.6°K (this is due mainly to the rapid decrease of the saturated vapor pressure). But a careful analysis of Eq. (200) shows that the terms depending on ρ_g are apparently completely negligible in this region. The attenuation, and therefore the Q value, are determined mainly by the properties of the substate.

To see this we assume that

$$q_{\mathrm{th}} h_g \ll 1 \tag{214}$$

and

$$q_{\mathrm{sub}} h_{\mathrm{sub}} \ll 1, \tag{215}$$

and then expand the hyperbolic tangents in (189) and (200) as a power series in these arguments, keeping only the first two terms. Assuming also that

$$q_{\mathrm{sub}}^2 \kappa_{\mathrm{sub}} h_{\mathrm{sub}}/B_1 \ll 1, \tag{216}$$

and using Eq. (65) for q_{sub}, we now get the following expression for the terms in the square brackets on the right-hand side of (200):

$$\frac{iB}{\omega} \cong \rho_{\mathrm{sub}} C_{\mathrm{sub}} h_{\mathrm{sub}} + \frac{i\omega\kappa_{\mathrm{sub}} h_{\mathrm{sub}}}{c_3^2} + \frac{i\omega\rho_{\mathrm{sub}}^2 C_{\mathrm{sub}}^2 h_{\mathrm{sub}}^2}{B_1} + \frac{i\omega\rho_{\mathrm{sub}}^2 C_{\mathrm{sub}}^2 h_{\mathrm{sub}}^3}{3\kappa_{\mathrm{sub}}}, \tag{217}$$

$$\frac{i\kappa_g q_{\mathrm{th}}}{\omega} \tanh \frac{q_{\mathrm{th}} h_g}{2} \cong \frac{1}{2}\rho_g C_p h_g + \frac{i\omega\rho_g h_g}{2c_3^2} + \frac{i\omega\rho_g^2 C_p^2 h_g^3}{24\kappa_g}. \tag{218}$$

The first terms of (217) and (218) are real and do not contribute to the attenuation. Of the remaining terms, only the ones in (217) could be large enough to make an important contribution to α. For a sufficiently thick substrate, a further simplification can occur in that only the last term of (217) is important. There is a range of values of h_{sub} for which this is true while nevertheless (215) and (216) still continue to hold. In that case, we would find a very strong temperature dependence for Q:

$$Q^{-1} \sim TC_{\mathrm{sub}}^2 \sim T^7. \tag{219}$$

This would lead us to expect a 7.5-fold increase in Q when we go from 0.8° to 0.6°K. In fact, an 8-fold increase has been observed (see Fig. 15).

For the experiment under discussion (Ratnam and Mochel, 1974) this simplification is not entirely justified, however, and one must really use all the terms of (217) in calculating Q. We have made such a calculation, and the results are shown in Fig. 16, together with the experimental results for comparison. There is some agreement in absolute values, though perhaps not in the dependence on temperature, between 0.5° and 0.65°K. But at higher temperatures the experimental results show an increasingly large deviation

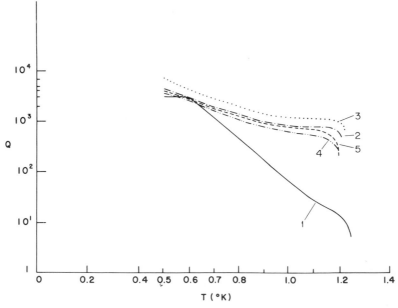

Fig. 16. Calculations of the Q value for the third sound resonator used by Ratnam and Mochel (1974). Its dimensions were $L = 1.79$ cm, $h_g = 4.26$ μm, $h_{\mathrm{sub}} = 100$ μm. P_{fill} was 552 mm Hg. Curves 4 and 5 show results for slightly different values of the substrate potential [the coefficient $\alpha m/k_B$ in Eq. (174) was taken to be 87° in 4 and 79° in 5]. Curves 2 and 3 were calculated for $\alpha m/k_B = 87°$, $h_{\mathrm{sub}} = 70\mu$, and $P_{\mathrm{fill}} = 528$ and 552 mm Hg, respectively. Curve 1 reproduces the experimental results for a pure He^4 film from Fig. 15.

from the calculation. The indication is again, as in the time-of-flight experiments, that other dissipation mechanisms must exist, not included in the hydrodynamic description of the helium film.

VII. Third Sound in Mixed He³–He⁴ Films

Recently, experiments were begun on third sound in superfluid films composed of a mixture of the two isotopes He³ and He⁴. Ratnam and Mochel (1974) have performed such experiments using the third sound resonators described in Section VI. They have measured both the resonance frequency (see Fig. 14) and the Q value (see Fig. 15) of the cavity as a function of temperature for a fixed amount of He⁴ inside the cell (about 5 atomic layers), and varying amounts of He³. The amount of He³ was varied up to almost the amount of He⁴. Downs and Kagiwada (1972) have also measured the velocity of third sound in unsaturated mixed films using the time-of-flight technique developed by Rudnick and co-workers, which was described briefly in Section IV (for a more complete description of this technique, see the review article by Atkins and Rudnick, 1970). For a fixed temperature (1.4°K), they measured u_3 both in a pure He⁴ film and in a mixture that contained a fixed percentage (about 17%) of He³ as a function of the He⁴ partial thickness (see Fig. 17).

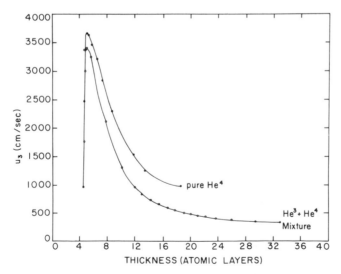

Fig. 17. Preliminary measurements of the third sound velocity u_3 for a pure He⁴ film and for a mixed He³–He⁴ film as a function of the partial thickness of He⁴. The He³ concentration in the mixed film was always around 17.4% of the total by number, and all measurements were made at $T = 1.400$°K. The full and empty circles represent measured points. The lines were drawn merely to aid the eye (Downs and Kagiwada, 1972).

VIII. Energy in Third Sound

The energy of a third sound wave is made up of three contributions: from the film, from the gas, and from the substrate.

To calculate the film contribution we start out by writing some basic thermodynamic relations for the superfluid helium film

$$dE_0 = T \, dS - P_A \, dA + \mu \, dM + (\mathbf{v}_n - \mathbf{v}_s) \cdot d\mathbf{J}_0, \tag{226}$$

$$-P_A A = E_0 - TS - \mu M - M_n (\mathbf{v}_n - \mathbf{v}_s)^2. \tag{227}$$

These are direct consequences of two-fluid hydrodynamics applied to a film (Khalatnikov, 1965). E_0 is the total energy in a frame of reference moving with the superfluid, $\mathbf{J}_0$ is the total momentum in that frame

$$\mathbf{J}_0 \equiv M_n(\mathbf{v}_n - \mathbf{v}_s), \tag{228}$$

M_n, M_s, and M are the normal, the superfluid, and the total masses, A is the area of the film, and P_A is the two-dimensional pressure acting on the film. Since we are dealing with a constant area film, we can write (226) in terms of quantities per unit area E_{0A}, S_A, M_A, J_{0A} as follows:

$$dE_{0A} = T \, dS_A + \mu \, dM_A + (\mathbf{v}_n - \mathbf{v}_s) \cdot d\mathbf{J}_{0A}. \tag{229}$$

The available energy ΔE in a nonequilibrium system (i.e. the energy that can be extracted to do work) is given by the difference between the instantaneous energy and the energy that the system would have if it were in equilibrium with its surroundings. We write this energy per unit area as follows:

$$\Delta E_A \equiv E_A(M_A, S_A, \mathbf{v}_s, \mathbf{v}_n) - E_A(\bar{M}_A, \bar{S}_A, 0, 0), \tag{230}$$

where the bar over $\bar{M}_A$, $\bar{S}_A$ signifies the ambient values of M_A, S_A. In a third sound wave the bar can signify alternatively an averaging over a single period of oscillation. The total energy E_A is related to E_{0A} by

$$E_A = E_{0A} + \mathbf{v}_s \cdot \mathbf{J}_{0A} + \tfrac{1}{2} M_A v_s^2. \tag{231}$$

Using this equation, and developing ΔE_{0A} in a Taylor series up to second order terms in $\Delta M_A \equiv M_A - \bar{M}_A$, etc., we find

$$\begin{aligned}
\Delta E_A &= T \, \Delta S_A + \mu \, \Delta M_A + \tfrac{1}{2}[\partial T/\partial S_A \, \Delta S_A + 2(\partial \mu/\partial S_A) \, \Delta S_A \, \Delta M_A \\
&\quad + (\partial \mu/\partial M_A) \, \Delta M_A^2] + (\mathbf{v}_n - \mathbf{v}_s) \cdot \mathbf{J}_{0A} + \mathbf{v}_s \cdot \mathbf{J}_{0A} + \tfrac{1}{2} M_A v_s^2 \\
&= T \, \Delta S_A + \mu \, \Delta M_A + \tfrac{1}{2}(\partial S_A/\partial T)_{MA} \, \Delta T^2 + \tfrac{1}{2}(\partial \mu/\partial M_A)_T \, \Delta M_A^2 \\
&\quad + \tfrac{1}{2} M_{nA} v_n^2 + \tfrac{1}{2} M_{sA} v_s^2.
\end{aligned} \tag{232}$$

We now calculate the average available energy by averaging ΔE_A over a single period, noting that

$$\overline{\Delta M_A} = \overline{\Delta S_A} = 0, \tag{233}$$

$$v_n \equiv 0, \tag{234}$$

$$M_{sA} \equiv h\bar{\rho}_s, \tag{235}$$

$$\Delta M_A = \Delta h(\partial M_A/\partial h)_T = \Delta h\bar{\rho}_f, \tag{236}$$

$$\left(\frac{\partial \mu}{\partial M_A}\right)_T = \left(\frac{\partial \mu}{\partial h}\right)_T \bigg/ \left(\frac{\partial M_A}{\partial h}\right)_T = \frac{f}{\bar{\rho}_f}, \tag{237}$$

$$(\partial S_A/\partial T)_M = h\bar{\rho}_f C_h/T. \tag{238}$$

We thus find

$$\overline{\Delta E_f} = \tfrac{1}{2}(h\bar{\rho}_f C_h/T)\,\overline{\Delta T_f^2} + \tfrac{1}{2}f\,\bar{\rho}_f\,\overline{\Delta h^2} + \tfrac{1}{2}h\bar{\rho}_s\,\overline{v_s^2}. \tag{239}$$

Noting that, because of the sinusoidal dependence on time, we get

$$\overline{\Delta T_f^2} = \tfrac{1}{2}T_f'^2; \text{ etc.}, \tag{240}$$

and using (126), (127), (152), and (153), we find

$$\overline{\Delta E_f} = \frac{1}{4}f\bar{\rho}_f|h'^2|\left\{\left|\frac{1}{J_{10}^2}\right|\frac{TC_h}{L}\frac{hf}{L} + 1 + \frac{\bar{\rho}_s}{\bar{\rho}_f|c_3^2|}\frac{hf}{L}\left(1 + \frac{T\bar{S}}{L}\right)^2\right\}. \tag{241}$$

In this equation the first term, which arises from $\overline{\Delta T_f^2}$, is negligible, while the two other terms are very nearly equal. Neglecting the difference between them, as well as the first term, we find for the energy per unit area of the film

$$\overline{\Delta E_f} = \tfrac{1}{2}|h'^2|\,\bar{\rho}_f f = \tfrac{1}{2}(L^2\bar{\rho}_f/f)|\,T_f'/T|^2|J_{10}^2|$$
$$= \tfrac{1}{2}h\bar{\rho}_s(\bar{\rho}_f/\rho_h)|v_{sx}^2|. \tag{242}$$

Similar considerations are applied to calculate the energy available from the helium gas. The available energy per unit volume is calculated as follows:

$$\Delta E_{gV} \equiv E_V(S_V, \rho_g, \mathbf{v}_g) - E_V(\bar{S}_V, \bar{\rho}_g)$$

$$= T\,\Delta S_V + \mu\,\Delta\rho_g + \frac{1}{2}\left(\frac{\partial T}{\partial S_V}\Delta S_V^2 + 2\frac{\partial T}{\partial\rho_g}\Delta S_V\,\Delta\rho_g + \frac{\partial\mu}{\partial\rho_g}\Delta\rho_g^2\right)$$

$$+ \frac{1}{2}\rho_g v_g^2$$

$$= T\,\Delta S_V + \mu\,\Delta\rho_g + \frac{1}{2}\left(\frac{\partial S_V}{\partial T}\right)_\rho\Delta T^2 + \frac{1}{2}\left(\frac{\partial\mu}{\partial\rho_g}\right)_T\Delta\rho_g^2 + \frac{1}{2}\rho_g v_g^2. \tag{243}$$

Averaging over one period of oscillation we find

$$\overline{\Delta E}_{gV} = \frac{\rho_g C_V}{2T}\overline{\Delta T_g^2} + \frac{1}{2\rho_g^2\kappa_T}\overline{\Delta\rho_g^2} + \frac{1}{2}\rho_g\overline{v_g^2}. \tag{244}$$

In our case, all the average quantities are in principle quadratic combinations of contributions from all three modes of the gas. Each of these contributions still has an exponential dependence on y, of the form

$$\exp[-y \, \mathrm{Re}(q_i + q_j)]. \tag{245}$$

One of the contributions, the acoustic mode, is even increasing in the y-direction though this increase is limited by the time at which the third sound wave started to propagate. Each contribution must be integrated over y to yield the energy per unit area of the film.

A careful consideration of the various contributions leads to the following, not unexpected results: In the thick film case the energy content of the gas is completely negligible. In the thin film case it is negligible as far as the acoustic and the viscous modes are concerned. The thermal mode makes the largest contribution to the energy content—this is consistent with the fact that it makes the only significant contribution to the dispersion equation. It contributes to the energy mainly through the first two terms of (244), i.e. the thermal and the potential energy terms. This contribution is obtained by substituting

$$\Delta\rho/\rho = -\Delta T/T \tag{246}$$

[see (103) and (137)] in (244), and multiplying the resulting ΔE_V by one half of the thermal length in the gas:

$$\tfrac{1}{2}l_{\mathrm{th}} = (\kappa_g/2\rho_g C_p \, \omega)^{1/2}. \tag{247}$$

We thus find

$$\overline{\Delta E}_{gA} = (5/8\sqrt{2}) \, |\, T_f'/T \,|^2 (\kappa_g \, \omega/\rho_g C_p \, c_3{}^2)^{1/2} P c_3/\omega, \tag{248}$$

where we have also used

$$\kappa_T = 1/P; \; C_v = \tfrac{3}{2}k_B/m. \tag{249}$$

The ratio of the energy in the gas to the energy in the film is, from (248) and (242), given by (note that in the thin film case $J_{10} = 1$)

$$\frac{\overline{\Delta E}_{gA}}{\overline{\Delta E}_f} = \frac{5}{4\sqrt{2}} \left(\frac{\kappa_g \, \omega}{\rho_g C_p \, c_3{}^2}\right)^{1/2} \frac{k_B \, T/m}{L} \frac{\rho_g}{\bar\rho_f} \frac{c_3}{h\omega} \frac{hf}{L} \sim \rho_g^{1/2} h^{-4} \omega^{-1/2}. \tag{250}$$

Numerically, this is still usually considerably less than 1, though it increases very fast as the film becomes thinner. For example, while it is only about 0.003 at $\omega = 10^3$, $T = 1.5°K$, and $h = 18$ atomic layers, it rises to 0.13 at $\omega = 10^4$, $T = 1.5°K$, and $h = 5.2$ layers, at which the onset of superfluidity occurs for that temperature.

For the energy of the substrate we get an expression analogous to the first term in (244):

$$\overline{\Delta E}_{\mathrm{sub,V}} = (\rho_{\mathrm{sub}} C_{\mathrm{sub}}/2T) \, \overline{\Delta T}_{\mathrm{sub}}^2. \tag{251}$$

As in the case of the gas, this only has any importance at all in the thin film case. As before, we use (134) to substitute for T_{sub} in terms of T_f, and we also have to multiply (251) by one half of the thermal diffusion length in the substrate

$$\tfrac{1}{2}l_{sub} = (\kappa_{sub}/2\rho_{sub}C_{sub}\,\omega)^{1/2} \tag{252}$$

in order to get the energy per unit area:

$$\overline{\Delta E}_{sub,A} = \frac{1}{4}\,\rho_{sub}\,TC_{sub}\left(\frac{\kappa_{sub}}{2\,\rho_{sub}C_{sub}\,\omega}\right)^{1/2}\left|1 - \frac{B}{B_1}\right|^2\left|\frac{T_f{}'}{T}\right|^2. \tag{253}$$

For an order of magnitude estimate of the ratio of this energy to the film energy we take

$$|1 - B/B_1| \cong 1 \tag{254}$$

and find

$$\frac{\overline{\Delta E}_{sub,A}}{\overline{\Delta E}_f} \simeq \frac{1}{2\sqrt{2}}\left(\frac{\kappa_{sub}}{\rho_{sub}C_{sub}\,c_3{}^2}\right)^{1/2}\frac{\rho_{sub}}{\bar{\rho}_f}\frac{TC_{sub}}{L}\frac{c_3}{\hbar\omega}\frac{hf}{L} \sim T^{3/2}h^{-4}\omega^{-1/2}. \tag{255}$$

Numerically this is again usually considerably less than 1: For $\omega = 10^3$, $T = 1.5°\text{K}$, and $h = 18$ atomic layers the ratio is about 0.004, rising to about 0.16 for $\omega = 10^4$, $T = 1.5°\text{K}$, and $h = 5.2$ layers.

We can summarize the results of this section by saying that most of the energy of a third sound wave usually resides in the height oscillations (potential energy) and the superfluid velocity oscillations (kinetic energy) of the helium film. The processes taking place in the gas and the substrate are only important because they provide the mechanisms for attenuation.

IX. The Normal Fluid Motion and Attenuation

In the previous sections we have always neglected the normal fluid motion in the film parallel to the substrate. We have indeed shown in Eq. (22) that this motion is very small due to the small thickness of the film. Nevertheless, one should still consider whether it makes any contribution to the attenuation of third sound, which is itself usually a small effect. This question has been looked into by Pollack (1966a,b). Similar considerations have also been made by Sanikidze *et al.* (1967) for the problem of fourth sound attenuation.

In discussing this problem, we begin by combining Eqs. (7) and (8) to get an equation of motion for v_{nx}

$$\rho_n\frac{\partial v_{nx}}{\partial t} = -\frac{\rho_n}{\rho_f}\frac{\partial P}{\partial x} - \rho_s S_f\frac{\partial T_f}{\partial x} + \eta_f\frac{\partial^2 v_{nx}}{\partial y^2} + \text{other dissipative terms.} \tag{256}$$

Noting that, when y is fixed, one can write either

$$d\mu = -S_f \, dT + (1/\rho_f) \, dP \tag{257}$$

or

$$d\mu = -\bar{S} \, dT + f \, dh, \tag{258}$$

we can rewrite Eq. (256) in the form

$$\frac{\partial v_{nx}}{\partial t} = -f \frac{\partial h}{\partial x} + \left[\bar{S} - S_f \left(1 + \frac{\rho_s}{\rho_n} \right) \right] \frac{\partial T_f}{\partial x} + \frac{\eta_f}{\rho_n} \frac{\partial^2 v_{nx}}{\partial y^2} + \text{other dissipative terms.} \tag{259}$$

If this equation is now integrated across the thickness of the film, we will still be left with a term $\partial v_{nx}/\partial y$. Both Pollack (1966a,b) and Sanikidze *et al.* (1967) have shown that one may make the following replacement

$$\eta_f \, \partial^2 v_{nx}/\partial y^2 \cong R v_{nx}|_{y=h}, \tag{260}$$

where $R v_{nx}$ is an effective dissipative force per unit volume opposing the motion v_{nx}. R was calculated for a flat film by Sanikidze *et al.* (1967) from two-fluid hydrodynamics, and found to be given by

$$R = 3\eta_f/h^2. \tag{261}$$

When the film thickness is less than the mean free path, we cannot describe this force by hydrodynamics. It is then a result of the collisions of elementary excitations (phonons) with the film boundaries. We will write this force in the form

$$R v_{nx} \cong \rho_n \, v_{nx}/\tau, \tag{262}$$

where τ is the time of flight of phonons across the film:

$$\tau \cong h/c_1 \tag{263}$$

where c_1 is the velocity of first sound in superfluid helium. In this case, therefore, we get

$$R = \rho_n/\tau \cong \rho_n c_1/h. \tag{264}$$

The other dissipative terms in (259) are unimportant and will be neglected.

We note that the term $-f \, \partial h/\partial x$ appears also in Eq. (61), and that it is approximately equal to $\dot{v}_{sx}$, since the other term, $\bar{S} \, \partial T_f/\partial x$, is small. From these observations, as well as the fact that $v_{nx} \ll v_{sx}$, we can write (259) in the form

$$0 = \dot{v}_{sx} + (R/\rho_n) v_{nx}, \tag{265}$$

from which we get

$$v_{nx} = (i\omega\rho_n/R) v_{sx}. \tag{266}$$

If we use (261) to substitute for R, we again obtain a result similar to (22).

We now calculate the contribution of the dissipative force R to the attenuation of third sound. Energy is dissipated by this force at the rate

$$\dot{E}_{\text{diss}} = R v_{\text{n}x}{}^2. \tag{267}$$

The contribution to the attenuation coefficient due to this process is therefore

$$\alpha = \frac{\bar{\dot{E}}_{\text{diss}}}{2u_3 \, \Delta E_{\text{f}}} = \frac{R \, |v_{\text{n}x}{}^2|}{2u_3 \bar{\rho}_{\text{s}} \, |v_{\text{s}x}^2|} = \frac{\omega^2 \rho_{\text{n}}{}^2}{2u_3 \, \bar{\rho}_{\text{s}} \, R}. \tag{268}$$

The two different expressions for R, (261) and (264), thus lead to

$$\alpha_{\text{hyd}} = \frac{h^2 \omega^2 \rho_{\text{n}}{}^2}{6 \eta_{\text{f}} \, \bar{\rho}_{\text{s}} \, u_3} \quad \text{in the hydrodynamic regime,} \tag{269}$$

$$\alpha_{\text{K}} = \frac{\tau \omega^2 \rho_{\text{n}}}{2u_3 \, \bar{\rho}_{\text{s}}} = \frac{h \omega^2 \rho_{\text{n}}}{2u_3 \, c_1 \, \bar{\rho}_{\text{s}}} \quad \text{in the nonhydrodynamic (Knudsen) regime.} \tag{270}$$

In both regimes these attenuation coefficients are much less than the ones we have calculated due to interactions of the film with its surroundings.

X. Microscopic Theories

The theories we have discussed up to now were all based on classical continuum physics—ordinary hydrodynamics for the gas, two-fluid hydrodynamics for the film, and thermal conduction theory for the substrate. Even simple experiments that measure only the velocity of third sound reveal, however, that these theories do not tell the whole story. In practice, it is found that there is an onset thickness, depending on the temperature, below which third sound is not observed (see Fig. 8). This onset thickness is in fairly good agreement with measurements of the onset thickness for dc superfluid flow (see Atkins and Rudnick 1970, Fig. 18). Since third sound is in fact a phenomenon of low frequency ac superfluid flow, it is natural to identify the two onsets. But even above the onset u_3 is less than the value one gets from the dispersion equations if one assumes that ρ_{s} has its bulk value everywhere in the film (see Fig. 8). We have anticipated this, which is why we have the average $\bar{\rho}_{\text{s}}$ appearing in our equations, and not $\rho_{\text{s,bulk}}$.

A. THE AVERAGE SUPERFLUID DENSITY

Part of the reduction of $\bar{\rho}_{\text{s}}$ below its bulk value can be understood as a result of the strong force field f due to the substrate. This causes the first atomic layer of the film to be so strongly bound to the substrate that it exhibits solid rather than fluid properties: Its specific heat looks like that of a two-dimensional solid (Brewer *et al.*, 1965). This decreases $\bar{\rho}_{\text{s}}$ below its bulk value. However, unpublished calculations made by M. Chester and by Bergman have shown that neither this effect nor the hydrostatic pressure effect on $\bar{\rho}_{\text{s}}$ are enough to explain the observed reduction in u_3.

Even when the film is fluid, we might expect surface effects to be important when the film is only a few atoms thick. Ginzburg and Pitaevskii (1958) have proposed a phenomenological theory which attempts to explain at the same time the properties of the lambda transition in bulk helium, and the properties of superfluid helium in constrained geometries. The theory assumed that the free energy of the superfluid can be expanded as a power series in the complex order parameter $\psi(\mathbf{x})$ which characterizes the superfluid state

$$F(T, \psi) = F_0(T) + \int [A(T)|\psi|^2 + B(T)|\psi|^4 + C(T)|\nabla\psi|^2]\, d\mathbf{x}. \quad (271)$$

Higher order terms are neglected. ρ_s is proportional to $|\psi|^2$. The equilibrium value of ψ is determined by minimizing F with respect to variations of $\psi(\mathbf{x})$. This leads to a nonlinear differential equation which ψ must satisfy

$$-C\nabla^2\psi + A\psi + 2B|\psi|^2\psi = 0. \quad (272)$$

ψ must also vanish at the walls.

In bulk systems, the gradient term can usually be neglected and the equilibrium value of $\psi(\mathbf{x})$ taken to be a constant. In fact, however, the gradient term determines the rate at which ψ changes near the walls. There is a characteristic "healing length" l, given by

$$l^2 = C(T)/A(T), \quad (273)$$

over which ψ must increase from zero at the walls to its fixed bulk value.

In a small system, such as a thin film where l is not small compared to the thickness h, ψ and hence ρ_s are not constant over a considerable part of the system. One can calculate from the theory the average value $\bar{\rho}_s$ as a function of h. $\bar{\rho}_s/\rho_{s,\mathrm{bulk}}$ depends only on the ratio $h/2l$, as shown in Fig. 18. At $h = \pi l$ $\bar{\rho}_s$ goes to zero, signifying inability to solve the differential equation (272) for ψ and also satisfy the boundary condition $\psi = 0$. This point is the onset thickness predicted by the theory.

In its original form, this theory included certain assumptions about the form of $A(T)$, $B(T)$, $C(T)$ which did not give good agreement with experimental results for the lambda transition in bulk helium. Josephson (1966) and Mamaladze (1967) showed that by a proper choice of these coefficients, agreement could be obtained. Using their choice we get the following expression for l:

$$l = \left[\frac{\hbar^2 T_\lambda \rho_{s,\mathrm{bulk}}/\rho}{2m^2 \Delta C_p(T_\lambda - T)^2}\right]^{1/2} \simeq 1.63 \left(\frac{T_\lambda}{T_\lambda - T}\right)^{2/3} \text{Å}. \quad (274)$$

Here T_λ is the lambda temperature ($T_\lambda = 2.172°\mathrm{K}$) and ΔC_p is the jump in the heat capacity per unit mass at constant pressure that is superimposed upon the symmetrical logarithmic singularity at T_λ. While the exact functional relationship between $\bar{\rho}_s/\rho_{s,\mathrm{bulk}}$ and $h/2l$ is fairly complicated and involves elliptic integrals, it has been shown (e.g., see Fraser, 1969, p. 33) that,

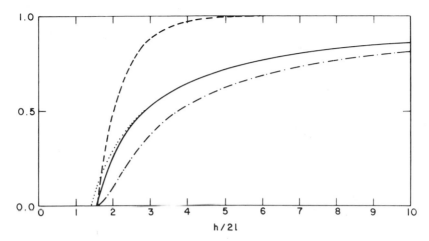

FIG. 18. The size effect in the Ginzburg–Pitaevskii theory. Plotted versus $h/2l$ are: —— $\bar{\rho}_s/\rho_{s,\text{bulk}}$; $\cdots$ the approximate form $\bar{\rho}_s/\rho_{s,\text{bulk}} = 1 - 2\sqrt{2l/h}$; – – – the maximum value of $\rho_s/\rho_{s,\text{bulk}}$; – · – · –$(F - F_0)/(F - F_0)_{\text{bulk}}$.

except for the region where $\bar{\rho}_s/\rho_{s,\text{bulk}}$ is small, a good approximation to this relation is the asymptotic formula (see Fig. 18)

$$\bar{\rho}_s/\rho_{s,\text{bulk}} = 1 - (D/h), \tag{275}$$

where

$$D = 2\sqrt{2}l. \tag{276}$$

One can include in this form also the correction due to the solid helium layer at the substrate by writing

$$D = h_s + 2\sqrt{2}l, \tag{277}$$

where h_s is the solid layer thickness.

A comparison of this theory with experiment (Kagiwada *et al.*, 1969), which we reproduce in Fig. 19 and Table II, shows that good quantitative agreement can be obtained for u_3 as a function of h if $l(T)$ is adjusted for each temperature, rather than being taken from Eq. (274). This is no surprise because (274) is only expected to be accurate for T close to T_λ. Empirically, it turns out that a fairly accurate representation for l over the entire region that was investigated is (see Revzen, 1969a,b)

$$l = B_0 T/\rho_{s,\text{bulk}}, \tag{278}$$

where

$$B_0 = 2.9 \times 10^{-9} \text{g cm}^{-2} \text{ deg}^{-1}. \tag{279}$$

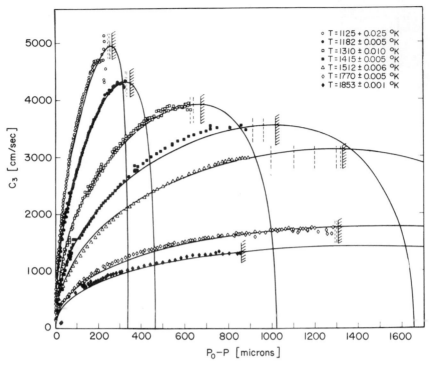

FIG. 19. Plots of u_3 versus $P_V - P$ at various temperatures. The full curves were calculated from the Ginzburg–Pitaevskii theory [i.e. Eq. (275)] with $D \equiv h_s + 2\sqrt{2}l$ chosen to give the best possible fit near the maximum of each curve. From Kagiwada *et al.* (1969), by permission of the American Institute of Physics and *Physical Review Letters*.

However, the superfluid onset does not occur when $h = h_s + \pi l$, but at a slightly greater thickness. Furthermore, while the theory predicts that $\bar{\rho}_s$, and therefore u_3, decrease continuously to zero as the onset is approached, it seems as though the onset occurs while u_3 is still quite large (close to its maximum value for the given temperature, in fact). dc measurements of superfluid flow show a similar behavior in that the critical velocity extrapolates to zero while $\bar{\rho}_s$ does not (Henkel *et al.*, 1968). From the last column of Table II it seems as though the value of $\bar{\rho}_s/\rho_{s,\text{bulk}}$ at the onset of third sound is approximately constant for different temperatures, being always around 0.36. It is worthwhile pointing out that it is not entirely clear whether $\bar{\rho}_s$ is in fact discontinuous at the onset. Another possibility is that, even though $\bar{\rho}_s$ itself goes to zero continuously, the onset that is in fact observed is due to the beginning of a strong attenuation of third sound. This is the prevalent view nowadays, since experiments have failed to detect any discontinuity in the volume (see Goodstein and Elgin, 1969) or a latent heat

TABLE II

SUPERFLUID PARAMETERS OF THIN HELIUM FILMS[a,b,c]

| $T(°K)$ | Experimental fit D | Healing lengths | | | Onset thickness h_c | $(h_c - 1)/l^d$ | Size effect at onset ρ_s/s, bulk |
		for $h_s = 1$	from Eq. (278)	from Eq. (274)			
1.125	2.79	0.63	0.63	0.76	3.84	4.51	0.38
1.192	3.04	0.72	0.67	0.79	4.29	4.91	0.39
1.205	3.05	0.73	0.68	0.80	4.34	4.84	0.37
1.310	3.22	0.79	0.76	0.86	4.41	4.48	0.35
1.415	3.41	0.85	0.84	0.934	4.71	4.42	0.36
1.495	3.61	0.92	0.92	1.01	5.10	4.45	0.37
1.512	3.67	0.94	0.94	1.03	5.18	4.45	0.36
1.586	3.89	1.02	1.03	1.11	5.73	4.59	0.38
1.778	4.99	1.41	1.39	1.45	7.35	4.57	0.36
1.852	5.53	1.60	1.61	1.67	9.27(8.25)	5.13(4.56)	0.44(0.37)
2.046	9.55	3.02	3.17	3.10	17.50(13.80)	5.21(4.04)	—

[a] From Rudnick and Fraser (1970), with permission of Plenum Publ. Co.

[b] Units of length are atomic layers (3.6 Å). The values in parentheses are for P/P_V at onset determined from a pool of other experiments.

[c] See Note added in proof at the end of this subsection.

[d] The parameter l is from Eq. (278).

(Frederikse, 1949; Evenson *et al.*, 1968) which were expected to accompany any jump in $\bar{\rho}_s$. We would like to point out, however, that $\bar{\rho}_s$ might still be discontinuous all by itself. An example of a model system where this occurs is the two-dimensional ideal Bose–Einstein gas which, at constant pressure, undergoes a phase transition into a condensed state. This transition involves no latent heat or volume discontinuity. But the occupation number of the single particle ground state jumps from essentially zero to essentially N—the total number of particles (Imry *et al.*, 1974; Gunther *et al.*, 1974).

In this connection, attempts have been made to explain the onset or cessation of superfluidity as being the result of a hydrodynamic instability which sets in at that point (Goodstein and Elgin, 1969; Goodstein, 1969). One result of the detailed hydrodynamic theory which we have described in Section II is to rule out completely such a possibility as far as the linearized equations are concerned.

A recent unpublished experiment has been performed by Kagiwada and Downs in which a great effort was made to take measurements in the region close to the onset thickness of third sound, where the attenuation is large and detection becomes very difficult. Using signal averaging techniques, Kagiwada succeeded in observing u_3 begin to descend towards zero as the onset was approached from above. The descent, though more rapid than expected from the theory, nevertheless looks quite continuous.

An attempt to modify the Ginzburg–Pitaevskii–Josephson–Mamaladze (GPJM) theory so as to get a superfluid transition of the first order in thin films was made by Amit (1968a,b,c). Amit argued that, in order to ensure stability of the solution $\psi \equiv 0$ at $T = T_\lambda$ in bulk helium, one should include a ψ^6 term in the expansion of the free energy [Eq. (271)]. His theory leads indeed to a first order superfluid transition in helium films. But the jump in $\bar{\rho}_s$ is accompanied by a jump in the thickness and by a latent heat, both of which have not been observed, as we mentioned before.

The GPJM theory is not a true microscopic theory. It is a phenomenological theory of the complex superfluid order parameter, similar to the Ginzburg–Landau (GL) theory of superconductivity, and in the spirit of Landau's general theory of second order phase transitions. Moreover, it has no rigorous microscopic justification similar to the one existing for the GL theory of superconductivity. It thus looks as though much more work on microscopic theories will have to be done in order to understand the onset of superfluidity and third sound.

Note added in proof: Because an improved value of α (the coefficient of the Van der Waals potential) is now available, thanks to Sabisky and Anderson (1973a,b), all the experimentally determined values of D have to be reduced by a factor .677. Obviously this will alter the healing lengths determined by Eq. (277) and appearing in the third column of Table II. It will also change the nature of the fit to experimental data of Eq. (278), necessitating at least a change in the empirical coefficient B_0. See also note in caption to Fig. 3.

B. The Onset of Superfluidity and Attenuation of Third Sound

A few attempts have been made to go beyond hydrodynamics in calculating the attenuation of third sound.

Revzen (1969a,b) has applied the Aslamazov-Larkin theory for fluctuations in thin superconductors to helium films. Revzen finds the following expression for the mobility μ of the superfluid fraction of the film

$$\mu = \mu_c \exp[4(h - h_c)/l], \qquad (280)$$

where μ_c and h_c are parameters of the theory. This is essentially equivalent to a coefficient of attenuation that depends exponentially on h. The onset is thus viewed as a point where the increasing attenuation becomes so large that third sound is undetectable in practice.

Putterman *et al.* (1971) have tried to correlate the anomalous attenuation with the appearance of quantum fluctuations in the film. They argue that analogously to the usual uncertainty principle for particles obeying quantum mechanics, one has the following uncertainty principle for helium films

$$m \, \Delta v_s \Delta h \simeq \hbar. \qquad (281)$$

On the left-hand side Δh is the uncertainty in the vertical position of a particle in the film and Δv_s is the corresponding uncertainty in the superfluid velocity, while on the right-hand side $\hbar$ is Planck's constant. They then argue that when $\Delta v_s \geqslant u_3$, a large attenuation of third sound is to be expected due to large fluctuations of a purely quantum nature. Wang and Rudnick (1972) have pointed out that together with the large fluctuations Δv_s one will have large fluctuations in h as well, which may lead to a local vanishing of $\bar{\rho}_s$ if h fluctuates to values below πl. The precise meaning of these ideas is not altogether clear since they do not arise from an underlying theory. They should be viewed, in our opinion, as preliminary ideas in the attempt to construct such a theory. Some interesting results nevertheless follow from (281): If one estimates Δh by $h/\sqrt{12}$ (this assumes a constant probability distribution for finding the particle at any position across the film), and replaces Δv_3 by u_3 in (281), one arrives at the following condition under which large attenuation due to large fluctuations in v_s is expected:

$$\hbar\sqrt{12}/m \geqslant u_3 h. \tag{282}$$

Comparison with experiment (Wang and Rudnick, 1972) shows that where (282) is not satisfied the attenuation is small and fairly well described by the continuum theory. At the onset, (282) is approximately satisfied as an equality. Furthermore, when one considers thick films, (282) is again satisfied, since u_3 then decreases very rapidly with thickness. This is precisely the region where the greatest discrepancy in the attenuation is found. The quantum fluctuations may thus be important not only for thin films, but for thick films as well!

Another idea, put forward by Rudnick (see the last paragraph of Atkins and Rudnick, 1970) is that nonlinearities in the propagation of third sound waves may contribute significantly to the attenuation. Such effects are well known in surface waves on an ordinary shallow liquid: The velocity at the crest is different from the velocity at the trough because: (a) the sound velocity varies with the total depth h and (b) the fluid is moving locally in the direction of propagation at the crest but in the opposite direction at the trough. In the case of third sound waves, the velocity decreases with increasing h for thick films but may increase with h when the size effects become important. The difference in sound velocity between crest and trough causes the wave to become highly nonsinusoidal in shape as it propagates, eventually developing either a leading or a trailing shock front—this is the well known phenomenon of wave breaking. Such behavior would give rise to attenuation. It is interesting to note that there is a film thickness for which the two velocity effects mentioned above, (a) and (b), exactly cancel and therefore nonlinearies do not develop. This occurs for rather thin films. Near that point one also observes experimentally that the attenuation has its minimum value, and is in good agreement with the hydrodynamic calculations (Putterman, 1974, Sect. 48).

Lastly, we will only mention the attempt that has been made by Chester and Maynard (1972) to construct a microscopic theory of third sound by quantizing the two-fluid equations of a helium film. The theory incorporates the macroscopic quantum uncertainty principle discussed previously. Quantitative results of the theory have not yet been published.

C. Summary

This section by its very nature cannot be summarized. There is clearly much more work to be done in developing a microscopic quantum theory for third sound and other properties of superfluid helium films. Some of it is being actively pursued at this time. The continuum theory and its comparison with experiments must serve as a guiding post in these attempts.

Glossary of Key Symbols and Phrases

In the second column we list, wherever possible, the equation or table where the symbol is defined, or where it appeared for the first time.

Symbol	Equation	Definition
A	88	Kinetic coefficient in the theory of evaporation.
B	69, 189	$1/B$ is the effective thermal resistance of the substrate
B_1	63	$1/B_1$ is the Kapitza thermal boundary resistance
C_p, C_v	99, 106	Heat capacity of the helium gas per unit mass at constant pressure, volume
C_{sub}	62	Heat capacity of the substrate per unit mass
C_h	40	Heat capacity of the helium film per unit mass
c	100	Complex velocity of sound in He gas
c_{01}	98	
c_{02}	99	Complex velocities of the three wave modes in He gas
c_{03}	100	
c_3	14	Complex velocity of third sound
c_1	263	Velocity of first sound in superfluid helium
D	275	Coefficient of $1/h$ in approximate formula for $\bar{\rho}_s$
f	3, 57	Force exerted by the substrate on a unit mass of helium at the gas–film interface
h	1	Thickness of the He film
h'	86	Amplitude of oscillations around h
h_s	277	Thickness of the layer of solid helium at the substrate
h_c	Table II, 280.	Thickness of the He film at onset of third sound
h_{sub}	67	Thickness of the substrate
$\mathbf{J}, J_x, J_y$	7, 7, 7	Mass current in the film and its components
J_M	33	Net mass flux from film to gas
J_g	44	Heat flux from film to gas
J_E	81	Net energy flux from film to gas
J_{sub}	41	Heat flux from film to substrate

Symbol	Equation	Definition
J_1	163	Numerical coefficient appearing in the thick film equations
J_{10}	162	Numerical coefficient appearing in the thick film equations
k	13	Complex wave vector of third sound
k_B	73	Boltzmann's constant
L	47	Latent heat of evaporation from film to gas per unit mass
L	204	Length of a third sound resonator
l	273	Healing length in the Ginzburg–Pitaevskii theory
l_g	34	Mean free path in the gas
l_{sub}	67	Thermal penetration depth in the substrate
l_η	23	Viscous penetration depth in liquid He
l_{vis}	102a	Viscous penetration depth in He gas
l_{th}	102b	Thermal penetration depth in He gas
M_1	98	The viscous mode in He gas
M_2	99	The thermal mode in He gas
M_3	100	The acoustic mode in He gas
q_{sub}	65	Thermal wave vector in the substrate
q_{vis}	98	Viscous wave vector in He gas
q_{th}	99	Thermal wave vector in He gas
q_{ac}	100	Acoustic wave vector in He gas
R	260	Coefficient of the effective drag force due to viscosity of the normal fluid
S_f	9	Entropy per unit mass of liquid He
S_g	42	Entropy per unit mass of the He gas
$\bar{S}$	39	Partial entropy per unit mass of the film
thick film	119	Film for which $\kappa_g \omega / \rho_g C_p c_3^2 \geqslant 1$
thin film	118	Film for which $\kappa_g \omega / \rho_g C_p c_3^2 \ll 1$
T	38	Equilibrium temperature
T_{sub}, T'_{sub}	41, 64	Instantaneous temperature of substrate, amplitude of its oscillations
T_g, T_g'	42, 86	Instantaneous temperature of gas, amplitude of its oscillations
T_f	9	Instantaneous temperature of the film
$\bar{T}_f, T_f'$	36, 86	Instantaneous temperature of the film averaged across the thickness, amplitude of the oscillations
T_{g2}, T_{g3}	103	Amplitudes of temperature oscillations in modes M_2, M_3 of the gas
u_3	5, 122	Real velocity of third sound
u_{30}	115	Approximate real velocity of third sound
v	113	$(\bar{\rho}_s / \rho_h)(v_{sx} / c_3)$

Symbol	Equation	Definition
$\mathbf{v}_s$, v_{sx}, v_{sy}	1, 8	Superfluid velocity and its components
$\mathbf{v}_n$, v_{nx}, v_{ny}	7	Normal fluid velocity and its components
$\mathbf{v}_g$, v_{gx}, v_{gy}	33, 42	Velocity of the He gas and its components
x	Fig. 1	Coordinate axis parallel to the third sound wave vector
y	Fig. 1	Coordinate axis perpendicular to the film
y_r	Fig. 1, 33	Position of the imaginary reference plane in the gas
α	123	Coefficient of attenuation of third sound
α	172	Constant appearing in the equation for f
γ	106	The ratio C_p/C_v for He gas
η_f, η_g	7, 71	Coefficient of shear viscosity of liquid, gas
ζ_1, ζ_2, ζ_3	7, 8	Coefficients of bulk viscosity of liquid He
κ_f, κ_g, κ_{sub}	9, 72, 62	Coefficient of thermal conductivity of liquid, gas, substrate
κ_T	244	Isothermal compressibility of He gas
μ, μ_g	2, 82	Chemical potential of helium liquid, gas
μ_f, μ_f'	56, 133	Chemical potential of helium averaged across the film, amplitude of its oscillations
ρ_f, ρ_h, $\bar{\rho}_f$	1, 31, 38	Mass density in the liquid film, at the surface, averaged across the film
ρ_s, ρ_n, $\bar{\rho}_s$	1, 11, 29	Superfluid, normal mass density, superfluid density averaged across the film
ρ_{sub}	62	Mass density of substrate
ρ_g, ρ_g'	33, 42, 103	Mass density of the gas, amplitude of its oscillations
$\omega/2\pi$	13	Frequency of third sound

Subscripts

g, f, sub		Signify quantities of the gas, film, and substrate, respectively
A, V	58, 229, 243	Signify quantities per unit area and volume, respectively

Acknowledgment

I would like to thank Professor I. Rudnick for a critical reading of the manuscript, for correcting some errors, and for making some very helpful suggestions which have been incorporated into it.

References

Amit, D. J. (1968a). *Phys. Lett.* **26A**, 448.
Amit, D. J. (1968b). *Phys. Lett.* **26A**, 466.
Amit, D. J. (1968c). Unpublished report.
Anderson, C. H., and Sabisky, E. S. (1970a). *Phys. Rev. Lett.* **24**, 1049.

Anderson, C. H., and Sabisky, E. S. (1970b). *J. Low Temp. Phys.* **3**, 235.

Andreev, A. F. (1966). *Sov. Phys.—JETP* **23**, 938.

Andreev, A. F., and Kompaneets, D. A. (1972). *Sov. Phys.—JETP* **34**, 1316.

Arkhipov, R. G. (1957). *Sov. Phys.—JETP* **6**, 90.

Atkins, K. R. (1957). *Physica (Utrecht)* **23**, 1143.

Atkins, K. R. (1959). *Phys. Rev.* **113**, 962.

Atkins, K. R., and Rudnick, I. (1970). *Progr. Low Temp. Phys.* **6**, 37.

Bergman, D. J. (1969). *Phys. Rev.* **188**, 370.

Bergman, D. J. (1971). *Phys. Rev. A* **3**, 2058.

Bergman, D. J. (1973). *Proc. Roy. Soc., Ser. A* **333**, 261.

Brewer, D. F., Symonds, A. J., and Thomson, A. L. (1965). *Phys. Rev. Lett.* **15**, 182.

Chester, M., and Maynard, R. (1972). *Phys. Rev. Lett.* **29**, 628.

Downs, J. L., and Kagiwada, R. S. (1972). *Bull. Am. Phys. Soc.* **17**, 19.

Dransfeld, K., Newell, J. A., and Wilks, J. (1958). *Proc. Roy. Soc., Ser. A* **243**, 500.

Evenson, A., Brewer, D. F., and Thomson, A. L. (1968). *Proc. Int. Conf. Low Temp. Phys., 11th, St. Andrews, Scot.* Vol. 1, p. 125.

Everitt, C. W. F., Atkins, K. R., and Denenstein, A. (1962). *Phys. Rev. Lett.* **8**, 161.

Everitt, C. W. F., Atkins, K. R., and Denenstein, A. (1964). *Phys. Rev.* **136**, A1494.

Fraser, J. C. (1969). Tech. Rep. No. 30. Off. Nav. Res., Washington, D.C.

Frederikse, H. P. R. (1949). *Physica (Utrecht)* **15**, 860.

Ginzburg, V. L., and Pitaevskii, L. P. (1958). *Sov. Phys.—JETP* **7**, 858.

Goodstein, D. L. (1969). *Phys. Rev.* **183**, 327.

Goodstein, D. L., and Elgin, R. L. (1969). *Phys. Rev. Lett.* **22**, 383.

Goodstein, D. L., and Saffman, P. G. (1971). *Proc. Roy. Soc., Ser. A* **325**, 447.

Gunther, L., Imry, Y., and Bergman, D. J. (1974). *J. Stat. Phys.* **10**, 299.

Henkel, R. P., Kukich, G., and Reppy, J. D. (1968). *Proc. Int. Conf. Low Temp. Phys., 11th, St. Andrews, Scot.* Vol. 1, p. 178.

Imry, Y., Bergman, D. J., and Gunther, L. (1974). *Proc. Int. Conf. Low Temp. Phys. 13th, Boulder, Colo.* Vol. 1, p. 80.

Josephson, B. D. (1966). *Phys. Lett.* **21**, 608.

Kagiwada, R. S., Fraser, J. C., Rudnick, I., and Bergman, D. J. (1969). *Phys. Rev. Lett.* **22**, 338.

Khalatnikov, I. M. (1965). "Introduction to the Theory of Superfluidity" (Engl. transl. by P. C. Hohenberg). Benjamin, New York.

Kuper, C. G. (1956a). *Physica (Utrecht)* **22**, 1291.

Kuper, C. G. (1956b). *Physica (Utrecht)* **24**, 1009.

McCormick, W. D., Goodstein, D. L., and Dash, J. G. (1968). *Phys. Rev.* **168**, 249.

Mamaladze, Y. G. (1967). *Sov. Phys.—JETP* **25**, 479.

Morse, P. M., and Feshbach, H. (1953). "Methods of Theoretical Physics." McGraw-Hill, New York.

Osborne, D. V. (1962). *Proc. Phys. Soc.* **80**, 103.

Pickar, K. A., and Atkins, K. R. (1969). *Phys. Rev.* **178**, 389.

Pollack, G. L. (1966a). *Phys. Rev.* **143**, 103.

Pollack, G. L. (1966b). *Phys. Rev.* **149**, 72.

Putterman, S. J. (1974). "Superfluid Hydrodynamics." North-Holland, Amsterdam, and American Elsevier, New York.

Putterman, S. J., Finkelstein, R., and Rudnick, I. (1971). *Phys. Rev. Lett.* **27**, 1697.

Ratnam, B., and Mochel, J. (1970a). *Phys. Rev. Lett.* **25**, 711.

Ratnam, B., and Mochel, J. (1970b). *J. Low Temp. Phys.* **3**, 239.

Ratnam, B., and Mochel, J. (1974). *Proc. Int. Conf. Low Temp. Phys., 13th, Boulder, Colo.* Vol. 1, p. 233.

Revzen, M. (1969a). *Phys. Rev. Lett.* **22**, 1102.

Revzen, M. (1969b). *Phys. Rev.* **185**, 337.

Rudnick, I., and Fraser, J. C. (1970). *J. Low Temp. Phys.* **3**, 225.

Rudnick, I., Kagiwada, R. S., Fraser, J. C., and Guyon, E. (1968). *Phys. Rev. Lett.* **20**, 430.

Sabisky, E. S., and Anderson, C. H. (1973a). *Phys. Rev. Lett.* **30**, 1122.

Sabisky, E. S., and Anderson, C. H. (1973b). *Phys. Rev. A* **7**, 790.

Sanikidze, D. G., Adamenko, I. I., and Kaganov, M. I. (1967). *Sov. Phys.—JETP* **25**, 383.

Scholtz, J. H., Mclean, E. O., and Rudnick, I. (1974). *Phys. Rev. Lett.* **32**, 147.

Wang, T. G., and Rudnick, I. (1972). *J. Low Temp. Phys.* **9**, 425.

Zinov'eva, K. N., and Bolarev, S. T. (1969). *Sov. Phys.—JETP* **29**, 585.

—2—

Physical Acoustics and the Method of Matched Asymptotic Expansions

M. B. LESSER

Institut CERAC,
Ecublens, Switzerland

and

D. G. CRIGHTON

Department of Applied Mathematical Studies,
University of Leeds, Leeds, England

  I. Introduction and Elementary Illustrations............................. 70
     A. Introduction 70
     B. Techniques of MAE through One-Dimensional Examples 76
     C. The Impedance Concept 89
     D. A Second-Order Model Equation 91
 II. Scattering and Diffraction Problems 93
     A. Introduction 93
     B. Scattering by a Hard Strip 94
     C. Higher Order Approximations 99
     D. Scattering by a Soft Strip 103
     E. Diffraction by a Thick Plate 106
     F. Three-Dimensional Problems 109
III. Acoustic Wave Guides 110
     A. Introduction 110
     B. The Long Wavelength Approximation and the Webster Horn Equation 111
     C. Radiation from an Open Slit in a Waveguide in the Long Wavelength
        Limit 114
     D. The Slowly Varying Guide in the Short Wavelength Limit 121

IV. Nonlinear Acoustics 125
    A. Introduction 125
    B. Development of Weak Shock Waves 126
    C. Shock Reflection from a Wall 132
    D. The Time-Harmonic Piston Problem 135
 V. Conclusions 143
    References 147

I. Introduction and Elementary Illustrations

A. Introduction

The main purpose of the present article is to show how the relatively new technique of "Matched Asymptotic Expansions" (MAE) can be applied to problems of interest in acoustic research. We hope in the course of the discussion to answer questions such as: What is MAE? What is its relevance to acoustics? What type of problems can be solved? and What physical insights does it provide? Also, sufficient information will be provided to enable the careful reader to apply the MAE technique to certain problems which are relevant to his research.

The technique of MAE is basically a refinement and extension of classical perturbation methods. Such methods typically make use of the solution to a known, relatively simple problem to obtain a useful approximation to a more complex situation. For example, the complete solution to a particular problem in acoustics might express the pressure p as a function of space, time, and some typical parameters such as a piston velocity amplitude and the sound speed in the medium,

$$p = p(x, t, \mathcal{U}, a).$$

The full solution in terms of arbitrary parameters might be unobtainable, or so complex as to be useless for interpretation—or even for computation. With perturbation theory one attempts to study such a situation by starting with a limiting case where the solution may be obtained in relatively simple closed analytical form, for example when $\mathcal{U}/a = \varepsilon \to 0$ in the above example. Using this solution the equations for the "full" problem are used to provide correction or perturbation terms so that the approximation to the above solution might take the form

$$p \cong p^{(0)}(x, t) + \mu_1(\varepsilon)p^{(1)}(x, t)$$

where $\mu_1(\varepsilon)$ is a function of the perturbation parameter ε which vanishes as $\varepsilon \to 0$. When such a form for the solution applies over the whole domain of independent variables of interest, x and t in the above example, the perturbation is termed *regular*. When the form fails to give meaningful results for some values of the independent variables, the perturbation is termed *singular*. The singular case is more the rule than the exception in practice, and MAE is one means of coping with this problem. Nonuniform perturbation expansions (singular perturbations) arise in an enormous variety of different ways, of course, and several basically distinct procedures have been developed for removing the nonuniformity in different types of problems. MAE is but one of these techniques, and forms the subject of this article because we believe it to be the one with the most immediate applicability in a number of areas within acoustics. References to work describing other singular perturbation techniques will be found in the concluding section.

The method of attack with MAE is to seek several perturbation series with differing domains of validity. In typical boundary value problems such as one encounters in acoustics, constants appear in the solution of the full problem which must be determined from boundary and initial conditions. However, when several perturbation series are employed for a problem the unknown constants for each solution cannot usually all be found from the boundary and initial conditions appropriate to that solution. This then leads to the problem of finding the remaining unknown constants in the separate series by utilizing the fact that each of the solutions is an approximation, in a certain sense, to the *same function*. The evaluation of the constants in this way is called matching; hence the name Matched Asymptotic Expansions.

Perhaps the most rapid appreciation of MAE can be obtained from a brief historical summary of how it developed. The origins of MAE are closely connected with acoustics' sister subject, fluid mechanics. The work of the great mathematicians and physicists of the eighteenth and nineteenth centuries had made theoretical fluid mechanics a subject of great beauty as well as a vehicle for the development of many of the significant techniques of modern applied mathematics. Paradoxically, however, despite their mathematical elegance, the methods were almost totally inadequate for predicting or understanding many observed flow situations. The crux of the problem faced by those wishing to bring the results of theoretical calculation into agreement with experimental findings lay in the treatment of fluid viscosity. When the ratio of inertial force to viscous stresses on a body of given characteristic length, as characterized by the famous nondimensional parameter known as the Reynolds number Re, is large or small, it appears reasonable to make some very helpful simplifications to the difficult hydrodynamic equations. As in many cases the fluid is air or water with relatively small viscosities, taking Re large and dropping the viscous terms from the equations would seem to be quite a reasonable procedure. The Re small case, where inertial terms are dropped, also appears plausible when bodies are small, and has even met with some success, for example in the famous Stokes drag formula for spheres. Somewhat astonishingly, however, both these reasonable approximations have produced results either totally at variance with experiment or completely inconsistent with plausibility. It is not our purpose to discuss Fluid Mechanics in detail here. However, as Physical Acoustics provides cases analogous with the small–large Re problems, some discussion is worthwhile. The reader interested in more of the details can consult Birkhoff (1960), Van Dyke (1964), or Kaplun (1967).

The problem faced in the large Re situation is that plausible reasoning leads to the nonphysical result known as d'Alembert's paradox—that a steadily translating rigid body suffers no drag in an infinite fluid. Attempts to correct this physically unreasonable result and to overcome other related problems met with little real success until the work of Prandtl (1904). With an inspired injection of physical reasoning, he saw both how to explain the lack of success and how to overcome the problem, at least in the important case of relatively streamlined bodies. Thus he observed that when the viscous terms are dropped from the equations of motion the order of the equations is lowered and fewer boundary conditions can be satisfied, i.e., dropping viscous terms implies dropping a boundary condition on the wall of the moving body. It seemed almost self-evident that no flow into a solid body could be allowed; hence this condition was retained, and the fluid was assumed to slip along the surface of the body. Prandtl observed, however, that at the surface the fluid must in general *not* be allowed to slip. His picture of the process Re → ∞ was that, in the region away from the body the viscosity plays a decreasingly important role as Re increases. However, near the body the gradient in velocity must be large to adjust to the no-slip condition on the surface. This in turn implies that even though 1/Re is small (in general implying small viscosity), the product of 1/Re and the velocity gradient terms can be quite comparable in magnitude with other terms. Hence he concluded that in such a "boundary layer," a different set of approximating equations—the "boundary layer equations"—should be applied. Thus the basic idea was established that two sets of approximating equations, one appropriate to the outer flow (the classic potential flow equations) and one for the inner flow (the boundary layer) were needed to provide an adequate calculational scheme for determining the viscous drag on streamlined bodies. The scheme and equations proposed by Prandtl have met with enormous success and are considered by many to have changed the whole face and role of Theoretical Fluid Mechanics.

The difficulties with small Re flows, on the other hand, proved to be a far more subtle problem, whose solution was not really found until recent times with the work of Kaplun (1957), Friedrichs (1955), Lagerström and Casten (1972), and others. The problem is exemplified in the so-called paradoxes of Stokes and Whitehead. Stokes (1851), dropping the inertial terms from the equations of motion, attempted to calculate the drag on a cylinder in uniform flow. He found that, because the fundamental solution to the resulting equation (the Stokes equation) is logarithmically singular at infinity, he could not produce a solution that would satisfy the uniform flow condition, i.e., the plausible assumption about viscous effects led to a problem with no mathematical solution. The three-dimensional problem of slow flow about a sphere, of course, does give the well-known Stokes drag formula, and all appeared well with the sphere until Whitehead (1889) tried to improve the result by the reasonable device of inserting the Stokes solution into the neglected terms, and treating the result as inhomogeneous terms in Stokes' equation. The result was found to lead again to a solution growing with distance from the body. The explanation for these difficulties is that as more and more of the fluid about the body is considered, the inertia effects become more and more important; hence one cannot expect the solution to retain its validity in this "outer" region. In the case of three dimensions, one is able to calculate the drag, as the body effect dies off with sufficient rapidity with distance to permit a meaningful first approximation. However, in two dimensions even this modest desire cannot be fulfilled. A partial solution was found by Oseen (1910,1913), who proposed an equation which gave a better representation of the outer flow. However, full understanding did not come until the work of Kaplun (1957), Proudman & Pearson (1957), and the full development of MAE. As with the large Re case, it was shown that two regions had to be considered, one near the body where inertia could be ignored, at least to first approximation, and an outer region where the body could be looked upon as a point singularity. While in this case no boundary condition is lost by reduction in the order of the governing equations, the problem of matching the solutions in the two regions proved to be very difficult.

It will be seen that the MAE method, as presently practiced, is derived from an abstraction of the above ideas. First, it consists of a formalization of the process used by Prandtl, which permits a logical development of correction terms in both the boundary layer region and the outer region. A discussion of this is given in the book by Van Dyke (1964). It should be noted that the higher order corrections to the drag are of great technical and engineering importance, yet fifty years passed between Prandtl's introduction of the boundary layer and the modern calculations. Thus, even in the technical area which gave birth to some of the basic ideas of MAE, the modern formalization has provided a significant advance. Second, it is becoming increasingly realized just how wide is the general applicability of the formalism to a large number of areas. As should become apparent, the formalism is both mathematical and physical in content. In fact, one of the very important features of MAE is the extent to which it ties together and helps to create physically significant conceptual structures. The classical example is how the boundary layer, and perfect (nonviscous) flow, corresponding to the inner and outer flow regions, join to provide a satisfying understanding of the complex real (viscous) fluid flow about bodies. The importance of MAE is thus in both the conceptual framework it provides for unifying mathematical models, and its effectiveness in resolving particular problems—sometimes with surprising simplicity.

In the application of MAE, dimensional reasoning and an understanding of physical scales and magnitudes in a problem play a very important role. As a foretaste of this, let us consider the small Re problem discussed above, and an equivalent acoustic problem—the scattering of sound by a body small compared with the wavelength of the incident radiation. From the viewpoint of MAE the problems are quite similar, while the latter problem makes a good introduction for the reader educated in physical acoustics, and hence will be considered in some detail below. A heuristic understanding of why the $Re \rightarrow 0$ problem calls for different approximations in two regions and why the "small" acoustic scatterer can be examined in a similar light can be gleaned from dimensional considerations. Both problems can be cast into forms where the variables are nondimensional. Thus, in the $Re \rightarrow 0$ problem the velocity can be given in units of the free stream velocity and the coordinates in units of body characteristic size. After this transformation, one is left with equations that contain only *one* parameter, Re, and the solution for any dependent variable (eg., pressure p) must be a function of the dimensionless or scaled coordinates and Re, i.e., $p = F(\mathbf{r}/l_B, Re)$. Examination of the parameter Re shows that it can be considered as the ratio of two lengths, as the characteristic body dimension l_B divided by the length found by appropriately grouping the parameters of viscosity, density, and free stream velocity, l_v (the so-called viscous length). When the body is small compared with the viscous length, $Re = l_B/l_v$ is small, viscous effects dominate, and inertia terms can be ignored, at least near the body surface.

Dropping the inertia terms is thus equivalent to taking

$$\lim_{Re \rightarrow 0} F(\mathbf{r}/l_B, Re)$$

in the solution for arbitrary Re. A little thought, however, shows that $Re \rightarrow 0$ can be characteristic of two very different physical situations. In one case, consider l_B fixed but $l_v \rightarrow \infty$; for example, a body of fixed size in a fluid of increased viscosity, giving a very large viscous length. Alternatively, suppose $l_B \rightarrow 0$ but hold l_v fixed. In the second case the body is very small and the free field dominates the physics. Clearly, the correct mathematical description must reflect this state of affairs. Thus, for the above limit to be unique, we must prescribe a relation $f(l_B, l_v, Re) = 0$ which holds as $Re \rightarrow 0$. The traditional Stokes problem is connected with the case $Re \rightarrow 0$, l_B fixed.

Now consider scattering of radiation of wavelength λ by a body of scale l. Again, the equation of interest (now the familiar Helmholtz equation) and the boundary condition can be put into suitably nondimensional form so that the scaled pressure is given by a function $F(\mathbf{r}/l, l/\lambda)$. If l is much smaller than λ, it seems plausible that an incompressible flow problem should provide an adequate description of the situation near the body. As the scattering problem is linear and has been studied for a number of years by a variety of techniques, such as integral equations, it is well known that the solution requires taking proper account of the outer radiation field. However, even if this were not so, the reasoning applied to the $Re \rightarrow 0$ problem applies here and, as we shall see, the problem provides a useful introduction to MAE methods in acoustics. The parameter l/λ can become small in two vastly different ways. If $l \rightarrow 0$ with λ constant, we should have a solution appropriate for a radiation region, but if $\lambda \rightarrow \infty$ with l constant, one expects a solution for an incompressible flow region.

Using MAE we will show how asymptotic expansions can be found directly from the differential equations in a term-by-term manner for the two regions, and how the linking of the two expansions ties up with some classical conceptual structures, such as scattering matrices and impedances.

A large part of what follows will in fact involve problems of linear physical acoustics. Aside from providing an excellent introduction, such problems will be shown to have considerable interest in themselves. Thus, in this case, we will take as basic the linearised compressible flow equations, and show how MAE can be used to arrive at various well-known theoretical acoustical constructs such as the Webster Horn equation, radiation end corrections, and others.

An important point to note is that in the MAE approach to a problem we choose a mathematical framework which contains *complete* descriptions of the physics of interest. Various simpler theories, and conceptual structures then arise as limiting cases, applicable to restricted spatial and temporal regions. Later in the article, for example, we will show how, starting with the Lighthill (1956) approximation to the Navier–Stokes equations, we can take a problem of nonlinear acoustics and by considering various limiting cases of the relevant parameters divide the problem into space–time regions where Burgers' equation, the linear wave equation and the heat conduction equation describe the asymptotic solution. Hence, as another side product, the MAE formalism provides logical derivations of a number of equations popular in acoustics, together with estimates of their validity and a very definite interpretation of their meaning in a larger context (e.g. Burgers' equation in relation to the linear wave equation and the Navier–Stokes equation).

We shall introduce the formalism by stages, starting with some very simple model examples of acoustical interest. After defining some of the notation, consideration will be given to the problem of linear wave propagation in a one-dimensional medium of varying sound speed. The well-known WKB method *can* be related to the MAE formalism. However, both for novelty and intrinsic interest, the opposite case, where the wavelength is *large* compared with the inhomogeneity scale, will be treated here. While the problem is simple, the asymptotic aspects are rich and provide an excellent foretaste of what follows in a fairly clear context. This problem will also be used to discuss the tricky problem of matching asymptotic series with different domains of validity. In some ways, matching is similar to finding connection formulas for turning point problems of the WKB type, though the context is considerably more general.

With this background, we will examine various problems for the wave equation where most of the parametric dependence arises from the geometrical situation and the boundary behavior, i.e., scattering problems, wave propagation in variously shaped ducts, radiation from cavities, and resonant systems. After this discussion of what might be called classical linear acoustics, we will enlarge our viewpoint and take the full Navier–Stokes equations of nonlinear viscous gas dynamics with allowed finite boundary motions as our basic model. After suitable normalizations based on particular situations, several theories important in nonlinear acoustics are developed, in particular those of weak shock formation, shock reflection from a wall, and the spatial evolution of an initially sinusoidal signal.

Matched Asymptotic Expansions are a tried and trusted technique in a number of areas of theoretical mechanics. Despite this and the fact that much of the technique's origins (Van Dyke, unpublished, 1971) lie in the great work of 19th century acousticians, the method has not been widely employed in modern acoustics research. For this reason the authors have worked out a number of new problems, not only for their expository value, but also for their current scientific interest.

We hope that this article will demonstrate the ability of MAE both to produce new and worthwhile results and to provide fresh insight into familiar results, and that readers will be thereby encouraged to apply these methods in acoustics as freely and widely as they have been in other branches of mechanics.

B. Techniques of MAE through One-Dimensional Examples

The main purpose of the present section is to provide a brief but useful introduction to the working techniques of MAE. To accomplish this task we shall first consider a model or analog problem of wave motion in a one-dimensional medium where waves can only propagate in the direction of increasing space coordinate. The problem chosen has the advantage that, while simple to solve exactly and easy to understand in its essential aspects, its asymptotic properties retain many of the essential technical difficulties found in more complicated problems.

Thus consider the *model* problem defined by the equation

$$\frac{\partial p}{\partial x} + \frac{1}{a(x)} \frac{\partial p}{\partial t} = 0, \tag{1}$$

where p is the "pressure," x and t space and time coordinates, and $a(x)$ the wave velocity, or "sound speed." In the rest of Section I, B we shall refer to these variables as if they were the pressure and sound speed of a gas. The reader should be aware that this use of language is only meant as a descriptive analogy. The algebra and computations associated with the model problem are simple enough for the reader to follow in all detail. Also, the situation described by Eq. (1) is close enough to an actual one-dimensional sound propagation problem to make the use of such language meaningful. We note in passing that a traditional way of introducing MAE is to treat the problem of a simple spring–mass–damper system, with the mass taken as the small, or perturbation, parameter. Such a treatment can be found, for example, in the book by Cole (1968). The present model is, however, felt to be more appropriate for the understanding of typical applications of MAE to acoustics.

In such a model the "physics" is present only in the choice of the function $a(x)$. Thus there are a number of situations of interest that can be examined by suitable choices of $a(x)$ and boundary conditions on p.

For the present discussion we consider a piston, oscillating harmonically at angular frequency ω, and located at $x = 0$. Waves of amplitude p_0 are radiated away from the piston through a region of varying sound speed, and out to regions of large x. One length scale in the problem is the wavelength $\lambda = 2\pi/k_0$, where $k_0 = \omega/a_0$ is the wavenumber and a_0 is the sound speed at $x = 0$. Another length scale can be introduced into the problem by our assumptions about $a(x)$. Thus we shall assume that $a(x)$ changes rapidly (has a large derivative) in a region of width l near the piston $(0 \leq x \lesssim l)$. However, when $x \gtrsim l$ it is assumed that $a(x)$ changes only slowly. These assumptions apply to the rate of change of a; a itself is assumed to be finite and bounded. We can think of l as being an inhomogeneity scale for sound speed variation in the vicinity of the piston. One would expect different kinds of results, depending on whether this inhomogeneity scale is larger than, smaller than, or comparable with the wavelength $2\pi/k_0$. For our demonstration of MAE we assume the wavelength to be large compared with the inhomogeneity scale, and thus choose as a small parameter $\varepsilon = 2\pi l/\lambda = k_0 l$. Even with these restrictions a number of qualitatively different situations can be generated by the detailed behavior of $a(x)$. As it is not our purpose to make an exhaustive study of all these situations, but merely to introduce MAE, we will confine ourselves to the case where $a(x)$ behaves algebraically as $x \to \infty$, as defined in Eq. (6) below.

In more formal terms our model is given by

$$p = p(x)e^{-i\omega t},$$
$$a_0/a = g(x/l),$$
$$(dp/dx) = ik_0 g(x/l)p, \tag{2}$$

where g varies only slowly for $x/l \gg 1$, and an asymptotic solution is sought for the case

$$p(0) = p_0$$
$$\varepsilon = k_0 l \to 0. \tag{3}$$

In keeping with the philosophy of the method, the first step is to scale the problem so that dependent and independent variables are $O(1)$ and the problem formulation is in a "universal" form with parameters appearing in suitable dimensionless groups. For Eq. (2) this is an easy task; thus, scale the pressure with the source pressure p_0, and scale length with $1/k_0$, which is proportional to the wavelength, to obtain

$$\bar{p} = p/p_0,$$
$$\bar{x} = k_0 x,$$
$$x/l = \bar{x}/k_0 l = \bar{x}/\varepsilon. \tag{4}$$

Hence

$$(d\bar{p}/d\bar{x}) - ig(\bar{x}/\varepsilon)\bar{p} = 0$$

where

$$\bar{p} \equiv \bar{p}(\bar{x},\ \varepsilon) \tag{5}$$

and

$$\bar{p}(0,\ \varepsilon) = 1.$$

Consistent with our assumptions about the behavior of g we assume that, as $\xi \to \infty$

$$g(\xi) \sim \beta + \sum_{n=1}^{\infty} \gamma_n / \xi^n. \tag{6}$$

This form fulfills the requirements that a vary rapidly near the piston at $x = 0$, and slowly in the region far away from the piston, the terms "near" and "far" being judged with reference to the wavelength $2\pi/k_0$. A concrete example of the kind of function envisaged would be

$$g(\xi) = 1 + (2\delta/\pi)\tan^{-1}\xi$$
$$\sim 1 + \delta - (2\delta/\pi\xi) + \cdots, \tag{7}$$

this implying an almost steplike change in wave speed in the vicinity of the source, followed by a slow change on the wavelength scale.

The problem of Eq. (5) is easy to solve exactly, and we find

$$\bar{p} = \exp[i\bar{\varphi}(\bar{x},\ \varepsilon)], \tag{8}$$

where the phase function is given by

$$\bar{\varphi} = \varepsilon \int_0^{x/\varepsilon} g(\xi)\ d\xi. \tag{9}$$

In the special case of Eq. (7)

$$\bar{\varphi} = \bar{x} + (2\delta/\pi)[\bar{x}\tan^{-1}(\bar{x}/\varepsilon) - \tfrac{1}{2}\varepsilon\ln(\varepsilon^2 + \bar{x}^2) + \varepsilon\ln\varepsilon]. \tag{10}$$

Having the full solution at our disposal, it is possible to make a number of observations about its asymptotic structure for small ε. However, it is instructive to try to appraise the structure directly from Eq. (5), this being the more natural situation in practice.

In formal terms we are faced with the problem of finding a useful approximation to a function $\bar{p}(x,\ \varepsilon)$ for small values of ε, where $\bar{p}$ is defined implicitly by Eq. (5). The common device for this problem is to seek an asymptotic or perturbation expansion of $\bar{p}$ of so-called Poincaré form, i.e. one assumes

$$\bar{p}(\bar{x},\ \varepsilon) \sim \sum_{n=0}^{N} \bar{p}^{(n)}(\bar{x})\mu_n(\varepsilon) + o(\mu_N(\varepsilon)) \tag{11}$$

where

$$\mu_0 = 1, \qquad \mu_{n+1} = o(\mu_n) \quad \text{as} \quad \varepsilon \to 0, \tag{12}$$

this meaning that $\mu_{n+1}/\mu_n \to 0$ as $\varepsilon \to 0$. Once the set of *gauge functions* $\{\mu_n\}$ is prescribed, the coefficients $\bar{p}^{(n)}$ can be uniquely defined by a limit process. We use the symbol $\overline{\lim}$ to denote the limit as $\varepsilon \to 0$ with $\bar{x}$ held constant (meaning in physical terms that we are regarding the limit as achieved by holding the wavelength constant and contracting the inhomogeneity scale l to zero). Then the coefficients in the Poincaré expansion of $\bar{p}(\bar{x},\,\varepsilon)$ with respect to the gauge functions $\{\mu_n\}$ are defined by the limits (which are assumed to exist)

$$\bar{p}^{(0)}(\bar{x}) = \overline{\lim}\ \bar{p}(\bar{x},\,\varepsilon), \tag{13}$$

$$\bar{p}^{(1)}(x) = \overline{\lim}\ \left[\frac{\bar{p}(\bar{x},\,\varepsilon) - \bar{p}^{(0)}(\bar{x})}{\mu_1(\varepsilon)}\right],$$

$$\bar{p}^{(2)}(\bar{x}) = \overline{\lim}\ \left[\frac{\bar{p}(\bar{x},\,\varepsilon) - \bar{p}^{(0)}(\bar{x}) - \mu_1(\varepsilon)\bar{p}^{(1)}(\bar{x})}{\mu_2(\varepsilon)}\right],$$

or, in general,

$$\bar{p}^{(N)}(\bar{x}) = \overline{\lim}\ \left\{\frac{\bar{p}(\bar{x},\,\varepsilon) - \sum_{n=0}^{N-1} \bar{p}^{(n)}(\bar{x})\mu(\varepsilon_n)}{\mu_N(\varepsilon)}\right\}. \tag{14}$$

It is convenient to express this process of *expanding* the function $\bar{p}(\bar{x},\,\varepsilon)$ by defining a formal expansion operator $E_N\{\mu_N\}$ for a *given* gauge sequence $\{\mu_N\}$ by

$$E_N\,\bar{p}(\bar{x},\,\varepsilon) = \sum_{n=0}^{N} \mu_n(\varepsilon)\bar{p}^{(n)}(\bar{x}),$$

where it is assumed that

$$\left|\bar{p}(\bar{x},\,\varepsilon) - E_N\,\bar{p}(\bar{x},\,\varepsilon)\right| = o(\mu_N(\varepsilon)), \tag{15}$$

that the limits of Eq. (14) exist, and that Eq. (15) holds for some set of points $\{\bar{x}\}$. It is evident that one can exercise a great deal of formalistic care in definitions like Eq. (15). However, except where deemed truly necessary, we shall take a heuristic approach, leaving unsaid fine details that would confuse the main points of the arguments. The reader interested in a careful step by step definition–proof discussion of MAE can consult the monograph of Eckhaus (1973) with profit.

If we insert a series of the type of Eq. (11) into Eq. (5) the question arises as to how to choose μ_n. In the region bounded away from $\bar{x} = 0$ for ε small, insertion of Eq. (6) into Eq. (5) gives

$$\frac{d\bar{p}}{d\bar{x}} - i\left\{\beta + \sum_{n=1}^{\infty} \frac{\varepsilon^n\gamma_n}{\bar{x}^n}\right\}\bar{p} = 0, \tag{16}$$

and at first glance it would appear that $\mu_n = \varepsilon^n$ is the proper choice of gauge function. The problem with this choice is that substitution into Eq. (16) of $\bar{p} = \bar{p}^{(0)} + \varepsilon\bar{p}^{(1)} + \cdots$ gives for $\bar{p}^{(1)}$ the inhomogeneous equation

$$(d\bar{p}^{(1)}/d\bar{x}) - i\beta\bar{p}^{(1)} = i\gamma_1\,\bar{p}^{(0)}/\bar{x} \tag{17}$$

where $\bar{p}^{(0)} = e^{i\beta x}$ for the condition $\bar{p}^{(0)}(0) = 1$. The solution of Eq. (17) will give a term proportional to $\ln\,\bar{x}$, and hence when $\bar{x} = O(\varepsilon)$, $\varepsilon\bar{p}^{(1)}$ will be

$O(\varepsilon \ln \varepsilon)$. This effect (referred to as "Switchback," Kaplun, 1967, p. 14), where one finds that terms of intermediate order must be inserted in an expansion, is a common one, and can be taken into account automatically, at least in principle, by assuming μ_n to be unknown. Another clue to the inconsistency of a straightforward ε^n type expansion is that the natural boundary condition on $\bar{p}^{(1)}$ would appear to be $\bar{p}^{(1)} = 0$. However, the presence of a term like $\ln \bar{x}$ which becomes infinite as $\bar{x} \to 0$ implies that $\bar{p}^{(1)}$ cannot be made to satisfy such a condition. The problem, of course, is in the approximation for $g(\bar{x}/\varepsilon)$, which is not valid in the domain $\bar{x} \sim \varepsilon$. This lack of validity, at least in a formal sense, is not serious enough to prevent us from finding a reasonable $\bar{p}^{(0)}$, although the $\ln \bar{x}$ terms seem to terminate the possibility of continuing the expansion process.

"Physical" reasoning suggests that we must introduce a type of *boundary layer* in the region of strong inhomogeity in the medium, i.e. for $\bar{x} = k_0 x = O(\varepsilon) = O(k_0 l)$, or in other words when $x/l = O(1)$. A formal device for doing this is to introduce the "stretching" transformation $\tilde{T}$: $\tilde{x} = \bar{x}/\varepsilon = x/l$, into the differential equation and boundary conditions. Applying $\tilde{T}$ to Eq. (5) gives

$$(d/d\tilde{x})\tilde{p}(\tilde{x},\ \varepsilon) - i\varepsilon g(\tilde{x})\tilde{p} = 0$$
$$\tilde{p}(0) \qquad\qquad\qquad = 1. \qquad\qquad (18)$$

From the above reasoning we expect Eq. (18) to provide a useful asymptotic series solution, at least for $\tilde{x} = O(1)$. Again, we can substitute a series of the form $E_n\tilde{p}(\tilde{x},\ \varepsilon)$ into the equation [note that now $\tilde{x}$ is fixed in the limit processes that define $\tilde{p}^{(n)}(\tilde{x})$]. However, as soon as $\mu_n = O(\varepsilon)$, *integration of the forcing function introduced from lower order terms will give a solution proportional to* $\tilde{x}$. In other words [as we shall see below, Eq. (28)] the expansion for $\tilde{p}$ will contain a term (constant) $\varepsilon\tilde{x}$.

This implies that as $\tilde{x}$ becomes increasingly large the second term of the expansion becomes larger than the first, although the terms were constructed on the basis that the second must be smaller than the first. The $\varepsilon\tilde{x}$ term above begins to have a magnitude comparable with the first term in the piston region solution when $\tilde{x} \sim 1/\varepsilon$.

One can now make the following observations: (1) the expansion in an asymptotic series of the problem for $\bar{x}$ fixed leads to results invalid when $\bar{x} = O(\varepsilon)$; (2) transformation $\tilde{T}$ leads to an (inner) asymptotic series invalid for $\tilde{x} = O(1/\varepsilon)$; (3) the region of validity of the (outer) series is in general outside the point of applicability of the boundary condition; (4) the choice of gauge functions $\mu_n(\varepsilon)$ is not *a priori* obvious, e.g. $\mu_n(\varepsilon) = \varepsilon^n$ leads to "switchback."

To resolve the difficulties we seek to determine *two separate series*, with the gauge functions $\mu_n(\varepsilon)$ to be found as part of the solution. For comparison with later work, the series

$$E_N\bar{p} = \sum_{n=0}^{N} \bar{p}^{(n)}\mu_n(\varepsilon) \qquad\qquad (19)$$

will be called the Helmholtz expansion, and $\tilde{x}$, etc. Helmholtz variables. The reason for this terminology is that the region (away from the piston) where this series applies is analogous to the region where the full Helmholtz or reduced wave equation applies in the scattering and waveguide problems of Sections II and III. In a similar way we shall use the terms, Laplace region and Laplace expansion, for the region near the piston where the variables $\tilde{p}$, $\tilde{x}$ are relevant.

It is assumed that the set of gauge functions can be found so that $\tilde{p}(\tilde{x}, \varepsilon)$ can also be expanded using them, with the proviso that $\tilde{p}^{(n)}$ or $\bar{p}^{(n)}$ may vanish identically for some n. We use the transformation symbol $\tilde{T}$ to denote the change from Helmholtz variables, $\tilde{x}$, to what we are calling Laplace variables, $\tilde{x}$,

$$E_N \tilde{p} = E_N \tilde{T}\bar{p}. \tag{20}$$

Because of the problem with boundary conditions and with determining μ_n we need some rule connecting, say, $E_N \tilde{p}$ and $E_M \bar{p}$ for as arbitrary an N and M as possible. Now $\bar{p}(\tilde{x}, \varepsilon)$ and $\tilde{p}(\tilde{x}, \varepsilon) = \tilde{T}\bar{p}(\tilde{x}, \varepsilon)$ are clearly just different ways of expressing the same function, the transformation $\tilde{T}$ providing a formal means of changing the form of the asymptotic expansion. Therefore, we have every reason to expect a relation between $E_N \tilde{p}$ and $E_M \bar{p}$, the real question being what form such a relation must take. For example, we might have

$$\lim_{\tilde{x} \to 0} \bar{p}^{(0)}(\tilde{x}) = \lim_{\tilde{x} \to \infty} \tilde{p}^{(0)}(\tilde{x}),$$

a simple instance being

$$\bar{p} = (1 + \tilde{x})^{-1}, \qquad \tilde{T}\bar{p} = (1 + \varepsilon\tilde{x})^{-1} \sim 1 - \varepsilon\tilde{x} + \cdots.$$

Now quite generally we can define a relation between the expansions as

$$(\tilde{T}E_N \tilde{T}E_M - E_M \tilde{T}E_N \tilde{T})\bar{p} = H_{N,M}(\tilde{x}, \varepsilon), \tag{21}$$

where $\tilde{T}$ denotes the transformation $\tilde{x} \to \tilde{x}$. For any given function $\bar{p}(\tilde{x}, \varepsilon)$ and gauge functions $\mu_n(\varepsilon)$, the " commutator matrix " $H_{N,M}$ can be readily and routinely calculated from this definition. The first term on the left of Eq. (21) is obtained by expanding $\bar{p}(\tilde{x}, \varepsilon)$ in terms of the $\mu_n(\varepsilon)$ according to the definition of Eq. (14), terms up to and including $\mu_M(\varepsilon)$ being retained. The result is then expressed in terms of the variable $\tilde{x} = \tilde{x}/\varepsilon$, and expanded again in terms of the $\mu_n(\varepsilon)$ with $\tilde{x}$ held fixed this time, and with the inclusion of all terms up to and including μ_N. This contribution is finally reexpressed in terms of the original variable $\tilde{x}$. For the second contribution, the function $\bar{p}(\tilde{x}, \varepsilon)$ is written at once as a function $\tilde{T}\bar{p}$ of $\tilde{x}$ and ε, and this is expanded for fixed $\tilde{x}$ through terms $O(\mu_N)$, the result then being written in terms of $\tilde{x}$ and further expanded through $O(\mu_M)$.

Usually, however, we do not know $\bar{p}(\tilde{x}. \varepsilon)$ exactly, though we do claim to be able to construct asymptotic series like $E_M \bar{p}$ and $E_N \tilde{T}\bar{p}$ by formally

the Laplace expansion proceeds as

$$\tilde{p} \sim 1 + \varepsilon \tilde{p}_K.$$

Note that the possibility of terms of order between 1 and ε in the Helmholtz expansion has been accounted for by not fixing K. Substituting into Eq. (5), after using $\tilde{T}$ and expanding, we have

$$d\tilde{p}^{(K)}/d\tilde{x} = ig(\tilde{x})\tilde{p}^{(0)} = ig(\tilde{x}),$$

and

$$\tilde{p}^{(K)}(0) = 0, \tag{28}$$

so that

$$\tilde{p}^{(K)} = i \int_0^{\tilde{x}} g(\hat{x})\, d\tilde{x}.$$

In applying the matching rule we shall need to know the singular behavior of Eq. (28) as $\tilde{x} \to \infty$. Using the assumed behavior of g for large values of the argument it is straightforward to see that

$$\tilde{p}^{(K)} \underset{\tilde{x} \to \infty}{\sim} \tilde{a}_{0,0}^{(K)} + \tilde{a}_{1,0}^{(K)} \tilde{x} + \tilde{a}_{0,1}^{(K)} \ln \tilde{x}. \tag{29}$$

As we will be most interested in the singular behavior of functions, it is useful to employ a uniform and suggestive notation. For almost all cases of interest forms such as

$$\tilde{p}^{(n)}(\tilde{x} \to 0) \sim \sum \tilde{a}_{j,k}^{(n)} \tilde{x}^j (\ln \tilde{x})^k \tag{30}$$

and

$$\tilde{p}^{(n)}(\tilde{x} \to \infty) \sim \sum \tilde{a}_{j,k}^{(n)} \tilde{x}^j (\ln \tilde{x})^k \tag{31}$$

suffice, where the sum is over nonzero $a_{j,k}^{(n)}$, and j and k may be negative.
    Inserting Eq. (29) into the rule

$$(\tilde{T} E_K \, \tilde{T} E_1 - E_1 \, \tilde{T} E_K \, \tilde{T})\tilde{p} = 0 \tag{32}$$

yields

$$\begin{aligned}
E_1 \, \tilde{T} E_K \, \tilde{T}\tilde{p} &= E_1 \, \tilde{T}[1 + \varepsilon(\tilde{a}_{0,0}^{(K)} + \tilde{a}_{1,0}^{(K)} \tilde{x} \\
&\quad + \tilde{a}_{0,1}^{(K)} \ln \tilde{x}) + \cdots] \\
&= E_1(1 + \varepsilon \ln \varepsilon \, \tilde{a}_{0,1}^{(K)} + \cdots).
\end{aligned}$$

Now $E_1 \tilde{p}$ will designate the series up to the first term after $\tilde{p}^{(0)}$. Hence, for matching to be possible $\mu_1 = \varepsilon \ln \varepsilon$, and in the chosen numbering system for gauge functions $K = 2$, i.e.,

$$\tilde{p} \sim \tilde{p}^{(0)}(\tilde{x}) + \varepsilon \ln \varepsilon \, \tilde{p}^{(1)}(\tilde{x}) + \varepsilon \tilde{p}^{(2)}(\tilde{x}) \tag{33}$$

and

$$\tilde{p} \sim \tilde{p}^{(0)}(\tilde{x}) + \varepsilon \ln \varepsilon \, \tilde{p}^{(1)}(\tilde{x}) + \varepsilon \, \tilde{p}^{(2)}(\tilde{x}), \tag{34}$$

where for this model problem it appears that $\tilde{p}^{(1)} = 0$.

After this preliminary work we retrace our steps, and insert Eqs. (33) and (34) into the problem in Helmholtz and Laplace coordinates, respectively. Thus we have

$$(d/d\bar{x})\bar{p}^{(0)} - i\beta\bar{p}^{(0)} = 0,$$
$$(d/d\bar{x})\bar{p}^{(1)} - i\beta\bar{p}^{(1)} = 0,$$
$$(d/d\bar{x})\bar{p}^{(2)} - i\beta\bar{p}^{(2)} = i\gamma_1\bar{p}^{(0)}/\bar{x}, \tag{35}$$

with the solutions

$$\bar{p}^{(0)} = \bar{A}_0\,e^{i\beta\bar{x}},$$
$$\bar{p}^{(1)} = \bar{A}_1\,e^{i\beta\bar{x}}, \tag{36}$$
$$p^{(2)} = \bar{A}_2\,e^{i\beta\bar{x}} + \bar{A}_0\,i\gamma_1\,e^{i\beta\bar{x}}\ln\bar{x},$$

and

$$d\tilde{p}^{(0)}/d\tilde{x} = d\tilde{p}^{(1)}/d\tilde{x} = 0,$$
$$d\tilde{p}^{(2)}/d\tilde{x} = i\tilde{p}^{(0)}g(\tilde{x}), \tag{37}$$

with the solutions (using the conditions $\tilde{p}^{(0)}(0) = 1$, $\tilde{p}(0)^{(n>0)} = 0$)

$$\tilde{p}^{(0)} = 1, \qquad \tilde{p}^{(1)} = 0, \tag{38}$$
$$\tilde{p}^{(2)} = i\int_0^{\tilde{x}} g(\tilde{x})\,d\tilde{x}.$$

The expression for $\tilde{p}^{(2)}$ can also be expressed in terms of the phase term Eq. (9) of the exact solution

$$\tilde{p}^{(2)} = i\bar{\varphi}(\tilde{x}, 1). \tag{39}$$

Substituting the expressions

$$\bar{p}^{(0)} \sim \bar{a}_{0,0}^{(0)} + \bar{a}_{1,0}^{(0)}\,\bar{x} + \cdots,$$
$$\bar{p}^{(1)} \sim \bar{a}_{0,0}^{(1)} + \bar{a}_{1,0}^{(1)}\,\bar{x} + \cdots, \tag{40}$$
$$\bar{p}^{(2)} \sim \bar{a}_{0,0}^{(2)} + \bar{a}_{1,0}^{(2)}\,\bar{x} + \bar{a}_{1,1}^{(2)}\,\bar{x}\ln\bar{x} + \cdots,$$

for $\bar{x} \to 0$ and

$$\tilde{p}^{(0)} = \tilde{a}_{0,0}^{(0)} = 1,$$
$$\tilde{p}^{(1)} = 0, \tag{41}$$
$$\tilde{p}^{(2)} \sim \tilde{a}_{0,0}^{(2)} + \tilde{a}_{1,0}^{(2)}\,\tilde{x} + \tilde{a}_{0,1}^{(2)}\ln\tilde{x},$$

for $\tilde{x} \to \infty$ into

$$(\bar{T}E_2\,\tilde{T}E_2 - E_2\,\bar{T}E_2\,\tilde{T})\bar{p} = 0 \tag{42}$$

gives

$$\bar{A}_0 = 1, \qquad \bar{A}_1 = i\gamma_1, \qquad \text{and} \qquad \bar{A}_2 = 0. \tag{43}$$

The matching rule also requires that $\tilde{a}_{1,0}^{(2)} = i\beta$ and $\tilde{a}_{0,1}^{(2)} = i\gamma_1$.

Before examining some of the interesting implications of these results, we now briefly examine the general matching rule Eq. (22), in the light of the specific example Eq. (10), with $\delta = \pi/2$. As we have the complete solution at our disposal for this case, we can verify Eq. (43) and perform an empirical test of Eq. (22). Thus, using Eq. (10) and the fact that in Laplace variables

$$\tilde{p} = \exp\left[i\varepsilon\bar{\varphi}(\tilde{x}, 1)\right], \tag{44}$$

we find, with

$$\bar{\varphi}(\tilde{x}, 1) = \tilde{x}(1 + \tan^{-1}\tilde{x}) - \tfrac{1}{2}\ln(1 + \tilde{x}^2), \tag{45}$$

and the definition

$$\bar{y} = (1 + \ln \bar{x}),$$

that the Laplace and Helmholtz expansions to $O(\varepsilon^2)$ are

$$E_5\,\tilde{T}\tilde{p} = 1 + \varepsilon i\bar{\varphi}(\tilde{x}, 1) + \tfrac{1}{2}\varepsilon^2[\bar{\varphi}(\tilde{x}, 1)]^2, \tag{46}$$

and

$$E_5\,\bar{p} = e^{i\beta\tilde{x}}[1 + i\varepsilon \ln \varepsilon - i\varepsilon\bar{y} - \tfrac{1}{2}\varepsilon^2(\ln \varepsilon)^2$$
$$+ \bar{y}\varepsilon^2 \ln \varepsilon - \tfrac{1}{2}\bar{y}^2\,\varepsilon^2]. \tag{47}$$

The argreement with the solutions Eqs. (36) and (38) is evident, if we note that in Eq. (43) $\gamma_1 = -1$ and that $\beta = 1 + \delta = 1 + \pi/2$.

To check the matching formula Eq. (22) it is convenient to use a slightly less explicit notation. Note that the principle expressed in Eq. (21) involves transformation of $E_N\,\tilde{T}E_M\,\bar{p}$ back to the coordinate $\bar{x} = \varepsilon\tilde{x}$. We can define two operators, one for the Helmholtz series, the other for the Laplace series, such that results are always expressed in the $\bar{x}$ coordinate. These two operators in the $\bar{x}$ space are

$$\bar{E}_M = E_M\,\tilde{T} \tag{48}$$

and

$$\bar{E}_N = \tilde{T}E_N\,\tilde{T}. \tag{49}$$

With this notation Eq. (21) reads

$$(\bar{E}_N\,\bar{E}_M - \bar{E}_M\,\bar{E}_N)\bar{p} = H_{N,M}(\bar{x}, \varepsilon). \tag{50}$$

Van Dyke (1964) proposed that $H_{N,M} = 0$ for *all* N, M. If Eqs. (46) and (47) are used in Eq. (50) we can calculate the matrices $\bar{E}_N\,\bar{E}_M\,\bar{p} = \bar{p}_{N,M}$ and $\bar{E}_N\,\bar{E}_M\,\bar{p} = \tilde{p}_{N,M}$. The "commutator" matrix $H_{N,M}$ has the structure for N, $M \le 5$,

$$
\begin{array}{c}
O(\ ) \\[4pt]
1 \\[6pt]
\varepsilon \ln \varepsilon \\[6pt]
\varepsilon \\[6pt]
\varepsilon^2(\ln \varepsilon)^2 \\[6pt]
\varepsilon^2 \ln \varepsilon \\[6pt]
\varepsilon^2
\end{array}
\quad
\begin{array}{ccccccc}
1 & \varepsilon \ln \varepsilon & \varepsilon & \varepsilon^2\,(\ln \varepsilon)^2 & \varepsilon^2 \ln \varepsilon & \varepsilon^2 \\
\left[\begin{array}{c}0\end{array}\right. & 0 & 0 & 0 & 0 & \left.\begin{array}{c}0\end{array}\right] \\
0 & * & 0 & 0 & 0 & 0 \\
0 & 0 & 0 & 0 & 0 & 0 \\
0 & 0 & 0 & * & * & 0 \\
0 & * & 0 & * & * & 0 \\
0 & 0 & 0 & 0 & 0 & 0
\end{array}
\tag{51}
$$

where the asterisks indicate values of N and M for which Van Dyke's rule fails. It is clear that Van Dyke's hypothesis is not generally correct. Fraenkel (1969) has made a careful study of the rule Eq. (22) and has compared it with other principles of matching. One of many conclusions reached was that Eq. (22) is probably the most convenient method for computations. It is usually only for simple problems that complete proofs of the correctness of asymptotic expansions can be found, and for this reason justification of an expansion is usually by physical arguments, the study of known special cases, and/or comparison with numerical and experimental results. It is thus worthwhile finding out under what conditions Eq. (22) may be correct.

Actually Fraenkel was able to achieve a great deal by showing that if the exact solution of the problem has a "reasonable" structure (which he shows to be the case in many interesting examples) then, under certain restrictions on N and M, $H_{N,M} = 0$. We shall return briefly to the problem of matching in the final section of this article. Here we now present a "practical" rule based on Theorem I, Assumption 3 of Fraenkel's paper. Under fairly general conditions, such as the existence of all appropriate limits, a *sufficient* condition for $H_{N,M}$ to vanish, for an arbitrary set of gauge functions μ_N, is that for some $\alpha > 0$

$$\lim_{\varepsilon \to 0} (\mu_{Q+1}/\mu_Q \, \varepsilon^\alpha) = 0 \tag{52}$$

where $Q = N$ or M. For example, in our test case

$$\lim_{\varepsilon \to 0} \frac{\mu_4}{\varepsilon^\alpha \mu_3} = \lim_{\varepsilon \to 0} \frac{\varepsilon^2 \ln \varepsilon}{\varepsilon^2 (\ln \varepsilon)^2 \varepsilon^\alpha} = \lim_{\varepsilon \to 0} \frac{1}{\varepsilon^\alpha \ln \varepsilon}$$

and clearly no $\alpha > 0$ exists for which this limit vanishes. The way to apply this criterion is to match in blocks, i.e. choose a subset of the gauge functions for which Eq. (52) holds. For the case

$$\{\mu_n\} = \{1, \, \varepsilon \ln \varepsilon, \, \varepsilon, \, \varepsilon^2 \ln^2 \varepsilon, \, \varepsilon^2 \ln \varepsilon, \, \varepsilon^2, \, \ldots\}$$

we choose

$$\{1, \, \varepsilon, \, \varepsilon^2, \, \varepsilon^3 \ldots\}$$

which, by Fraenkel's theorem, leads us to match using N, M as indicated in the matrix

	1	$\varepsilon \ln \varepsilon$	ε	$\varepsilon^2 \ln^2 \varepsilon$	$\varepsilon^2 \ln \varepsilon$	ε^2	$\varepsilon^3 \ln^3 \varepsilon$
1	0	—	0	—	—	0	—
$\varepsilon \ln \varepsilon$	—	—	—	—	—	—	—
ε	0	—	0	—	—	0	—
$\varepsilon^2 \ln^2 \varepsilon$	—	—	—	—	—	—	—
$\varepsilon^2 \ln \varepsilon$	—	—	—	—	—	—	—
ε^2	0	—	0	—	—	0	—
$\varepsilon^3 \ln^3 \varepsilon$	—	—	—	—	—	—	—

i.e.,

$$H_{0,0} = H_{2,2} = H_{0,2} = H_{2,0} = H_{3,3} = H_{3,0} = \cdots = 0. \tag{53}$$

The fact that other elements of the matrix (51) vanish must be taken as peculiar to the special example chosen. The possibility that the criterion (52) cannot be satisfied for *any* N, M also occurs in practice, for example in the problem of scattering from a " soft " cylinder. However, we shall defer further discussion of this case until Section II, where we shall see that it is advantageous, and indeed vital, to choose the gauge functions in such a way that the rule Eq. (22) does hold.

Before leaving the problem of matching (to be taken up again in the final section) it is important to consider another point of view on how the matching process works, and which has considerable intuitive appeal despite its computational awkwardness. This is the idea that the matching of two asymptotic series is possible because for a fixed $\varepsilon \to 0$, both series have a common domain of validity to some given order in ε, and hence may be compared. One technique for performing this comparison is to pose an " intermediate " expansion in terms of intermediate variables such as $x^* = \bar{x}/\varepsilon^\alpha$, $1 > \alpha > 0$, to express both asymptotic series in terms of $x^*(\varepsilon)$, and examine the difference as $\varepsilon \to 0$ to determine unknown constants. The book by Cole (1968) gives many examples of this process.

The idea of an intermediate region where both expansions are valid leads to the idea of combining the expansions to produce a so-called " composite expansion," which has the property of being asymptotically equivalent to each expansion in its respective domain of validity. One popular way of accomplishing this is to add the series to given order and to subtract off the common part. Alternatively, one may multiply the two series and divide the product by the common part, though zero divisor problems can occur with this type of composition (see Schneider, 1973).

In terms of the operators $\bar{E}_M$, $\tilde{E}_N$, a composite expansion (not unique) in the space $\bar{x}$ which is valid to $O(\mu_M)$ in the " inner " region and $O(\mu_N)$ in the outer is given by the composite operator

$$E_{M,N} = \bar{E}_M + \tilde{E}_N - \bar{E}_M \tilde{E}_N. \tag{54}$$

Note that

$$\bar{E}_M E_{M,N} \bar{p} = \bar{E}_M \{\bar{E}_M + \tilde{E}_N - \bar{E}_M \tilde{E}_N\}\bar{p} = \bar{E}_M \bar{p}, \tag{55}$$

and, using

$$H_{M,N} = 0,$$
$$\tilde{E}_N E_{M,N} \bar{p} = \tilde{E}_N \bar{p}. \tag{56}$$

Thus the operator $E_{M,N}$ gives a single smooth function in the variable $\bar{x}$ which has asymptotic equivalence with both expansions. For our example problem

$$E_{2,2} \bar{p} = e^{i\beta \bar{x}}[1 + i\varepsilon \ln \varepsilon - i(1 + \ln \bar{x})\varepsilon] + 1$$
$$+ i\varepsilon\{(\bar{x}/\varepsilon)[1 + \tan^{-1}(\bar{x}/\varepsilon)] - \tfrac{1}{2} \ln[1 + (\bar{x}^2/\varepsilon^2)]\}$$
$$-[1 + i\beta \bar{x} + i\varepsilon \ln \varepsilon - i\varepsilon(1 + \ln \bar{x})].$$

C. The Impedance Concept

A major conceptual advantage of MAE is that it provides formal definitions of various useful conceptual structures. For example, the general idea of impedance is frequently related to making a division of a problem, one part of which is then represented as an impedance to another part. Again we illustrate this with our sample problem. In an impedance approach we think of the "piston" in the region of strong sound speed variation as being represented by a "generator" and a "source impedance" which drive a "transmission line" represented by Eq. (1b). Thus generator and source impedance represent boundary conditions for Eq. (16). If one-dimensional sound waves in a gas were being described by our model, the velocity would be given by

$$\bar{u} = -i\, d\bar{p}/d\bar{x} \tag{57}$$

where $\bar{u} = uZ_0/p_0$, $Z_0 = \rho_0 a_0$ being the wave impedance.

We label the applied driving pressure at $\bar{x} = 0$, the effective pressure, and the source impedance with subscripts I, E, and S, respectively. Then, referring to Fig. 1 for sign conventions, we have

$$\bar{p}_{\mathrm{I}} = \bar{p}_{\mathrm{E}} + \bar{Z}_{\mathrm{S}}\,\bar{u}_{\mathrm{E}}. \tag{58}$$

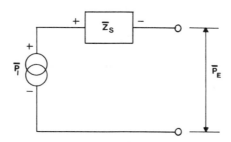

Fig. 1. Equivalent circuit for source.

In our perturbation procedure *all the source* $\bar{p}_{\mathrm{I}}$ was taken to drive the first term $\bar{p}^{(0)}$ in the expansion for $\bar{p}$. Therefore, using Eq. (57) in Eq. (58) and the expansion for $\bar{p}$, we find, with $\bar{Z}_{\mathrm{S}} = \bar{Z}_{\mathrm{S}}^{(0)} + \varepsilon \ln \varepsilon \bar{Z}_{\mathrm{S}}^{(1)} + \cdots$,

$$\bar{p}_{\mathrm{I}} = \lim_{\bar{x}\to 0} (\bar{p}^{(0)} - i\bar{Z}^{(0)}\, d\bar{p}^{(0)}/d\bar{x}) \tag{59}$$

and if we now substitute the series (40) for $\bar{p}^{(0)}$ as $\bar{x} \to 0$, this gives

$$\bar{p}_{\mathrm{I}} = \bar{a}_{0,0}^{(0)} - i\bar{Z}_{\mathrm{S}}^{(0)}\,\bar{a}_{1,0}^{(0)}. \tag{60}$$

The matching $(\tilde{E}_0\bar{E}_0 - \bar{E}_0\tilde{E}_0)\bar{p} = 0$ gives $\bar{a}_{0,0}^{(0)} = \tilde{a}_{0,0}^{(0)}$, and as $\tilde{a}_{0,0}^{(0)}$ is the constant term in the Laplace region, i.e., the driving pressure $\bar{p}_{\mathrm{I}}$, we find

$$\bar{Z}_{\mathrm{S}}^{(0)} = 0. \tag{61}$$

The next term is $O(\varepsilon \ln \varepsilon)$, and Eq. (58) gives

$$\lim_{\bar{x} \to 0} [i \, \bar{Z}_{\mathrm{S}}^{(1)}(d\bar{p}^{(0)}/d\bar{x}) - \bar{p}^{(1)}] = 0. \tag{62}$$

Then substitution of the small $\bar{x}$ form of $\bar{p}^{(1)}$ from Eq. (40) gives

$$\bar{a}_{0,0}^{(1)} = i\bar{Z}_{\mathrm{S}}^{(1)}\bar{a}_{1,0}^{(0)}. \tag{63}$$

The coefficients $\bar{a}_{0,0}^{(1)}$ and $\bar{a}_{1,0}^{(0)}$ are found from the matching

$$(\bar{E}_2 \bar{E}_2 - \tilde{E}_2 \bar{E}_2)\bar{p} = 0, \tag{64}$$

and can be expressed in terms of the coefficients $\tilde{a}$ in the large $\bar{x}$ series for $\tilde{p}^{(n)}$, Eq. (41), to give

$$\bar{Z}_{\mathrm{S}}^{(1)} = i\tilde{a}_{0,1}^{(2)}/\tilde{a}_{1,0}^{(2)}. \tag{65}$$

The same line of attack, using

$$\lim_{\bar{x} \to 0} (\bar{p}^{(2)} - i \, \bar{Z}^{(2)} \, d\bar{p}^{(0)}/d\bar{x}) = 0, \tag{66}$$

gives $\bar{Z}^{(2)}$. However, it is at this point that the true singular behavior of the outer approximation series comes into play, in the fact that as $\bar{x} \to 0$, $\bar{p}^{(2)} \sim \bar{a}_{0,1}^{(2)} \ln \bar{x}$. This indicates that $\bar{Z}^{(2)}$ must display the same behavior, if the impedance is to replicate the true outer solution. We adopt the formal device that

$$F(0) = a\{\ln x\}$$

means

$$F(x) \sim a \ln x \tag{67}$$

as

$$x \to 0,$$

and with the matching Eq. (64) this gives

$$\bar{Z}_{\mathrm{S}}^{(2)} = -i(\tilde{a}_{0,1}^{(2)}/\tilde{a}_{1,0}^{(2)})\{\ln \bar{x}\}. \tag{68}$$

Determination of the inner coefficients $\tilde{a}^{(2)}$ requires solution of the inner or Laplace region problem to $O(\varepsilon^2)$. For our example Eq. (46)

$$\tilde{p}^{(0)} = \bar{p}_{\mathrm{I}}, \qquad \tilde{p}^{(1)} = 0$$

and

$$\tilde{p}^{(2)} \sim i\beta\tilde{x} - i \ln \tilde{x}$$

($\bar{p}_{\mathrm{I}} = 1$ with our normalization), and hence we find

$$\bar{Z}_{\mathrm{S}} = \varepsilon(1 + \ln \varepsilon)(i/\beta)(1 - \{\ln \bar{x}\}) + O(\varepsilon^2 \ln^2 \varepsilon). \tag{69}$$

Using Eq. (69) with the impedance definition Eq. (58) and the $O(\varepsilon)$ truncated version of Eq. (16),

$$d\bar{p}/d\bar{x} - i[\beta - (\varepsilon/\bar{x})]\bar{p} = 0, \tag{70}$$

one can solve the problem in the outer *Helmholtz region* to $O(\varepsilon)$. The main point is that the large $\tilde{x}$ behavior of the inner problem sequence can be used to define the effective impedance of the source region in a quite definite manner which can be carried out to as high an order of approximation as needed. We shall come across a number of examples in which MAE can be used to verify the accuracy and validity as well as compute the needed parameters in such idealized acoustic constructs. Section II, E deals with the "end correction" for a thick plate, for example, while an equivalent "scattering matrix" is discussed in Section III, C.

D. A Second-Order Model Equation

Before leaving the domain of elementary one-dimensional examples, let us examine a problem closer to the type that one meets in practice, i.e. a problem with two directions of wave motion possible, so that instead of Eq. (2) as our starting point, we have a second-order equation. Thus, after suitable normalization and removal of the explicit time dependence by Fourier analysis, our equivalent of Eq. (5) takes the form

$$d^2\bar{p}/d\tilde{x}^2 + g(\tilde{x}/\varepsilon)\bar{p} = 0,$$
$$g(\xi) = \beta^2 + \alpha e^{-\gamma\xi},$$
$$\bar{p}(0) = 1, \tag{71}$$
$$\bar{p}(\tilde{x} \to \infty) \sim e^{i\beta\tilde{x}}$$

To fix ideas, the reader might think of this as a simple model expressing, for example, the variation in sound speed in a bonding layer between a crystal transducer and a target medium with nondimensional wavenumber β, the limit $\varepsilon \to 0$ expressing the fact that the layer thickness is small compared with the typical wavelength. In accord with this physical situation, we restrict $\tilde{x}$ to positive values.

The operation $\varepsilon \to 0$ on Eq. (71) with $\tilde{x}$ fixed gives that $g(\tilde{x}/\varepsilon) = \beta^2 +$ terms of exponentially small order. Therefore the $\tilde{x}$ dependence of $\bar{p}$ for correction terms of algebraic order, that also meet the outgoing or radiation condition, must be of the form

$$\bar{p}^{(L)} = \bar{A}_L e^{i\beta\tilde{x}} \tag{72}$$

and hence,

$$E_L \bar{p} = E_L \bar{T}\bar{p} = \left(\sum_{n=0}^{L} \mu_n(\varepsilon)\bar{A}_L\right)e^{i\beta\tilde{x}}. \tag{73}$$

Clearly, such a representation as Eq. (73) is not capable of providing an asymptotic representation of p "near" the piston surface, where the exponential dependence of the varying sound speed becomes important, i.e., when $\bar{x} = O(\varepsilon)$. In this Laplace region, over which phase changes are small, the suitable variable is again $\tilde{x} = \bar{x}/\varepsilon$, and Eqs. (71) take the form

$$d^2\tilde{p}/d\tilde{x}^2 + \varepsilon^2 g(\tilde{x})\tilde{p} = 0,$$
$$\tilde{p}(0) = 1, \tag{74}$$
$$p(\tilde{x} \to \infty): \text{ form given by matching.}$$

The appearance of the term ε^2 in Eq. (74) might give the impression that the gauge functions $\mu_K(\varepsilon)$ for the problem might proceed as $1, \varepsilon^2, \varepsilon^4, \ldots$. We shall adopt a somewhat cautious point of view and leave the general form of μ_K undetermined. When $\varepsilon \to 0$ in Eq. (74), we are clearly left with a meaningful problem. It is also evident that the effect of $g(\tilde{x})$ will first become significant for a term $O(\varepsilon^2)$. Therefore we assume, as a tentative hypothesis, the form

$$E_2\tilde{T}\tilde{p} = E_2\tilde{p} = \tilde{p}^{(0)} + \mu_1(\varepsilon)\tilde{p}^{(1)} + \varepsilon^2\tilde{p}^{(2)}. \tag{75}$$

If Eq. (75) does not prove to have enough structure between terms $O(1)$ and $O(\varepsilon^2)$, we shall have to relabel the terms and add more gauge functions. (In the next section we will study diffraction by a strip, where we are confronted with this type of problem and show how it can be dealt with in a systematic way.) Inserting Eq. (75) into Eq. (74) and thus developing equations for $\tilde{p}^{(0)}$, $\tilde{p}^{(1)}$, and $\tilde{p}^{(2)}$ we find

$$\tilde{p}^{(0)} = 1 + \tilde{B}_0\tilde{x}, \qquad \tilde{p}^{(1)} = \tilde{B}_1\tilde{x},$$
$$\tilde{p}^{(2)} = (\alpha/\gamma^2)(1 - e^{-\gamma\tilde{x}}) + [\tilde{B}_2 - (\alpha/\gamma)]\tilde{x} - \tfrac{1}{2}\beta^2\tilde{x}^2, \tag{76}$$

where we have used the boundary conditions $\tilde{p}^{(0)}(0) = 1$, $\tilde{p}^{(n>0)}(0) = 0$. The reader will note that $\tilde{B}_0$, $\tilde{B}_1$, and $\tilde{B}_2$ remain to be determined by the matching principle, and that the so-called *eigensolutions* (general solutions to the homogeneous problem) $\tilde{B}_0\tilde{x}$, $\tilde{B}_1\tilde{x}$, and $\tilde{B}_2\tilde{x}$ are singular as $x \to \infty$.

Under our assumptions about the sequence μ_n, the outer solution to $O(\varepsilon^2)$ is given by Eq. (73) as

$$E_2\bar{p} = (\bar{A}_0 + \mu_1\bar{A}_1 + \varepsilon^2\bar{A}_2)e^{i\beta\bar{x}}. \tag{77}$$

With the assumed gauge function defining E and with $\bar{E} = E\bar{T}$, $\tilde{E} = \tilde{T}E\tilde{T}$, the matching principle has the form

$$(\bar{E}_2\tilde{E}_2\tilde{E}_2 - \bar{E}_2)\tilde{p} = 0. \tag{78}$$

For this simple problem the reader can easily confirm that, for Eq. (78) to hold, it is required that

$$\mu_1 = \varepsilon, \qquad \bar{A}_0 = 1,$$
$$\bar{A}_1 = 0, \qquad \bar{A}_2 = \alpha/\gamma^2, \tag{79}$$

and that

$$\tilde{B}_0 = 0, \qquad \tilde{B}_1 = i\beta, \qquad \tilde{B}_2 = \alpha/\gamma. \tag{80}$$

Therefore

$$E_2 \bar{p} = [1 + (\alpha/\gamma^2)\varepsilon^2]e^{i\beta\tilde{x}}, \tag{81}$$

and

$$E_2 \tilde{p} = 1 + i\beta\tilde{x}\varepsilon + \varepsilon^2[(\alpha/\gamma^2)(1 - e^{-\gamma\tilde{x}}) - \tfrac{1}{2}\beta^2\tilde{x}^2]. \tag{82}$$

Finally, we can form a *composite* expansion by using the formula (54). Thus

$$E_{2,2}\bar{p} = e^{i\beta\tilde{x}}[1 + (\varepsilon^2\alpha/\gamma^2)(1 - \exp\{-[i\beta + (\gamma/\varepsilon)]\tilde{x}\})], \tag{83}$$

a result which should be relatively simple to interpret and which provides a *uniformly* valid approximation in $0 \leqslant \tilde{x} < \infty$ to $O(\varepsilon^2)$.

We could continue with a number of one-dimensional wave propagation problems of increasing complexity, including ones that fall into the class of WKB turning point problems (see, for example, Harper *et al.*, 1971; Nayfeh, 1973, Chapter 7; Murray, 1974, Chapter 6; O'Malley, 1974). However, it is more instructive for the purpose of this article to examine some classical diffraction problems, and problems relating to resonant cavities and wave-guides. For some informative MAE treatments of one-dimensional linear problems, the reader can consult the second chapter of Cole (1968). We shall, however, return to one-dimensional wave propagation in our examination of nonlinear acoustics.

The present section contains all the formalism needed to understand other sections, which can now be read independently.

II. Scattering and Diffraction Problems

A. Introduction

In this section we examine some acoustic scattering and diffraction problems from the MAE viewpoint. We start by considering the scattering of a plane wave by a strip of width $2a$, using MAE to find a uniformly valid asymptotic solution in the low frequency limit $\varepsilon = k_0 a \to 0$. (We shall some-times refer to ε, essentially the ratio of obstacle size to wavelength, as the Helmholtz number.) Without having to carry the expansion too far, we find that this problem exemplifies several interesting and important features. Firstly, whether the strip is hard or soft, there are a large number of eigen-solutions to both the inner (Laplace) and outer (Helmholtz) problems, depending upon conditions which can be tolerated at infinity and at the strip edges. Secondly, logarithmic gauge functions and the "switchback" phenomenon arise early on, in the second approximation for the hard strip, in fact. Thirdly, an inhomogeneous Laplace equation arises in the second inner approximation, and in this simple case the reader can see in detail how to deal with this kind of situation. Fourthly, if the strip is soft the asymp-totic sequence of gauge functions appears to proceed very slowly, in inverse powers of $(\ln \varepsilon)$, giving a situation analogous to that arising in plane viscous flow at low Reynolds number (Van Dyke, 1964, p. 161). We shall see that a naive application of the Matching Rule Eq. (22) is quite unsatisfactory, the rule working for some values of (N, M) and failing for others. However, a simple change of the gauge function $(\ln \varepsilon)^{-1}$ to $(\ln \varepsilon + K)^{-1}$ effectively allows the whole series in inverse powers of $(\ln \varepsilon)$ to be summed, leaving only more rapidly decreasing algebraic functions of ε, and allows the matching to be performed with no difficulty.

We then go on to summarize some results obtained elsewhere for a more complicated problem, involving the diffraction of a plane wave by a semiinfinite screen of small, but finite, thickness. For this problem the approximate MAE method leads to simple, exact, closed-form results for some quantities, in contrast to the approximate purely numerical results which have been given previously from an "exact" solution of the problem. The article concludes with a section which quotes some results for three-dimensional scattering problems which may be constructively attacked using MAE. The great difficulties encountered in applying MAE, as much as any other method, to three-dimensional scattering problems are also emphasized.

B. Scattering by a Hard Strip

We take the simplest case, in which the plane wave, with potential $\phi^i = \exp(-ik_0 y - i\omega t)$, is at normal incidence to the hard strip occupying $y = 0$, $|x| < a$; see Fig. 2a. Then, suppressing the time factor $\exp(-i\omega t)$, $(\omega = a_0 k_0)$, the complete problem for the scattered potential is

$$(\nabla^2 + k_0{}^2)\phi = 0, \tag{84}$$

$$(\partial/\partial y)[\phi + \exp(-ik_0 y)] = 0 \qquad (y = 0, \quad |x| < a),$$

which implies that ϕ is odd in y and hence that

$$\phi = 0 \qquad (y = 0, \quad |x| > a). \tag{85}$$

In addition, ϕ must satisfy a radiation condition at infinity, and ϕ must satisfy some edge conditions of the kind

$$\phi = O(1) \qquad \rho\nabla\phi = o(1) \tag{86}$$

as $\rho = [(x \pm a)^2 + y^2]^{1/2} \to 0$. These conditions imply that the pressure must be finite and the velocity no more than integrably singular at the edges. By well-known theorems in diffraction theory, the problem defined by Eqs. (84)–(86) has a unique solution (e.g., Jones, 1964).

Outer or Helmholtz variables are defined as

$$\bar{x} = k_0 x, \qquad \bar{y} = k_0 y,$$

$$\phi(x, y) \equiv \bar{\phi}(\bar{x}, \bar{y}), \tag{87}$$

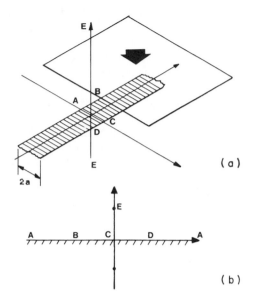

FIG. 2. Scattering by a strip and conformal mapping of Eq. (97).

and we assume an outer expansion

$$\bar{\phi}(\bar{x}, \bar{y}) \sim \sum_{n=0}^{\infty} \mu_n(\varepsilon) \bar{\phi}^{(n)}(\bar{x}, \bar{y}).$$

In outer variables, the complete problem is

$$
\begin{aligned}
(\bar{\nabla}^2 + 1)\bar{\phi} &= 0, \\
(\partial/\partial\bar{y})\bar{\phi} &= i \qquad (\bar{y} = 0, \quad |\bar{x}| < \varepsilon), \\
\bar{\phi} &= 0 \qquad (\bar{y} = 0, \quad |\bar{x}| > \varepsilon).
\end{aligned}
\tag{88}
$$

When the outer expansion is substituted into this problem, the condition on $\partial\bar{\phi}/\partial\bar{y}$ is irrelevant and all we can enforce is the condition on $\bar{\phi}$, together with the radiation condition. Thus

$$
\begin{aligned}
(\bar{\nabla}^2 + 1)\bar{\phi}^{(n)} &= 0, \\
\bar{\phi}^{(n)} &= 0 \qquad (\bar{y} = 0, \quad |\bar{x}| \neq 0),
\end{aligned}
\tag{89}
$$

i.e. $\bar{\phi}^{(n)}$ is an outer eigensolution for all n. The most general (generalized) function satisfying Eq. (89) can easily be shown to be

$$\bar{\phi}^{(n)} = \sum_m \bar{A}_{nm} \lambda_m(\bar{r}, \theta) \tag{90}$$

where $\lambda_m(\bar{r}, \theta) = H_m^{(1)}(\bar{r}) \sin m\theta$ $(m = 1, 2, \ldots)$ and $\bar{x} = \bar{r} \cos \theta$, $\bar{y} = \bar{r} \sin \theta$, $-\pi < \theta \leqslant +\pi$.

Now the outer and inner solutions do not both necessarily start with the same order term, so that, according to our scheme for numbering the gauge functions, we should not prejudice the issue by assuming that the first term of $\bar{\phi}$ is in fact $O(\mu_0)$, i.e., the inner solution may start off with higher order terms than the outer solution. We therefore assume merely that to leading order

$$\bar{\phi} \sim \mu_R(\varepsilon)\bar{\phi}^{(R)} = \mu_R \sum_m \bar{A}_{Rm} \lambda_m(\bar{r}, \theta). \tag{91}$$

According to the principle of Minimum Singularity (Van Dyke, 1964, p. 53), it is a matter of practical experience that only the least singular term of Eq. (91) is capable of matching a suitably well-behaved inner solution, and therefore

$$\bar{\phi} \sim \mu_R \bar{A}_{R1} H_1^{(1)}(\bar{r}) \sin \theta \tag{92}$$

is the leading term, denoted by $E_R\bar{\phi}(\bar{x}, \bar{y}, \varepsilon)$ of the outer expansion. Terms corresponding to $\bar{A}_{Rm}$ with $m > 1$ can be included if the reader is suspicious of the Minimum Singularity Principle, but then matching with an inner solution will indeed be found to show that $\bar{A}_{Rm} = 0$ for $m > 1$, *provided the inner solution satisfies the edge conditions.*

Now the appropriate inner variables are evidently

$$(\tilde{x}, \tilde{y}) = \tilde{T}(\bar{x}, \bar{y}) = (\bar{x}/\varepsilon, \bar{y}/\varepsilon)$$

and

$$\tilde{\phi}(\tilde{x}, \tilde{y}, \varepsilon) = \tilde{T}\bar{\phi}(\bar{x}, \bar{y}, \varepsilon). \tag{93}$$

The gauge functions are assumed to be chosen so that $\tilde{\phi}$ may also be expanded in them,

$$\tilde{\phi}(\tilde{x}, \tilde{y}, \varepsilon) \sim \sum_{n=0}^{\infty} \mu_n(\varepsilon)\tilde{\phi}^{(n)}(\tilde{x}, \tilde{y}).$$

The complete problem in inner variables is

$$
\begin{aligned}
(\tilde{\nabla}^2 + \varepsilon^2)\tilde{\phi} &= 0, \\
\partial\tilde{\phi}/\partial\tilde{y} &= i\varepsilon \qquad (\tilde{y} = 0, \quad |\tilde{x}| < 1), \\
\tilde{\phi} &= 0 \qquad (\tilde{y} = 0, \quad |\tilde{x}| > 1),
\end{aligned}
\tag{94}
$$

where the edge conditions are to be enforced, but the radiation condition must be relinquished in favor of matching with the outer solution. When the inner expansion is inserted into Eq. (94), we find that the leading term must vanish identically unless its gauge function is appropriately chosen. If we require $\tilde{\phi}^{(0)}$ to be nonzero, then there is only one possible choice for $\mu_0(\varepsilon)$—the *distinguished limit* (Cole, 1968, p. 10) $\mu_0(\varepsilon) = \varepsilon$, which preserves just enough structure in the differential equation and boundary conditions to start the expansion immediately with a nontrivial term. Then

$$
\begin{aligned}
\tilde{\nabla}^2\tilde{\phi}^{(0)} &= 0, \\
\partial\tilde{\phi}^{(0)}/\partial\tilde{y} &= i \qquad (\tilde{y} = 0, \quad |\tilde{x}| < 1), \\
\tilde{\phi}^{(0)} &= 0 \qquad (\tilde{y} = 0, \quad |\tilde{x}| > 1),
\end{aligned}
\tag{95}
$$

which is satisfied by $\tilde{\phi}^{(0)} = i\tilde{y} + f_0$, where f_0 is any inner eigensolution, i.e., a solution of

$$\tilde{\nabla}^2 f_0 = 0,$$
$$\partial f_0 / \partial \tilde{y} = 0 \qquad (\tilde{y} = 0, \quad |\tilde{x}| < 1), \qquad (96)$$
$$f_0 = 0 \qquad (\tilde{y} = 0, \quad |\tilde{x}| > 1).$$

The most general inner eigensolution can be found by conformal mapping. We use the imaginary unit j in this context, to avoid confusion with the i arising from the time dependence. Then the mapping

$$w = u + jv = j\left(\frac{\tilde{z} - 1}{\tilde{z} + 1}\right)^{1/2}, \qquad (97)$$

with $\tilde{z} = \tilde{x} + j\tilde{y}$, takes the segment $|\tilde{x}| < 1$, $\tilde{y} = 0$ into the whole line $v = 0$, and the flow domain into $v > 0$, as depicted in Fig. 2b. The line $\tilde{y} = 0$, $|\tilde{x}| > 1$, on which $f_0 = 0$, maps into the imaginary axis $u = 0$. The eigensolution f_0 is the real part (with respect to j) of a complex potential $\Omega(w)$ which has constant imaginary part on $v = 0$, and zero real part on $u = 0$, $v > 0$.

Now, in order to match the outer solutions, the potentials $\tilde{\phi}^{(n)}$ must all be small as $|\tilde{z}| \to \infty$. Because of particular solutions like the $i\tilde{y}$ in $\tilde{\phi}^{(0)}$, the eigensolutions need not be small, but can be algebraically large at infinity in the $\tilde{z}$-plane. Correspondingly, in the w-plane the eigensolutions can have a pole of finite order at $w = j$. In general, they may also have half-power singularities at the edges of the strip, which in the w-plane allows the complex potential to have algebraic growth as $|w| \to \infty$, and to have an algebraic singularity as $w \to 0$. These arguments apply, of course, only to the flow domain $v > 0$. However, since f_0 is the real part of an Ω which has zero imaginary part on $v = 0$, Ω is analytically continuable to $v < 0$, and may have a pole of finite order at $w = -j$ (identical with that at $w = +j$) and poles of finite order at $w = \infty$, $w = 0$. The most general form of Ω is thus

$$\Omega(w) = \frac{w^n}{(1 + w^2)^m} \qquad (98)$$

(times any constant which is real with respect to j), m amd n being any positive or negative integers. This gives

$$\Omega(\tilde{z}) = (j^n / 2^m)(\tilde{z} - 1)^{n/2}(\tilde{z} + 1)^{m - n/2}, \qquad (99)$$

which clearly has all the stated properties.

In the present case of normal incidence, symmetry between $\tilde{z} - 1$ and $\tilde{z} + 1$ requires $m = n$. The resulting function Ω then has zero imaginary part on $\tilde{y} = 0$, $|\tilde{x}| < 1$, but has zero real part on $\tilde{y} = 0$, $|\tilde{x}| > 1$ only when $m = \pm 1$, ± 3, ± 5, $\ldots$. Thus the most general symmetric inner eigenfunction is

$$\Lambda_m = \text{Re}_j[(j/2)^{2m-1}(\tilde{z}^2 - 1)^{m-1/2}]_{(m = \ldots, -2, -1, 0, +1, +2, \ldots)}. \qquad (100)$$

The eigensolutions which fulfill the edge condition have $m \geqslant 1$; those which vanish at infinity have $m \leqslant 0$. As $|\tilde{z}| \to \infty$,

$$\Lambda_m \sim \mathrm{Re}_j[(j/2)^{2m-1}\tilde{z}^{2m-1}f_m(\tilde{z})], \tag{101}$$

$$f_m(\tilde{z}) \sim 1 - \left(m - \frac{1}{2}\right)\frac{1}{\tilde{z}^2} + \frac{(m - \frac{1}{2})(m - \frac{3}{2})}{2!}\frac{1}{\tilde{z}^4} + \cdots.$$

Now consider the matching principle

$$(\tilde{E}_N \bar{E}_M - \bar{E}_M \tilde{E}_N)\bar{\phi} = 0, \qquad \bar{T}E_N \tilde{T} = \bar{E}_N, \qquad E_M \bar{T} = \bar{E}_M, \tag{102}$$

where $\tilde{T}$ denotes the transformation from $\bar{x} \to \tilde{x}$, $\bar{T}$ that from $\tilde{x} \to \bar{x}$, and the E_N are the partial expansion operators through $O[\mu_N(\varepsilon)]$. Take $M = R$ and then

$$\tilde{T}E_R\bar{\phi} = \mu_R(\varepsilon)\bar{A}_{R1}H_1^{(1)}(\varepsilon\tilde{r})\sin\theta$$
$$\sim (1/\varepsilon)\mu_R(\varepsilon)\bar{A}_{R1}(-2i/\pi\tilde{r})\sin\theta$$

to leading order. For this to match the inner expansion starting with $\mu_0(\varepsilon)\phi^{(0)} \equiv \varepsilon\phi^{(0)}$, we must have

$$\mu_R(\varepsilon) = \varepsilon^2.$$

Thus the matching rule is to be applied with $M = R$ (where $\mu_R = \varepsilon^2$) and $N = 0$ ($\mu_0 = \varepsilon$), and we have

$$\bar{T}E_0\,\tilde{T}E_R\bar{\phi} = \varepsilon^2\bar{A}_{R1}(-2i/\pi\tilde{r})\sin\theta. \tag{103}$$

To match this, $\bar{\phi}^{(0)}$ must vanish as $|\tilde{z}| \to \infty$, and yet must satisfy the edge conditions, so that the only possibility is that

$$\bar{\phi}^{(0)} = i\tilde{y} + \tilde{a}_{01}\Lambda_1(\tilde{z}). \tag{104}$$

This gives

$$E_0\,\tilde{T}\bar{\phi} = \varepsilon[i\tilde{y} + \tilde{a}_{01}\Lambda_1(\tilde{z})],$$
$$\bar{T}E_0\,\tilde{T}\bar{\phi} = \varepsilon[(i\bar{y}/\varepsilon) + \tilde{a}_{01}\Lambda_1(\tilde{z}/\varepsilon)],$$
$$E_R\,\bar{T}E_0\,\tilde{T}\bar{\phi} = \varepsilon\{(i\bar{y}/\varepsilon) + \tilde{a}_{01}\,\mathrm{Re}_j[(j\tilde{z}/2\varepsilon) - (j\varepsilon/4\tilde{z})]\}$$
$$= i\bar{y} - \tfrac{1}{2}\tilde{a}_{01}\bar{y} - \tfrac{1}{4}\tilde{a}_{01}\varepsilon^2(\sin\theta)/\bar{r}, \tag{105}$$

and matching Eq. (105) with Eq. (103) then yields

$$\tilde{a}_{01} = 2i, \qquad \bar{A}_{R1} = \pi/4, \tag{106}$$

so that

$$\bar{\phi} = \varepsilon^2(\pi/4)H_1^{(1)}(\bar{r})\sin\theta + o(\varepsilon^2),$$
$$\bar{\phi} = \varepsilon[i\tilde{y} + 2i\Lambda_1(\tilde{z})] + o(\varepsilon). \tag{107}$$

C. HIGHER ORDER APPROXIMATIONS

We continue to higher order by observing that the expansion Eq. (105) proceeds with a term $O(\varepsilon^4)$,

$$E_S \bar{T} E_0 T\bar{\phi} = -\frac{i\varepsilon^2}{2\bar{r}} \sin \theta - \frac{i\varepsilon^4}{8\bar{r}^3} \sin 3\theta. \tag{108}$$

This suggests that

$$\bar{\phi} = \mu_R(\varepsilon)\bar{\phi}^{(R)} + \mu_S(\varepsilon)\bar{\phi}^{(S)} + o(\mu_S),$$

where $\mu_S(\varepsilon) \equiv \varepsilon^4$ and

$$\bar{\phi}^{(S)} = \sum_{m=1}^{K} \bar{A}_{Sm} H_m^{(1)}(\bar{r}) \sin m\theta.$$

For matching with an inner solution which is at most $O(\varepsilon)$, the inner expansion of $\varepsilon^4 \bar{\phi}^{(S)}(\varepsilon\tilde{x})$ must be at most $O(\varepsilon)$, and therefore K cannot exceed 3, so that

$$\bar{\phi}^{(S)} = \sum_{m=1}^{3} \bar{A}_{Sm} H_m^{(1)}(\bar{r}) \sin m\theta. \tag{109}$$

The coefficient $\bar{A}_{S3}$ can be immediately determined by using the matching rule with $M = S$, $N = 0$,

$$(\bar{E}_0 \bar{E}_S - \bar{E}_S \bar{E}_0)\bar{\phi} = 0. \tag{110}$$

The second member of this equation is given by Eq. (108) above, while the first member is

$$-\frac{i\varepsilon^2}{2\bar{r}} \sin \theta - \frac{i\varepsilon^4 16\bar{A}_{S3}}{\pi} \frac{\sin 3\theta}{\bar{r}^3}, \tag{111}$$

and therefore

$$\bar{A}_{S3} = \pi/128. \tag{112}$$

Now the next term, after $O(\varepsilon)$, of the inner expansion of the outer series as far as $O(\mu_S(\varepsilon))$ is in fact $O(\varepsilon^2)$ arising from the $H_2^{(1)}(\bar{r})$ term in Eq. (109)

$$E_R \bar{T} E_S \bar{\phi} = E_0 \bar{T} E_S \bar{\phi} - \frac{4i}{\pi} \bar{A}_{S2} \varepsilon^2 \frac{\sin 2\theta}{\bar{r}^2}. \tag{113}$$

The inner expansion therefore continues with a term $O(\mu_R) = O(\varepsilon^2)$

$$\tilde{\phi} \sim \varepsilon\tilde{\phi}^{(0)} + \mu_R(\varepsilon)\tilde{\phi}^{(R)} + o(\mu_R),$$

where $\tilde{\phi}^{(R)}$ is an eigensolution which must satisfy the edge conditions *and* vanish as $|\tilde{z}| \to \infty$ in order to match Eq. (113). There is no eigensolution with this property, and therefore

$$\bar{A}_{S2} = 0. \tag{114}$$

We may now simplify our notation a little by writing

$$\mu_R(\varepsilon) \equiv \mu_1(\varepsilon) = \varepsilon^2. \tag{115}$$

Continuing the inner expansion of the outer series, we find that the next terms beyond $O(\varepsilon^2)$ are $O(\varepsilon^3 \ln \varepsilon)$ and $O(\varepsilon^3)$,

$$E_Q \tilde{T} E_S \tilde{\phi} = E_0 \tilde{T} E_S \tilde{\phi} - \frac{2i}{\pi} \bar{A}_{S1} \varepsilon^3 \frac{\sin \theta}{\tilde{r}}$$

$$- \frac{i\varepsilon^3}{64} \frac{\sin 3\theta}{\tilde{r}} + \frac{\pi\varepsilon^3}{4} \sin \theta \left\{ \frac{i\tilde{r}}{\pi} (\ln \tilde{r} + \ln \varepsilon) \right.$$

$$\left. + \frac{1}{2} \tilde{r} \left[1 - \frac{i}{\pi} (1 - 2\gamma + 2 \ln 2) \right] \right\} \tag{116}$$

where $\gamma = 0.5772 \ldots$ is the Euler constant. This suggests an inner expansion

$$\tilde{\phi} \sim \varepsilon \tilde{\phi}^{(0)} + \mu_P(\varepsilon) \tilde{\phi}^{(P)} + \mu_Q(\varepsilon) \tilde{\phi}^{(Q)}, \tag{117}$$

where

$$\mu_P(\varepsilon) \equiv \varepsilon^3 \ln \varepsilon, \qquad \mu_Q(\varepsilon) \equiv \varepsilon^3,$$

and we treat μ_P and μ_Q as both of the same algebraic order, $O(\varepsilon^3)$, in order to avoid the kind of possible errors mentioned in the Introduction. $\tilde{\phi}^{(P)}$ is an eigensolution which we consider later. The problem for $\tilde{\phi}^{(Q)}$ involves a Poisson equation,

$$\tilde{\nabla}^2 \tilde{\phi}^{(Q)} = -\tilde{\phi}^{(0)}, \tag{118}$$

where

$$\phi^{(0)} = i\tilde{y} + i \operatorname{Re}_j j(\tilde{z}^2 - 1)^{1/2}.$$

$\tilde{\phi}^{(Q)}$ must also satisfy the boundary conditions $\tilde{\phi}^{(Q)} = 0$ on $\tilde{y} = 0$, $|\tilde{x}| > 1$, $\partial \tilde{\phi}^{(Q)}/\partial \tilde{y} = 0$ on $\tilde{y} = 0$, $|\tilde{x}| < 1$, must vanish at infinity, and must satisfy the edge conditions.

Write

$$\tilde{\phi}^{(Q)} = (-i\tilde{y}^3/3!) - i \operatorname{Re}_j(j\psi), \tag{119}$$

and then

$$\tilde{\nabla}^2 \psi = (\tilde{z}^2 - 1)^{1/2},$$

or

$$\partial^2 \psi / \partial \tilde{z} \, \partial \tilde{z}^* = \tfrac{1}{4}(\tilde{z}^2 - 1)^{1/2}$$

where

$$\tilde{z}^* = \tilde{x} - j\tilde{y}.$$

Thus

$$\psi = \tfrac{1}{8}\tilde{z}\tilde{z}^*(\tilde{z}^2 - 1)^{1/2} - \tfrac{1}{8}\tilde{z}^* \cosh^{-1} \tilde{z}, \tag{120}$$

plus any function of $\tilde{z}$ alone. The first term in ψ does satisfy the boundary conditions on $\tilde{\phi}^{(Q)}$, as does the $\tilde{y}^3$ term in Eq. (119). The second contribution to ψ can be made to satisfy the boundary conditions if a term

$$\tfrac{1}{8}\tilde{z}\cosh^{-1}\tilde{z}$$

is added to ψ. The complete expression for $\tilde{\phi}^{(Q)}$ also satisfies the edge conditions, but is not small at infinity. We must therefore add on eigensolutions which satisfy the edge conditions, and grow no more rapidly than $\tilde{z}^3$ at infinity, to give

$$\tilde{\phi}^{(Q)} = (i\tilde{y}^3/3!) - i\,\mathrm{Re}_j[\tfrac{1}{8}j|\tilde{z}|^2(\tilde{z}^2-1)^{1/2}$$
$$- \tfrac{1}{8}j(\tilde{z}*-\tilde{z})\cosh^{-1}\tilde{z}] + \tilde{a}_{Q1}\Lambda_1(\tilde{z}) + \tilde{a}_{Q2}\Lambda_2(\tilde{z}). \tag{121}$$

For the eigensolution $\tilde{\phi}^{(P)}$ a term growing as rapidly as $\tilde{z}^3$ is not permitted, but there is no reason to suppose that the less rapidly growing eigensolution Λ_1 will not be present, so that we take

$$\tilde{\phi}^{(P)} = \tilde{a}_{P1}\Lambda_1(\tilde{z}). \tag{122}$$

Thus

$$E_Q\,\tilde{T}\tilde{\phi} = E_0\,\tilde{T}\tilde{\phi} + \varepsilon^3\ln\varepsilon\,\tilde{\phi}^{(P)} + \varepsilon^3\tilde{\phi}^{(Q)}$$

and

$$E_S\,\tilde{T}E_Q\,\tilde{T}\tilde{\phi} = E_S\,\tilde{T}E_0\,\tilde{T}\tilde{\phi}$$
$$+ \varepsilon^3 a_{Q2}\left[\frac{1}{8}\left(\frac{3\tilde{x}^2\tilde{y}}{\varepsilon^3} - \frac{\tilde{y}^3}{\varepsilon^3}\right) - \frac{3}{16}\frac{\tilde{y}}{\varepsilon} - \frac{3}{64}\frac{\varepsilon\sin\theta}{\tilde{r}}\right]$$
$$+ \varepsilon^3\tilde{a}_{Q1}\left(-\frac{1}{2}\frac{\tilde{y}}{\varepsilon} - \frac{1}{4}\frac{\varepsilon\sin\theta}{\tilde{r}}\right) - \frac{i\tilde{y}^3}{6}$$
$$- i\varepsilon^3\left(-\frac{\tilde{r}^3\sin\theta}{8\varepsilon^3} - \frac{\tilde{r}\sin\theta}{16\varepsilon} - \frac{\varepsilon}{64\tilde{r}}(3\sin\theta - 4\sin^3\theta)\right.$$
$$\left.- \frac{1}{4}\frac{\tilde{r}}{\varepsilon}\sin\theta(\ln 2 + \ln\tilde{r} - \ln\varepsilon) + \frac{\varepsilon\sin\theta}{16\tilde{r}}(1 - 2\sin^2\theta)\right)$$
$$+ \varepsilon^3\ln\varepsilon\,\tilde{a}_{P1}\left(-\frac{\tilde{y}}{2\varepsilon} - \frac{\varepsilon\sin\theta}{4\tilde{r}}\right). \tag{123}$$

This expression must match the result of applying $\tilde{T}$ to Eq. (116), namely,

$$\tilde{T}E_0\,\tilde{T}E_S\tilde{\phi} = \frac{2i}{\pi}\,\bar{A}_{S1}\varepsilon^4\frac{\sin\theta}{\tilde{r}} - \frac{i\varepsilon^4}{64\tilde{r}}(3\sin\theta - 4\sin^3\theta)$$
$$+ \frac{\pi\varepsilon^3}{4}\sin\theta\left\{\frac{i\tilde{r}}{\varepsilon}\ln\tilde{r} + \frac{\tilde{r}}{2\varepsilon}\left[-\frac{i}{\pi}(1 - 2\gamma + 2\ln 2)\right]\right\}. \tag{124}$$

The first terms in Eqs. (123) and (124) agree because of the matching which has already been carried out. Matching of the remaining terms gives the following results:

$$\tilde{a}_{Q2} = -\tfrac{1}{3}i$$
$$\tilde{a}_{Q1} = -\tfrac{1}{4}\pi + \tfrac{1}{2}i - \tfrac{1}{2}i\gamma + i \ln 2,$$
$$\tilde{a}_{P1} = -\tfrac{1}{2}i,$$
$$\bar{A}_{S1} = \tfrac{1}{2}\pi(5/64 + i\pi/16 - \gamma/8 + \tfrac{1}{4} \ln 2), \qquad (125)$$

and at the same time provides a partial check on the working in that these choices, determined from a particular set of matchings, then automatically ensure the matching of other terms. There is, however, a term $(i/8)\varepsilon^4 \ln \varepsilon \, \bar{r}^{-1} \sin \theta$ left unmatched in Eq. (123). This deficiency indicates that we should add a "Switchback" term

$$-(\pi/16)\varepsilon^4 \ln \varepsilon \, H_1^{(1)}(\bar{r}) \sin \theta \qquad (126)$$

to the outer solution, for this supplies only the contribution

$$(i/8)\varepsilon^4 \ln \varepsilon \, \bar{r}^{-1} \sin \theta$$

to Eq. (124), and therefore matches the extra term in Eq. (123) *without* violating any of the other matchings we have performed.

Thus the inner solution proceeds with the series

$$\tilde{\phi} = \{\varepsilon, \, \varepsilon^3 \ln \varepsilon, \, \varepsilon^3, \, \ldots\},$$

the outer with

$$\bar{\phi} = \{\varepsilon^2, \, \varepsilon^4 \ln \varepsilon, \, \varepsilon^4, \, \ldots\},$$

and we have now determined the outer series completely to $O(\varepsilon^4)$, and the inner to $O(\varepsilon^3)$.

As a check on the results obtained, we can calculate the total power in the radiated wave,

$$P = (\rho_0 \omega^2/a_0) \lim_{r \to \infty} \int_0^{2\pi} |\phi|^2 \, rd\theta$$

and we find

$$P = (2\rho_0 \omega^2/a_0)\varepsilon^4(\pi^2/16)[1 + (5\varepsilon^2/16)(1 - 8\delta/5) + o(\varepsilon^2)] \qquad (127)$$

where

$$\delta = \gamma + \ln \varepsilon - 2 \ln 2.$$

This agrees exactly with results given by Bowman *et al.* (1969, p. 210). It is clear, moreover, that the MAE calculation presented here can be carried out to any order in ε, using precisely the methods which have been needed to derive Eq. (127). Further, that while some of the many classical methods which have been derived to attack this problem may quickly yield a number of terms in the expansion for particular quantities, such as the pressure on the strip, or the far-field scattering coefficient, there is no method superior to the MAE method for providing a detailed and readily interpretable picture of the entire field. Usually it is convenient to have that picture in the form of separate inner and outer expansions, as we are normally interested in either the pressure field on the scattering surface or in the distant radiation field. Composite expansions can, however, be formed if desired (see Van Dyke, 1964, p. 94). Thus, from Eqs. (107) a composite approximation, valid to $O(\varepsilon^2)$ for $\bar{r} = O(1)$ and to $O(\varepsilon)$ for $\tilde{r} = O(1)$, may be formed either by additive composition, giving

$$\phi_a \sim \varepsilon^2 (\pi/4) \sin \theta [H_1^{(1)}(\bar{r}) + (2i/\pi\bar{r})] + \varepsilon [i\tilde{y} + 2i\Lambda_1(\tilde{z})], \qquad (128)$$

or by multiplicative composition, which gives the somewhat neater result

$$\phi_m \sim (\pi\bar{r}/2i)\varepsilon H_1^{(1)}(\bar{r})[i\tilde{y} + 2i\Lambda_1(\tilde{z})]. \qquad (129)$$

D. Scattering by a Soft Strip

Turning now to the soft strip, the boundary condition $\phi + e^{-ik_0 y} = 0$ on $(y = 0, |x| < a)$ implies that ϕ is even in y, so that $\partial\phi/\partial y = 0$ for $(y = 0, |x| > a)$. All terms of the outer expansion

$$\bar{\phi}(\bar{x}, \bar{y}, \varepsilon) \sim \sum g_n(\varepsilon)\bar{\phi}^{(n)}(\bar{x}, \bar{y}) \qquad (130)$$

are outer eigensolutions, with the general form

$$\bar{\phi}^{(N)} = \sum_{m=0} \bar{A}_{Nm} H_m^{(1)}(\bar{r}) \cos m\theta. \qquad (131)$$

For the leading order term, $\bar{\phi}^{(1)}$ say, the Minimum Singularity Principle, or matching, shows that only a monopole is present, with

$$\bar{\phi}^{(1)} = \bar{A}_{10} H_0^{(1)}(\bar{r}), \qquad (132)$$

and to leading order

$$g_1(\varepsilon)\bar{\phi}^{(1)}(\varepsilon\bar{r}) \sim g_1(\varepsilon)(2i\bar{A}_{10}/\pi) \ln \varepsilon. \qquad (133)$$

Assuming an inner expansion

$$\tilde{\phi}(\tilde{x}, \tilde{y}, \varepsilon) \sim \sum g_n(\varepsilon)\tilde{\phi}^{(n)}(\tilde{x}, \tilde{y}), \qquad (134)$$

we see that the leading term will be nonzero (implying $\bar{A}_{10} \neq 0$) only if the leading gauge function is

$$g_0(\varepsilon) = 1,$$

and that then

$$\tilde{\phi}^{(0)} = -1. \qquad (135)$$

If we now take $g_1(\varepsilon) = 1/\ln \varepsilon$, and tentatively apply the rule

$$\bar{T}E_0 \tilde{T}E_1\bar{\phi} = E_1\bar{T}E_0 \tilde{T}\bar{\phi},$$

we have Eq. (133) for the left side, -1 for the right, and hence we get

$$\bar{A}_{10} = \pi i/2. \tag{136}$$

Proceeding beyond Eq. (133) we have

$$E_1 \tilde{T} E_1 \bar{\phi} = -(1/\ln \varepsilon)(\ln \varepsilon + \ln \tilde{r} - \ln 2 + \gamma - \tfrac{1}{2}\pi i), \tag{137}$$

$$\bar{T} E_1 \tilde{T} E_1 \bar{\phi} = -(1/\ln \varepsilon)(\ln \tfrac{1}{2}\tilde{r} + \gamma - \tfrac{1}{2}\pi i). \tag{138}$$

The first of these equations suggests that

$$\tilde{\phi} = -1 + (1/\ln \varepsilon)\tilde{\phi}^{(1)} + \cdots, \tag{139}$$

in which $\tilde{\phi}^{(1)}$ is an inner eigensolution, a harmonic function with $\tilde{\phi}^{(1)} = 0$ on $\tilde{y} = 0$, $|\tilde{x}| < 1$ and with $\partial \tilde{\phi}^{(1)}/\partial \tilde{y} = 0$ on $\tilde{y} = 0$, $|\tilde{x}| > 1$. The general form of this eigensolution can be found using the mapping Eq. (97) again. For the present case of a soft body, however, it is more convenient to use the mapping

$$\tilde{z} = \tfrac{1}{2}[\tilde{w} + (1/\tilde{w})]. \tag{140}$$

Since this sends the strip in the $\tilde{z}$-plane into the unit circle in the $\tilde{w}$-plane, it is obvious that the general inner eigensolution is a linear combination of $\mathrm{Re}_j \ln \tilde{w}$ and $\mathrm{Re}_j(\tilde{w}^n - \tilde{w}^{-n})$. The source term $\ln \tilde{w}$ is evidently required for $\tilde{\phi}^{(1)}$ so that we assume

$$\tilde{\phi}^{(1)} = \tilde{a}_1 \, \mathrm{Re}_j \ln[\tilde{z} + (\tilde{z}^2 - 1)^{1/2}]. \tag{141}$$

There are now two possibilities for determining $\tilde{a}_1$. Firstly, since $\tilde{\phi}$ contains no $O(1)$ term, we can use the matching rule

$$\bar{T} E_1 \tilde{T} E_0 \tilde{\phi} = E_0 \, \bar{T} E_1 \tilde{T} \tilde{\phi}, \tag{142}$$

in which the left side is zero, while

$$E_1 \tilde{T} \tilde{\phi} = -1 + (1/\ln \varepsilon)\tilde{a}_1 \, \mathrm{Re}_j \ln[\tilde{z} + (\tilde{z}^2 - 1)^{1/2}] \tag{143}$$

so that

$$E_0 \, \bar{T} E_1 \tilde{T} \tilde{\phi} -- 1 - \tilde{a}_1. \tag{144}$$

Thus

$$\tilde{a}_1 = -1.$$

Alternatively, using the rule

$$\bar{T} E_1 \tilde{T} E_1 \bar{\phi} = E_1 \bar{T} E_1 \tilde{T} \tilde{\phi}, \tag{145}$$

the left side is given by Eq. (138) above, while from Eq. (143) the right side is equal to

$$-1 - \tilde{a}_1 + (1/\ln \varepsilon)\tilde{a}_1 \ln 2\tilde{r}. \tag{146}$$

Clearly there is no choice of $\tilde{a}_1$ which makes Eq. (138) identical with Eq. (146), though the choice $\tilde{a}_1 = -1$ does at least have the merit of matching the $O(1)$ terms and the terms proportional to $\ln \tilde{r}/\ln \varepsilon$.

The fact that the matching nearly works leads us to suspect that a slight change in the gauge functions may yield expansions which have the structure necessary for matching to work.

Fraenkel (1969, p. 217) has discussed this kind of situation, showing by example that three cases may arise, depending on the gauge functions. With the "most appropriate" choices the principle Eq. (22) holds; with a less appropriate choice the principle may or may not hold, while with other choices there is no matching principle at all. At present we are in the third position. However, experience of other problems involving inverse powers of $\ln \varepsilon$ (for example, plane flow at low Reynolds number, Van Dyke, 1964, p. 161), consideration of, say, the scattering of a plane acoustic wave by a soft circular cylinder, for which the total potential is given exactly by

$$\phi_t = \exp(ik_0 r \cos \theta) - \sum_{n=-\infty}^{+\infty} i^n e^{in\theta} \frac{J_n(k_0 a)}{H_n^{(1)}(k_0 a)} H_n^{(1)}(k_0 r), \qquad (147)$$

and consideration of the fact that mere redefinition of ε by a constant factor involves the addition of a constant to $\ln \varepsilon$, all lead to the idea of trying

$$g_0(\varepsilon) = 1, \qquad g_1(\varepsilon) = 1/(\ln \varepsilon + K). \qquad (148)$$

Then, with notation as before,

$$E_1 \bar{\phi} = [1/(\ln \varepsilon + K)]\bar{A}_{10} H_0^{(1)}(\bar{r}), \qquad (149)$$

$$E_1 \tilde{T} E_1 \bar{\phi} = \frac{2i\bar{A}_{10}}{\pi} + \frac{2i\bar{A}_{10}}{\pi} \frac{(\ln \tilde{r} - \ln 2 + \gamma - \frac{1}{2}\pi i - K)}{(\ln \varepsilon + K)},$$

while

$$E_1 \tilde{T} \bar{\phi} = -1 + [1/(\ln \varepsilon + K)]\tilde{a}_1 \text{ Re}_j \ln[\tilde{z} + (\tilde{z} - 1)^{1/2}] \qquad (150)$$

and

$$E_1 \tilde{T} E_1 \tilde{T} \bar{\phi} = -1 + [\tilde{a}_1/(\ln \varepsilon + K)][-(\ln \varepsilon + K) + K + \ln 2\tilde{r}].$$

Now the rules

$$\bar{T} E_0 \tilde{T} E_1 \bar{\phi} = E_1 \bar{T} E_0 \tilde{T} \bar{\phi}$$

and

$$\bar{T} E_1 \tilde{T} E_0 \bar{\phi} = E_0 \bar{T} E_1 \tilde{T} \bar{\phi}$$

lead to the same choice of $\bar{A}_{10}$ and $\tilde{a}_1$ as previously, namely $\bar{A}_{10} = \pi i/2$, $\tilde{a}_1 = -1$. However, the rule Eq. (145) which failed before now requires

$$-(K + \ln 2\tilde{r}) = (2i\bar{A}_{10}/\pi)(\ln \tilde{r} - \ln 2 + \gamma - \frac{1}{2}\pi i)$$

which is satisfied if

$$K = \gamma - \frac{1}{2}\pi i - \ln 4. \qquad (151)$$

Thus the matching is satisfactorily accomplished, and leads quite naturally to the introduction of the parameter

$$K + \ln \varepsilon \equiv \ln(\tfrac{1}{4} k_0 a) + \gamma - \tfrac{1}{2} \pi i,$$

whose origin in other approximate treatments of this kind of problem is normally rather obscure. Further, what appeared earlier to take the form of a very slow expansion in inverse powers of $\ln \varepsilon$,

$$\bar\phi = (1/\ln \varepsilon)\bar\phi^{(1)} + [1/(\ln \varepsilon)^2]\bar\phi^{(2)} + \cdots, \tag{152}$$

is now seen to arise simply from the binomial expansion of $(\ln \varepsilon + K)^{-1}$. We have in fact summed the logarithmic series, and the second term in the *new* outer expansion is of algebraic order in ε. To see this, expand Eq. (150) to the next term beyond $g_1(\varepsilon)$,

$$E_2 \bar{T} E_1 \tilde{T} \bar\phi = - \frac{(K + \ln 2\bar{r})}{(\ln \varepsilon + K)} + \frac{\varepsilon^2}{(\ln \varepsilon + K)} \frac{\cos 2\theta}{4\bar{r}^2}.$$

This suggests that the outer expansion takes the form

$$\bar\phi \sim [1/(\ln \varepsilon + K)]\bar\phi^{(1)} + [\varepsilon^2/(\ln \varepsilon + K)]\bar\phi^{(2)} + \varepsilon^2 \bar\phi^{(3)} + \cdots \tag{153}$$

where we should again treat the last two terms simply as both $O(\varepsilon^2)$. Note, however, that there is no reason why a switchback effect should not occur, forcing us to introduce an earlier term with gauge function $\varepsilon^2(\ln \varepsilon + K)$, for example. In fact that does not occur at this stage, though it does at $O(\varepsilon^6)$, where the gauge functions $\varepsilon^6(\ln \varepsilon + K)^2$, $\varepsilon^6(\ln \varepsilon + K)$, ε^6, $\varepsilon^6(\ln \varepsilon + K)^{-1}$, and $\varepsilon^6(\ln \varepsilon + K)^{-2}$ are all needed (cf. Bowman *et al.*, 1969, p. 190). Pursuit of higher order terms for this problem is an arduous business, though in principle it can be accomplished to any order using only the methods which have already been demonstrated in this article. The interested reader is recommended to look at the somewhat simpler problem of scattering by a soft circular cylinder, for which the exact solution Eq. (147) will serve as a ready check on the working. Harper (1969) has examined this problem in detail from the MAE viewpoint, showing how the time-harmonic results can be used in a Fourier synthesis to describe transient pulse scattering by a soft cylinder.

E. Diffraction by a Thick Plate

As a demonstration of the full power of the MAE technique we examine briefly the problem of the diffraction of a plane wave by a semiinfinite rigid plate of small but finite thickness $2a$, the expansion parameter being the Helmholtz number $\varepsilon = k_0 a$ (Crighton and Leppington, 1973). Although the outer solutions here are much more complicated than the eigensolutions of Sections II,B and C above, the principle is just the same; an outer wavefield with boundary conditions appropriate to a zero-thickness half-plane is matched to an inner incompressible field describing the details of the flow around the blunt edge of the plate.

This problem has been attacked by other authors using a formally exact extension of the Wiener–Hopf method. In particular, Jones (1953) gives a contour integral solution, valid for arbitrary ε, involving an infinite sequence of parameters which are themselves the solutions of an infinite system of linear algebraic equations with coefficients dependent on ε. For $\varepsilon \ll 1$ this system can be truncated to a small closed finite system, and the parameters have purely numerical values which can be found approximately. In the limit $\varepsilon \to 0$, the parameters have a simple interpretation in terms of the Fourier coefficients of the pressure across the end face of the plate, and a simple interpretation of the effect of finite thickness on the distant diffracted sound field can also be given.

The specific results obtained by Jones are that the first few parameters have the approximate values

$$A_1 = 0.3481, \qquad A_3 = 0.0543,$$
$$A_5 = 0.0229, \qquad A_7 = 0.0130, \dots \tag{154}$$

and that to $O(\varepsilon)$ the distant field of the thick plate is identical with that of a parallel plate duct of width $2a$ whose plates are longer than the plate faces by an amount

$$L - 0.22a, \tag{155}$$

apart from an obvious monopole effect associated with plane wave propagation down the duct.

It is perhaps remarkable that the approximate MAE method improves on these results by providing *exact closed form expressions*. Moreover, the MAE method can be developed much further than the "exact" method, though admittedly the process is very laborious, and in this problem it illustrates in a very striking manner the danger of failing to include all logarithmic terms with those of the same algebraic order in performing the matching.

Suppose the plate occupies $x < 0$, $|y| < a$ with a plane incident wave $\exp[-ik_0(x \cos \theta_0 + y \sin \theta_0)]$. Then the scattered field satisfies

$$(\overline{\nabla}^2 + k_0{}^2)\phi = 0,$$
$$\partial \overline{\phi}/\partial y = ik_0 \sin \theta_0 \exp(-ik_0 x \cos \theta_0 \mp i\varepsilon \sin \theta_0) \qquad (y = \pm a, \quad x < 0),$$
$$\partial \overline{\phi}/\partial x = ik_0 \cos \theta_0 \exp(-ik_0 y \sin \theta_0) \qquad (x = 0, \quad -a < y < +a),$$

with no more than an integrable singularity in $\nabla \phi$ at the corners, and with a radiation condition at infinity. Defining outer coordinates as $\bar{x} = k_0 x$, $\bar{y} = k_0 y$, it is found that the outer expansion takes the form

$$\overline{\phi}(\bar{x}, \bar{y}; \varepsilon) \sim \overline{\phi}^{(0)} + \varepsilon \ln \varepsilon \, \overline{\phi}^{(1)} + \varepsilon \, \overline{\phi}^{(2)}$$
$$+ \varepsilon^2 \ln^2 \varepsilon \, \overline{\phi}^{(3)} + \varepsilon^2 \ln \varepsilon \, \overline{\phi}^{(4)} + \varepsilon^2 \overline{\phi}^{(5)} + O(\varepsilon^3), \tag{156}$$

while in inner coordinates $\tilde{x} = \bar{x}/\varepsilon$, $\tilde{y} = \bar{y}/\varepsilon$,

$$\tilde{\phi}(\tilde{x}, \tilde{y}; \varepsilon) \sim \varepsilon^{1/2}\tilde{\phi}^{(0)} + \varepsilon\,\tilde{\phi}^{(1)} + \varepsilon^{3/2}\ln\varepsilon\,\tilde{\phi}^{(2)}$$
$$+ \varepsilon^{3/2}\tilde{\phi}^{(3)} + \varepsilon^2\ln\varepsilon\,\tilde{\phi}^{(4)} + \varepsilon^2\tilde{\phi}^{(5)} + O(\varepsilon^{5/2}), \qquad (157)$$

(using a notation more convenient here than our usual one). The first outer solution $\bar{\phi}^{(0)}$ is Sommerfeld's solution for the zero-thickness problem,

$$\bar{\phi}^{(0)} = \tfrac{1}{2}\exp[-i\bar{r}\cos(\theta + \theta_0)] - \tfrac{1}{2}\exp[-i\bar{r}\cos(\theta - \theta_0)]$$
$$-\tfrac{1}{2}\exp[-i\bar{r}\cos(\theta + \theta_0)]\,\mathrm{erf}\,\{e^{-\pi i/4}(2\bar{r})^{1/2}\cos[(\theta + \theta_0)/2]\}$$
$$-\tfrac{1}{2}\exp[-i\bar{r}\cos(\theta - \theta_0)]\,\mathrm{erf}\,\{-e^{-\pi i/4}(2\bar{r})^{1/2}\cos[(\theta - \theta_0)/2]\}. \qquad (158)$$

This fails to describe the field correctly *only* within a circular neighborhood of radius $O(a)$ around the plate edge, and it is therefore appropriate to stretch both of the coordinates in the same way. We find then that the inner behavior of $\bar{\phi}^{(0)}$ is

$$\bar{\phi}^{(0)}(\varepsilon\tilde{r}, \theta) \sim \varepsilon^{1/2}(2^{3/2}e^{-\pi i/4}/\pi^{1/2})\tilde{r}^{1/2}\sin(\theta/2)\sin(\theta_0/2). \qquad (159)$$

The first inner solution $\tilde{\phi}^{(0)}$ is harmonic, with zero normal derivative over the whole of the thick plate ($\tilde{x} < 0$, $|\tilde{y}| < 1$), and with behavior $\tilde{r}^{1/2}\sin(\theta/2)$ as $\tilde{r} \to \infty$ in order to match Eq. (159). It is proportional to the real part of a complex potential $\Omega(\tilde{w}) = \tilde{w}$, where

$$-\tfrac{1}{2}\pi(\tilde{z} + i) = \tilde{w}(\tilde{w}^2 - 1)^{1/2} - \ln[\tilde{w} + (\tilde{w}^2 - 1)^{1/2}] \qquad (160)$$

maps the flow domain into $\mathrm{Im}\,\tilde{w} > 0$, the surface of the plate into $\mathrm{Im}\,\tilde{w} = 0$, and the corners into $\tilde{w} = \pm 1$.

Now although Eq. (160) defines $\tilde{w}$ implicitly in terms of $\tilde{z}$, the Fourier coefficients of the potential across the end face, $\tilde{x} = 0$, $-1 < \tilde{y} < +1$, can be found explicitly, and the result for the quantities A_{2n+1} is

$$A_{2n+1} = \frac{1}{2(2n+1)}\left[J_n\left(n + \frac{1}{2}\right) - J_{n+1}\left(n + \frac{1}{2}\right)\right], \qquad (161)$$

the first few values of which agree with those in Eq. (154) to within the numerical accuracy of those results. By continuing the outer expansion through $O(\varepsilon)$ and matching it to the inner expansion through $O(\varepsilon)$, one can also show that the so-called "end correction" L is given exactly by

$$L = a(\ln 2)/\pi, \qquad (162)$$

which agrees numerically with Eq. (155). Further, as if the numerical agreement were not a sufficiently convincing demonstration of the validity of the MAE approach, it was shown by Crighton and Leppington (1973) that Eq. (161) does indeed constitute the exact solution of the full infinite system of equations whose approximate numerical solution is given by Eq. (154), and that a certain infinite series involving the A_{2n+1} can be summed exactly to produce the result Eq. (162) for the end correction.

In the calculation of higher order terms an interesting phenomenon presents itself. If one attempts to treat logarithmic and algebraic gauge functions independently, then the outer $O(\varepsilon^2 \ln^2 \varepsilon)$ term can be found and matched to the inner solution in an apparently perfectly satisfactory way— that is, the right functional forms arise from inner and outer expansions, and there are just enough matching equations to determine all the unknowns uniquely. However, one finds that the asymptotic form of $\bar{\phi}^{(3)}$ is then not symmetric in the angles θ, θ_0, as is required at all orders by the Reciprocal Theorem, permitting the interchange of source and receiver. Without this additional check one would not suspect any error in $\bar{\phi}^{(3)}$. Of course, at even higher order some inconsistency would be bound to arise, but then one might not be able to carry the calculations through to a sufficiently high order for the error to be revealed.

This then serves as another warning as to the importance of assessing orders of magnitude on algebraic grounds alone [if $\bar{\phi}^{(3)}$, $\phi^{(4)}$ and $\phi^{(5)}$ are all treated as $O(\varepsilon^2)$ and taken together in a " block " matching then it is found that, as $\bar{r} \to \infty$, these functions are indeed each symmetric in θ and θ_0], and of the danger in regarding consistency alone as a sufficient justification for either the form of an expansion or for the form of a matching rule.

F. THREE-DIMENSIONAL PROBLEMS

Three-dimensional scattering problems at low Helmholtz number can in principle be treated in just the same way as the problems already discussed here. In practice, however, we are limited by the relative scarcity of solutions to the inner problem in three dimensions, unless the scattering body coincides with a coordinate surface in one of the coordinate systems in which the Laplace equation is separable. The reader will find it a straightforward matter, as suggested by Kanwal (1967), for example, to derive several terms in the inner and outer series for the scattering of a plane wave $\phi^i = \exp(-ik_0 r \cos \theta)$ by a sphere $r = a$, in the limit $\varepsilon = k_0 a \to 0$. The results should be compared with those obtained by applying the inner and outer limit processes to the exact solutions

$$\phi = -\sum_{n=0}^{\infty} (-i)^n (2n + 1) \frac{j_n(\varepsilon)}{h_n^{(1)}(\varepsilon)} h_n^{(1)}(k_0 r) P_n(\cos \theta), \qquad (163)$$

$$\phi = -\sum_{n=0}^{\infty} (-i)^n (2n + 1) \frac{j_n'(\varepsilon)}{h_n^{(1)'}(\varepsilon)} h_n^{(1)}(k_0 r) P_n(\cos \theta), \qquad (164)$$

valid, respectively, for the soft sphere and the hard sphere, with ϕ denoting the scattered potential. Note that the solution Eq. (163) indicates that the difficulties experienced in Section II,D are likely to arise only in two-dimensional scattering problems with soft surfaces.

A more taxing problem in three dimensions is created by the scattering of an axially incident plane wave $\phi^i = \exp ik_0 z$ by a paraboloid of revolution $\eta = \eta_1$, the paraboloidal coordinates being defined by $x = 2(\xi \eta)^{1/2} \cos \alpha$,

$y = 2(\xi\eta)^{1/2} \sin \alpha$, $z = \xi - \eta$. η_1 may be interpreted as one-half of the radius of curvature at the nose, and in the low frequency limit $\varepsilon = k_0\,\eta_1 \to 0$. Again, the results should be compared with those derived from the exact solutions

$$\phi = - \left[\frac{\frac{1}{2}\pi - \mathrm{Si}(2k_0\,\eta) + i\,\mathrm{Ci}(2k_0\,\eta)}{\frac{1}{2}\pi - \mathrm{Si}(2k_0\,\eta_1) + i\,\mathrm{Ci}(2k_0\,\eta_1)} \right] e^{ik_0 z}, \tag{165}$$

$$\phi = \left(\frac{\frac{1}{2}\pi - \mathrm{Si}(2k_0\,\eta) + i\,\mathrm{Ci}(2k_0\,\eta)}{(e^{2ik_0\eta_1}/k_0\,\eta_1) - [\frac{1}{2}\pi - \mathrm{Si}(2k_0\,\eta_1) + i\,\mathrm{Ci}(2k_0\,\eta_1)]} \right) e^{ik_0 z}, \tag{166}$$

for the soft and hard surface, respectively. These solutions are given by Bowman *et al.* (1969, pp. 602 and 611). The sine and cosine integral functions are defined as

$$\mathrm{Si}(x) = \int_0^x (\sin t)/t\,dt, \qquad \mathrm{Ci}(x) = - \int_x^\infty (\cos t)/t\,dt.$$

A vastly more difficult problem is created by the three-dimensional analog of the problem of Section II,E—that of the scattering by a semiinfinite circular rod in the low frequency limit. Matched Expansions would seem the ideal technique for attacking this problem, yet it is impossible to make any progress short of solving the entire problem. The first-order inner flow is simply that of uniform incompressible streaming past the rod $-\infty < \tilde{x} < 0$, $\tilde{r} \leq 1$, which at present can be found only with the aid of Jones' (1953, 1955) modified Wiener–Hopf method, which, as in Section II,E, reduces the problem to that of numerical inversion of an infinite matrix—just the kind of problem we hope to circumvent by using MAE. Moreover, technical difficulties often arise in the solution of purely incompressible flow problems by the Wiener–Hopf method, and what is normally taken is the limit $k_0 \to 0$ of the compressible problem, so that even the first inner flow can only be found at present by determining the "exact" solution by the method we hoped to avoid. This sort of difficulty sets a real limitation on the *usefulness* of purely analytical applications of MAE in three-dimensional scattering problems. Though the analytical difficulties associated with inner problems in three dimensions are quite severe, there is however a real possibility that matching can be very useful in a practical sense when combined with numerical or experimental data. A good example of this viewpoint can be seen in the work of Landahl *et al.* (1971), where near field pressure data taken in a wind tunnel are matched to analytical outer fields to yield sonic boom signatures.

III. Acoustic Waveguides

A. INTRODUCTION

Problems of acoustic waves in guides provide another example of how MAE unifies and extends classical ideas. As in the section on diffraction problems, attention is confined to linear problems. No attempt to review the long history of this class of problems and the various *ad hoc* schemes developed to deal with them will be given. However, we shall simply indicate how scaling ideas and recognition of singular regions can be used to develop, extend, and render more precise the notions used to deal with such problems.

As a first step we show how considerations of scale lead to a formal scheme which gives, for the equation obeyed by the lowest order term, the Webster Horn Equation (for a discussion of this equation, see Eisner, 1964). Our derivation is an extension, to the time dependent three-dimensional case, of the derivation given in Lesser and Lewis (1972a). The derivation gives clear conditions for the applicability of the scheme which applies when the scale of the guide variation is comparable to, and the guide diameter is small in relation to a typical wavelength.

If these conditions break down in some region of the guide, coordinate stretching leads to an equivalent static problem whose solution matches the Webster Horn expansion. In effect this provides a formalization of the ideas of Rayleigh (1945, Art. 264), which have been applied to a number of problems, as described, for example, by Morse and Ingard (1968, Chapter 9).

To indicate the advantage of the formal MAE approach, we investigate the singularity introduced into the Webster expansion by a small slit in the guide wall. This forces us to introduce a third expansion, valid in the region exterior to the guide, which describes the radiation from the slit. The problem thus formulated is easily solved by asymptotic expansions, though an exact solution would present great difficulties. We are also given an opportunity to show how the matching formula can be used to calculate a scattering matrix for the slit.

The section closes with a brief discussion of the case where the guide width varies slowly on the scale of the average width and the wavelength is of the order of the width. As well as being important in its own right, this problem demonstrates the ready applicability of matching methods in contexts where other singular perturbation techniques (such as multiple scaling or averaging methods; see, for example, Nayfeh, 1973) have almost invariably been used in the literature.

B. THE LONG WAVELENGTH APPROXIMATION AND THE WEBSTER HORN EQUATION

The Webster Horn equation has played an important role in duct acoustics and it is of some interest to see how it fits into a more complete expansion scheme which would permit the estimation of errors and the calculation of higher order terms. The equation is expected to give a reasonable representation of the plane wave mode traveling in a guide where the guide width is small compared to the wavelength and the guide cross sectional area varies slowly with respect to the average guide diameter. Thus we shall scale the acoustic equations and boundary conditions in accordance with these ideas. The situation under consideration is shown in Fig. 3, in which H is the typical sectional dimension and L is the longitudinal length scale, so that the guide shape is given by

$$r = Hh(x/L, \theta), \tag{167}$$

where h is dimensionless and

$$\partial h/\partial(x/L) = f(x/L, \theta) = O(1),$$
$$\partial h/\partial \theta = g(x/L, \theta) = O(1), \tag{168}$$

as

$$H/L = \varepsilon \to 0.$$

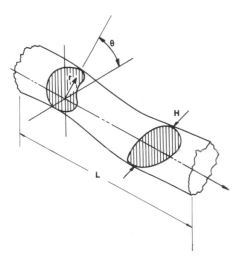

FIG. 3. Duct of slowly varying cross section.

If the duct walls are hard, the condition of no flow through the wall gives that

$$v_1 + Hgv_2/r = \varepsilon fu \tag{169}$$

on

$$r = Hh,$$

where $\mathbf{q} = (u, v_1, v_2)$ are the x, r, and θ velocity components. The acoustic equations are

$$\rho_0(\partial \mathbf{q}/\partial t) + \nabla p = 0 \tag{170}$$

and

$$(\partial p/\partial t) + \rho_0 a_0^2 (\nabla \cdot \mathbf{q}) = 0. \tag{171}$$

The concept of wavelength is introduced by considering that a typical pulse has a time scale T and that the Helmholtz number

$$k = L/(a_0 T) = O(1). \tag{172}$$

If we consider the guide to be driven by a piston at large negative x with a mean velocity U the pressure scale will be $\rho_0 a_0 U$. Also the boundary relation (169) indicates that v_1/u and v_2/u are $O(\varepsilon)$. The profile variation function and the requirements on its derivatives, (168), indicate that L and H are the appropriate longitudinal and cross sectional length scales. Therefore we define the "Webster variables" as

$$\bar{t} = t/T, \qquad \bar{x} = x/L, \qquad \bar{r} = r/H$$
$$\bar{u} = u/U, \qquad \bar{v}_1 = v_1/\varepsilon U, \qquad \bar{v}_2 = v_2/\varepsilon U, \tag{173}$$
$$\bar{p} = p/(\rho_0 a_0 U).$$

If $\nabla^1 = (\partial/\partial\bar{r},\ (1/\bar{r})\ \partial/\partial\bar{\theta})$ is defined as the cross sectional gradient and $\mathbf{q} = (\bar{v}_1,\ \bar{v}_2)$ is the cross sectional velocity vector, the equations in Webster variables are

$$k(\partial\bar{u}/\partial\bar{t}) + (\partial\bar{p}/\partial\bar{x}) = 0, \tag{174}$$

$$\varepsilon^2 k(\partial\bar{\mathbf{q}}/\partial\bar{t}) + \nabla^1\bar{p} = 0, \tag{175}$$

and

$$k(\partial\bar{p}/\partial\bar{t}) + (\partial\bar{u}/\partial\bar{x}) + \nabla^1 \cdot \bar{\mathbf{q}} = 0. \tag{176}$$

If $\mathbf{n} = (1,\ g/r)$ is the component of a normal vector in a cross sectional plane, the boundary condition (169) is

$$(\mathbf{n} \cdot \bar{\mathbf{q}}) = f\bar{u} \qquad \text{on} \quad \bar{y} = h. \tag{177}$$

We define the leading term in the "Webster expansion" by

$$\lim_{\varepsilon \to 0} \bar{p}(\bar{x},\ \bar{r},\ \theta,\ \bar{t},\ \varepsilon) = \bar{p}^{(0)}(\bar{x},\ \bar{r},\ \theta,\ \bar{t}). \tag{178}$$

Application of this limit to Eqs. (174)–(177) gives

$$\nabla^1\bar{p}^{(0)} = 0, \tag{179}$$

and hence

$$\bar{p}^{(0)} = F(\bar{x},\ \bar{t}). \tag{180}$$

Taking time derivatives and some simple algebraic manipulation gives

$$\nabla^1 \cdot \mathbf{w} = (1/k)(\partial^2 F/\partial\bar{x}^2 - k^2\ \partial^2 F/\partial\bar{t}^2) \tag{181}$$

and

$$\mathbf{n} \cdot \mathbf{w} = -(1/k)\ f\ \partial F/\partial x \qquad \text{on} \quad \bar{r} = h, \tag{182}$$

where the transverse acceleration is

$$\mathbf{w} = (\partial/\partial\bar{t})\bar{\mathbf{q}}^{(0)}.$$

Now integrate $\nabla^1 \cdot \mathbf{w}$ over a cross section and apply the Gauss theorem to obtain

$$\iint_{A_{\bar{x}}} \nabla^1 \cdot \mathbf{w}\ dA_x = \int_{\partial A_{\bar{x}}} (\mathbf{w} \cdot \mathbf{n})\ d\sigma. \tag{183}$$

with the boundary conditions

$$\bar{v} = 0 \qquad \text{on} \quad \bar{y} = 0, \tag{191}$$

$$\bar{v} = 0 \qquad \text{on} \quad \bar{y} = 1, \quad |\bar{x}| > \sigma\varepsilon. \tag{192}$$

We assume general excitation conditions in the duct, i.e.,

$$\bar{p} \sim \alpha_1 e^{ik\bar{x}} \qquad \text{as} \quad \bar{x} \to -\infty, \quad |\bar{y}| < 1, \tag{193}$$

$$\bar{p} \sim \alpha_2 e^{-ik\bar{x}} \qquad \text{as} \quad \bar{x} \to +\infty, \quad |\bar{y}| < 1. \tag{194}$$

This will result in reflected waves given by

$$\bar{p} \sim \beta_1 e^{-ik\bar{x}} \qquad \text{as} \quad \bar{x} \to -\infty, \quad |\bar{y}| < 1, \tag{195}$$

$$\bar{p} \sim \beta_2 e^{ik\bar{x}} \qquad \text{as} \quad \bar{x} \to +\infty, \quad |\bar{y}| < 1. \tag{196}$$

The reflected wave amplitudes will of course be related to the incident wave amplitudes by the duct geometry. This relation is given by the scattering matrix S, defined by

$$\begin{pmatrix} \beta_1 \\ \beta_2 \end{pmatrix} = S \begin{pmatrix} \alpha_1 \\ \alpha_2 \end{pmatrix}. \tag{197}$$

From the reciprocity theorem $S_{12} = S_{21}$, and from the symmetry of the present problem $S_{11} = S_{22}$.

One of our tasks is to calculate an asymptotic representation for S,

$$S \sim \sum \mu_n(\varepsilon) S^{(n)}. \tag{198}$$

As radiation takes place from the slit, S will not be unitary and another task is to calculate and specify the characteristics of the radiation field. Because the slit width is small on the wavelength scale, we will be able to specify the radiation field in terms of multipoles located at the center of the slit.

We shall designate the Webster series valid for $\bar{x} < -\varepsilon\sigma$ by

$$\bar{p}_1 = \sum \mu_n \bar{p}_1^{(n)} \tag{199}$$

and the series valid for $\bar{x} > \varepsilon\sigma$ by

$$\bar{p}_2 = \sum \mu_n \bar{p}_2^{(n)}. \tag{200}$$

We assume that the excitation conditions are such that the leading gauge function $\mu_0 = 1$. Because the duct width is fixed, Eqs. (188)–(190) show that for any n

$$\bar{p}_{1,2}^{(n)} = \alpha_{1,2}^{(n)} e^{\pm ik\bar{x}} + \beta_{1,2}^{(n)} e^{\pm ik\bar{x}}. \tag{201}$$

As the limit expansion based on Webster variables fails near the slit, $\bar{p}_1$ and $\bar{p}_2$ cannot be directly related to one another. To treat the singular behavior of the Webster expansion near $\tilde{x} = 0$ we introduce the Laplace variables

$$\tilde{T}(\tilde{x}, \tilde{y}) = (\varepsilon\tilde{x}, \tilde{y}) \tag{202}$$

and

$$\tilde{T}(\tilde{v}) = \tilde{v}/\varepsilon.$$

The equations in Laplace variables take the form:

$$-\varepsilon ik\tilde{u} + \tilde{p}_{\tilde{x}} = 0, \tag{203}$$

$$-\varepsilon ik\tilde{v} + \tilde{p}_{\tilde{y}} = 0, \tag{204}$$

$$-\varepsilon ik\tilde{p} + \tilde{u}_{\tilde{x}} + \tilde{v}_{\tilde{y}} = 0. \tag{205}$$

Designate terms with integral powers of ε such as $\varepsilon^n\tilde{p}^{(m)}$ by $\varepsilon^n\tilde{p}_{(n)}$. These are terms related to gauge functions directly expected from the form of Eqs. (203)–(205). In addition, we can expect "switchback" terms of intermediate orders. Thus

$$\tilde{p} = \tilde{p}^{(0)} + \mu_1\tilde{p}^{(1)} + \cdots + \varepsilon\tilde{p}_{(1)} + \cdots. \tag{206}$$

Also define the "up to ε^n" expansion operator by

$$E_{(n)}\tilde{p} = \tilde{p}^{(0)} + \cdots + \mu_m(\varepsilon)\tilde{p}^{(m)} + \varepsilon^n\tilde{p}^{(m+1)}. \tag{207}$$

Formal insertion of $E_{(1)}\tilde{p}$ into Eqs. (203)–(205) shows that, with $\tilde{\mathbf{q}} = (\tilde{u}, \tilde{v})$,

$$\tilde{p}^{(0)} = \tilde{C}_0^{(0)} = \text{constant},$$

$$\tilde{\mathbf{q}}^{(0)} = +(ik)^{-1}\tilde{\nabla}\tilde{p}_{(1)}, \qquad \tilde{\mathbf{q}}_{(1)} = +(ik)^{-1}\tilde{\nabla}\tilde{p}_{(2)}, \tag{208}$$

$$\tilde{\nabla}^2\tilde{p}_{(1)} = 0, \qquad \tilde{\nabla}^2\tilde{p}_{(2)} = -ik\tilde{p}^{(0)}.$$

To determine solutions to these equations we need conditions as

$$\tilde{x} \to \pm\infty \qquad \text{in} \quad |\tilde{y}| < 1$$

and as

$$\tilde{r} = (\tilde{x}^2 + \tilde{y}^2)^{1/2} \to \infty, \qquad |\tilde{y}| > 1.$$

The former are given by matching with the two Webster region series for $\tilde{p}_1$ and $\tilde{p}_2$. The latter present a problem, as the source solutions of Laplace's equation in two dimensions will be singular as $\tilde{r} \to \infty$, and we have a situation similar to that existing in our treatment of the scattering problem. To account for the singular behavior as $\tilde{r} \to \infty$ it is thus natural to try a variable transformation that will lead to a limit process expansion whose terms satisfy the Helmholtz equation.

The most convenient choice is given by

$$\hat{T}(\tilde{x}, \tilde{y}) = (\hat{x}/\varepsilon, \hat{y}/\varepsilon + 1), \tag{209}$$

so that

$$\hat{\nabla}^2 \hat{p} + k^2 \hat{p} = 0, \tag{210}$$

$$\hat{\mathbf{q}} = (ik)^{-1} \hat{\nabla}\hat{p}. \tag{211}$$

In the limit $\varepsilon \to 0$, with $(\hat{x}, \hat{y})$ fixed, the Laplace region shrinks, crudely speaking, to the point $\hat{r} = 0$. The solution to (210) satisfying the wall boundary condition and the outgoing wave condition will thus be

$$\hat{p} = \sum \mu_n(\varepsilon) \hat{p}^{(n)},$$

$$\hat{p}^{(n)} = \sum_{m=0}^{\infty} \hat{C}_m^{(n)} H_m^{(1)}(k\hat{r}) \cos m\hat{\theta}, \tag{212}$$

$$\hat{r} = (\hat{x}^2 + \hat{y}^2)^{1/2}, \qquad \hat{\theta} = \tan^{-1}(\hat{y}/\hat{x}).$$

As we have considered this type of situation in our treatment of scattering problems we will not carry all the terms given in (212), and will give only heuristic arguments for the terms retained. The reader can verify that the terms ignored cannot be suitably matched, i.e. can be rejected on the basis of the matching principle.

Thus as $\varepsilon \to 0$ the slit vanishes and we would expect no $O(1)$ radiation field; in fact, as the slit area is proportional to ε, we expect the first nonzero term in $\hat{p}$ to be $E_{(1)}\hat{p} = \varepsilon \hat{p}_{(1)}$. Because of the behavior of $H_m^{(1)}(\rho)$ as $\rho \to 0$, namely

$$H_0^{(1)}(\rho) \sim \ln \rho, \qquad H_m^{(1)}(\rho) \sim \rho^{-m},$$

the only eigensolution in $\hat{p}^{(1)}$ that will match with the Laplace region in (208) is $H_0^{(1)}(k\hat{r})$. Therefore, we take

$$E_{(1)}\hat{p} = \hat{C}_{(1)0} H_0^{(1)}(k\hat{r}). \tag{213}$$

We could continue with the general approach, not labeling terms in the expansion until we have filled in "switchback" terms as in our treatment of scattering. However, it is reasonably clear that the logarithmic behavior of $H_0^{(1)}$ will lead to the presence of $\ln \varepsilon$ terms in the gauge functions, and either by trial and error or by a deductive procedure it is found that

$$\{\mu_n\} = \{1, \, \varepsilon \ln \varepsilon, \, \varepsilon, \, \ldots\} \tag{214}$$

and from now on we shall number terms accordingly, so that

$$E_{(1)}\hat{p} = E_2 \hat{p} = \hat{C}_0^{(2)} H_0^{(1)}(k\hat{r}). \tag{215}$$

In the Laplace region we have

$$E_2 \tilde{p} = \tilde{C}_0^{(0)} + \varepsilon \ln \varepsilon \, \tilde{C}_0^{(1)} + \varepsilon \tilde{p}^{(2)}(\tilde{x}, \tilde{y})$$

where $\tilde{\nabla}^2 \tilde{p}^{(2)} = 0$. To resolve this problem we make use of conformal mapping, again using j for the complex variable to avoid confusion with the i of the time dependence. Thus with $\tilde{z} = \tilde{x} + j\tilde{y}$, $w = \xi + j\eta$ the mapping

$$\tilde{z} = j + w + (1/\pi) \ln(w + b) - (1/\pi) \ln(w - b) \tag{216}$$

with

$$w_0 = \{b[b + (2/\pi)]\}^{1/2} \tag{217}$$

and

$$\sigma = w_0 + (1/\pi) \ln (w_0 + b) - (1/\pi) \ln (w_0 - b), \tag{218}$$

takes the upper half $\tilde{z}$ plane into the upper half w plane, with the guide sides along the ξ axis as indicated in Fig. 4b.

At the points B and F we can add eigensolutions that satisfy the edge and matching conditions, while at C and E we must place source singularities corresponding to the behavior required by matching with the Webster regions. Near the edges [the mapping (216) shows for $\tilde{z} \to \sigma + j$ that $w - w_0 \sim (z - \sigma - j)^{1/2}$], a term such as $(w - w_0)^m$ corresponds to $(z - \sigma - j)^{m/2}$, so that the edge conditions require $m = 2, 3, 4, \ldots$. Matching eliminates all such eigenfunctions from $\tilde{p}^{(2)}$ as can be readily verified. Therefore, we have

$$\tilde{p}^{(2)} = \mathrm{Re}_j[\tilde{C}_0^{(2)} + \tilde{C}_1^{(0)} \ln (w + b) + \tilde{C}_2^{(2)} \ln (w - b)]. \tag{219}$$

As Eq. (216) gives $w(\tilde{z})$ implicitly, matching requires the derivation of the asymptotic behavior of Eq. (216) as $w \to \pm b$, for which $\mathrm{Re}_j \tilde{z} \to \pm \infty$ in $0 \le \tilde{y} < 1$, and as $|w| \to \pm \infty$, for which $\tilde{z} \to \infty, \tilde{y} > 1$. By a relatively straightforward calculation it can be established that as

$$\tilde{x} \to -\infty, \qquad 0 < \tilde{y} < 1,$$

$$w = -b + 2b \exp[\pi(\tilde{z} + b)] + O\{\exp[2\pi(\tilde{z} + b)]\}, \tag{220}$$

as

$$\tilde{x} \to +\infty, \qquad 0 < \tilde{y} < 1,$$

$$w = b - 2b \exp[\pi(b - \tilde{z})] + O\{\exp[2\pi(b - \tilde{z})]\}, \tag{221}$$

and as

$$\tilde{r} \to \infty, \qquad \tilde{y} > 1,$$

$$w = \tilde{z} - j - \frac{2b}{\pi(\tilde{z} - j)} + O\left(\frac{1}{\tilde{z}^2}\right). \tag{222}$$

The Webster expansions

$$E_2 \bar{p}_1 = \bar{p}_1^{(0)} + \varepsilon \ln \varepsilon \; \bar{p}_1^{(1)} + \varepsilon \bar{p}_1^{(2)}, \qquad \tilde{x} < 0$$

and

$$E_2 \bar{p}_2 = \bar{p}_2^{(0)} + \varepsilon \ln \varepsilon \; \bar{p}_2^{(1)} + \varepsilon \bar{p}_2^{(2)}, \qquad \tilde{x} > 0$$

are matched to the Laplace expansion by

$$(\bar{T} E_2 \; \tilde{T} E_2 - E_2 \; \bar{T} E_2 \; \tilde{T})\bar{p} = 0, \tag{223}$$

while the Helmholtz expansion is matched by using

$$(\hat{T} E_2 \; \tilde{T} E_2 - E_2 \; \hat{T} E_2 \; \tilde{T})\hat{p} = 0. \tag{224}$$

As the procedure is now familiar we simply give the results as read off from a straightforward application of Eqs. (223) and (224);

$$\alpha_1^{(0)} + \beta_1^{(0)} = \tilde{C}_0^{(0)}, \qquad \text{(Matching of } E_2\bar{p}_1 \text{ with } E_2\tilde{p})$$
$$\alpha_1^{(0)} - \beta_1^{(0)} = (\pi/ik)\tilde{C}_1^{(2)},$$
$$\alpha_1^{(1)} + \beta_1^{(1)} = \tilde{C}_0^{(1)}, \qquad\qquad\qquad\qquad (225)$$
$$\alpha_1^{(2)} + \beta_1^{(2)} = \tilde{C}_0^{(2)} + \pi b\tilde{C}_1^{(2)} + (\tilde{C}_1^{(2)} + \tilde{C}_2^{(2)})\ln 2b,$$

$$\alpha_2^{(0)} + \beta_2^{(0)} = \tilde{C}_0^{(0)}, \qquad \text{(Matching of } E_2\bar{p}_2 \text{ with } E_2\tilde{p})$$
$$\alpha_2^{(0)} - \beta_2^{(0)} = (\pi/ik)\tilde{C}_2^{(2)},$$
$$\alpha_2^{(1)} + \beta_2^{(1)} = \tilde{C}_0^{(1)}, \qquad\qquad\qquad\qquad (226)$$
$$\alpha_2^{(2)} + \beta_2^{(2)} = \tilde{C}_0^{(2)} + \pi b\tilde{C}_2^{(2)} + (\tilde{C}_1^{(2)} + \tilde{C}_2^{(2)})\ln 2b,$$

$$\tilde{C}_0^{(0)} = 0, \qquad \text{(Matching of } E_2\hat{p} \text{ with } E_2\tilde{p})$$
$$\tilde{C}_0^{(1)} - (\tilde{C}_1^{(2)} + \tilde{C}_2^{(2)}) = 0,$$
$$\tilde{C}_0^{(2)} = \Lambda\hat{C}_0^{(2)}, \qquad\qquad\qquad\qquad (227)$$
$$\tilde{C}_1^{(2)} + \tilde{C}_2^{(2)} = (2i/\pi)\hat{C}^{(2)},$$

where $\Lambda = 1 + (2i/\pi)(\gamma + \ln k/2)$; $\gamma =$ Euler Constant. Thus matching gives us 12 equations for the 18 constants α, β, $\tilde{C}$, $\hat{C}$. Hence we can solve for 12 of the constants in terms of a suitably chosen set of 6. We choose

$$\alpha = (\alpha_1^{(0)},\ \alpha_2^{(0)},\ \alpha_1^{(1)},\ \alpha_2^{(1)},\ \alpha_1^{(2)},\ \alpha_2^{(2)})$$

as our known constants representing the amplitudes of the incoming waves from $\pm\infty$ in the strip. From Eq. (197) or (198) we now seek the scattering matrix $S(\varepsilon)$. Thus manipulating Eqs. (225)–(227) we find, using Eq. (198), that

$$E_2 S = S^{(0)} + \varepsilon \ln \varepsilon\, S^{(1)} + \varepsilon S^{(2)}, \qquad\qquad (228)$$

$$S^{(0)} = -I = - \begin{pmatrix} 1 & 0 \\ 0 & 1 \end{pmatrix}, \qquad\qquad (229)$$

$$S^{(1)} = (2ik/\pi)I, \qquad\qquad (230)$$

$$S^{(2)} = \begin{pmatrix} r + 2ikb & r \\ r & r + 2ikb \end{pmatrix}, \qquad\qquad (231)$$

where

$$r = k\Lambda + (2ik/\pi)\ln 2b. \qquad\qquad (232)$$

As a partial check on our work we note that the components of S satisfy the symmetry properties required by the problem. The source strength of the radiation term is

$$\hat{C}_0^{(2)} = k(\alpha_1^{(0)} + \alpha_2^{(0)}) \qquad\qquad (233)$$

and the constants for the Laplace region solution are

$$\tilde{C}_0^{(0)} = 0,$$
$$\tilde{C}_0^{(1)} = (2ik/\pi)(\alpha_1^{(0)} + \alpha_2^{(0)}),$$
$$\tilde{C}_0^{(2)} = k\Lambda(\alpha_1^{(0)} + \alpha_2^{(0)}), \tag{234}$$
$$\tilde{C}_1^{(2)} = (2ik/\pi)\alpha_1^{(0)},$$
$$\tilde{C}_2^{(2)} = (2ik/\pi)\alpha_2^{(0)}.$$

The dependence of b on σ is given by (217) and (218). The calculation of higher order terms involves careful attention to eigenfunctions arising from the edge. However, $\tilde{p}_{(2)}$ still only involves the solution of a homogeneous equation. In addition, enough information already is contained in $E_2\tilde{p}$ to calculate some higher order radiation terms.

　　Another problem closely connected with waveguides, the calculation of eigenfunctions and eigenvalues in a closed cavity, can also be dealt with by MAE. This application can be found in Lesser and Lewis (1974), where the effect of a hard scatterer, small on the wavelength scale, is treated by the MAE method. Other problems which may be treated in this fashion are easily found, and solved, for example, propagation past an iris in a parallel plate waveguide, radiation from the end of a parallel plate waveguide with either a small flange or an infinite flange fitted, and propagation past a T-junction in a parallel plate waveguide.

D. The Slowly Varying Guide in the Short Wavelength Limit

　　A common problem of great practical interest in waveguide theory is the calculation of waveguide modes when the parameters defining the guide are variable. For example, the guide may be curved, have variable cross sectional area, and contain material whose sound speed varies. In Section III,B we examined one case where a perturbation treatment proved useful, that where the guide width variation scale L was large compared with the guide width H, and $1/k_0$, the wave or pulse length scale was large compared with H, but approximately the same as L. In terms of the parameter $\varepsilon = H/L$ this implies $\varepsilon \ll 1$ and $k_0 L = O(1)$, where k_0 is the dimensional wavenumber. A brief treatment will now be given of a case where the guide is *not* narrow compared with the wavelength, and thus we now assume that

$$k_0 H = O(1) \qquad \text{as} \quad \varepsilon \to 0. \tag{235}$$

The formal problem to be solved is now

$$\nabla^2 p + k_0^2 p = 0 \tag{236}$$

in

$$0 < x < \infty, \qquad 0 < y < Hh(x/L),$$

with boundary conditions appropriate to hard walls given by

$$p_y(x, Hh, \varepsilon) = (H/L)h'(x/L)p_x(x, Hh, \varepsilon),$$
$$p_y(x, 0, \varepsilon) = 0, \tag{237}$$

and

$$h' = O(1).$$

Also, for the sake of completeness assume that

$$p(0, y, \varepsilon) = p_0 f(y/H), \tag{238}$$

that as $x \to \infty$, $h \to 1$, and that only waves traveling in the direction of increasing x are allowable. The obvious scaling is to let

$$\bar{y} = y/H, \qquad \bar{x} = x/L,$$
$$\bar{p} = p/p_0, \qquad k = k_0 H \tag{239}$$

so that the problem expressed in dimensionless variables takes the form

$$\varepsilon^2 \bar{p}_{\bar{x}\bar{x}} + \bar{p}_{\bar{y}\bar{y}} + k^2 \bar{p} = 0$$

on

$$\bar{y} = h, \qquad \bar{p}_{\bar{y}} = \varepsilon^2 h'(\bar{x})\bar{p}_{\bar{x}} \tag{240}$$

and on

$$\bar{y} = 0, \qquad \bar{p}_{\bar{y}} = 0.$$

If $\varepsilon \to 0$ in Eq. (240) as written we arrive at a nonsensical conclusion, namely,

$$\bar{p}_{\bar{y}\bar{y}}^{(0)} + k^2 \bar{p}^{(0)} = 0,$$
$$\bar{p}^{(0)} \sim \cos k\bar{y} \quad \text{or} \quad \sin k\bar{y}, \tag{241}$$
$$\bar{p}_{\bar{y}}^{(0)} = 0 \quad \text{on} \quad \bar{y} = 0, h,$$

and as k is considered given and fixed in the limit process, the problem stated by Eq. (241) has no solution unless $kh(\bar{x}) = n\pi$, n an integer. As h is considered to vary with $\bar{x}$ this condition can not be expected to hold.

One apparent way of resolving the problem is to stretch $\bar{x}$, i.e., to let

$$\tilde{x} = \bar{x}/\varepsilon = x/H, \qquad \tilde{y} = \bar{y}, \tag{242}$$

which leads to the problem

$$\tilde{p}_{\tilde{x}\tilde{x}} + \tilde{p}_{\tilde{y}\tilde{y}} + k^2 \tilde{p} = 0$$

and on

$$\hat{y} = h(\varepsilon \tilde{x}) \tag{243}$$
$$\tilde{p}_{\tilde{y}} = \varepsilon h'(\varepsilon \tilde{x})\tilde{p}_{\tilde{x}}.$$

An examination of Eq. (243) in the limit $\varepsilon \to 0$ shows that this reduces, for the lowest order term, to the case of a waveguide of constant width valid in the *local* region about $\bar{x} = 0$. As h' ceases to be defined at this point (see Fig. 5), one would expect in general a need for such a local expansion about such a point. However, a little thought indicates that such an expansion must eventually become nonuniform in higher order. The reason for this is that the guide shape is treated by a power series expansion of the upper wall condition as given by Eq. (243). Thus an expansion based on Eq. (242) is a local or inner expansion valid for regions where the scaling of Eq. (239) is inappropriate or $h' \neq O(1)$.

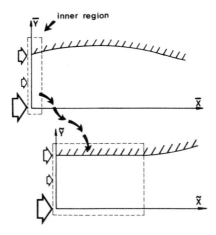

Fig. 5. Inner region needed for uniform expansion of waveguide problem.

The problem with Eq. (240) is that straightforward application of $\lim \varepsilon \to 0$ with $\bar{x}$ fixed neglects the rapid phase variations expected in $\bar{p}$. One way of overcoming this while remaining in the context of MAE is to introduce as new dependent variables

$$\bar{p} = e^{i\bar{\psi}/\varepsilon} \tag{244}$$

for the outer region and

$$\tilde{p} = e^{i\tilde{\psi}} \tag{245}$$

for the inner region. This leads to nonlinear equations for $\bar{\psi}$ and $\tilde{\psi}$, and the MAE method can be applied to obtain suitable matched expansions for $\bar{\psi} = \varepsilon\tilde{\psi}$. The algebra in such an approach is quite tedious, and for simplicity we shall adopt a simpler technique appropriate to obtaining the leading term in the outer region. Thus, assume the form

$$\bar{p} = \bar{P}(\bar{x}, \bar{y}, \varepsilon) \, \exp[i\Omega(\bar{x})/\varepsilon]. \tag{246}$$

Insertion of Eq. (246) into Eq. (240) gives

$$\bar{P}_{\bar{y}\bar{y}} + \bar{\beta}^2(\bar{x})\bar{P} + \varepsilon(2i\Omega'\bar{P}_{\bar{x}} + i\Omega''\bar{P}) + \varepsilon^2\bar{P}_{\bar{x}\bar{x}} = 0, \tag{247}$$

where

$$\bar{\beta}^2 = k^2 - (\Omega')^2 \tag{248}$$

and ()′ indicates, as always, differentiation with respect to the argument. If

$$\bar{P}^{(0)} = \lim_{\varepsilon \to 0} \bar{P}, \qquad \bar{P}^{(1)} = \lim_{\varepsilon \to 0} (\bar{P} - \bar{P}^{(0)})/\varepsilon$$

(i.e., assuming the gauge functions are 1 and ε), then

$$L_{\bar{\beta}}\bar{P}^{(0)} \equiv [(\partial^2/\partial\bar{y}^2) + \bar{\beta}^2]\bar{P}^{(0)} = 0, \tag{249}$$

$$L_{\bar{\beta}}\bar{P}^{(1)} = -2i\Omega'\bar{P}_{\bar{x}}^{(0)} - \Omega''\bar{P}^{(0)}. \tag{250}$$

The boundary conditions are now

$$\bar{P}_{\bar{y}}^{(0)} = 0 \qquad \text{on} \quad \bar{y} = 0 \quad \text{and} \quad \bar{y} = h, \tag{251}$$

and

$$\begin{aligned}
\bar{P}_{\bar{y}}^{(1)} &= 0 \qquad \text{on} \quad \bar{y} = 0; \\
\bar{P}_{\bar{y}}^{(1)} &= i\Omega'h'\bar{P}^{(0)} \qquad \text{on} \quad y = h.
\end{aligned} \tag{252}$$

First consider the solution of Eq. (249). Only derivatives with respect to $\bar{y}$ are contained in $L_{\bar{\beta}}$. Therefore

$$\bar{P}_{\bar{x}}^{(0)} = \bar{A}_{\bar{\beta}}(\bar{x}) \cos \bar{\beta}\bar{y} \tag{253}$$

and

$$\bar{\beta}(\bar{x})h(\bar{x}) = n\pi, \qquad n = 0, 1, 2, \ldots, \tag{254}$$

where

$$(\Omega_{\beta}')^2 = (k^2 - \bar{\beta}^2). \tag{255}$$

The implications of Eqs. (253)–(255) should be clear to readers familiar with waveguide phenomena; thus

$$\begin{aligned}
\bar{p}_{\bar{\beta}} = \bar{p}_n \sim {}&\exp\left[+\int^{\cdot}\frac{i}{\varepsilon}\left(k^2 - \frac{n^2\pi^2}{h^2(\bar{x})}\right)^{1/2} d\bar{x}\right]\bar{A}_n(\bar{x}) \cos(n\pi\bar{y}/h) \\
&+ \exp\left[-\int^{\cdot}\frac{i}{\varepsilon}\left(k^2 - \frac{n^2\pi^2}{h^2(\bar{x})}\right)^{1/2} d\bar{x}\right]\bar{B}_n(\bar{x}) \cos(n\pi\bar{y}/h)
\end{aligned} \tag{256}$$

represents a *mode* given by the condition (254) (hence the replacement of $\bar{\beta}$ by n as a subscript).

If $k^2 > n^2\pi^2/h^2(\bar{x})$, the radical in (256) is real and we can expect the mode to propagate. Also the phase speed of a mode varies with $\bar{x}$, i.e., with the changing guide width.

The reader will note that the mode amplitudes $\bar{A}_n$ for waves traveling to the right and $\bar{B}_n$ for those to the left are not determined. *Two procedures* are needed to find them; (1) matching with an inner solution, such as one valid for the condition (238) or for a source or a region where $h' \neq O(1)$ (e.g., a step) and (2) consideration of the solvability of higher order problems.

Thus Eq. (249) and the boundary condition (251) define an eigenvalue problem for the operator $L_{\tilde{\beta}}$. This means that the forcing terms in Eqs. (250) and (252) must be orthogonal to the eigenfunctions of the (fortunately) self-adjoint operator $L_{\tilde{\beta}}$ (see, e.g., Stakgold, 1967). The consistency condition is that

$$\int_0^h (\bar{P}_n^{(0)} L_n \ \bar{P}_n^{(1)} - \bar{P}_n^{(1)} L_n \ \bar{P}_n^{(0)}) \, d\bar{y} = [(\bar{P}_n^{(0)} \bar{P}_{ny}^{(1)} - \bar{P}_n^{(1)} \bar{P}_{n\bar{y}}^{(0)})]_0^h, \qquad (257)$$

which leads to the results

$$\bar{A}_n = \bar{C}_n / h(\Omega_n')^{1/2}, \qquad n > 0 \qquad (258)$$

and

$$\bar{A}_0 = \bar{C}_0 / (hk)^{1/2}, \qquad n = 0. \qquad (259)$$

As would be expected, Eq. (258) fails to yield a valid approximation when a mode changes from the traveling to the evanescent form, i.e., when $\Omega_n = 0$. These regions would require a treatment using a local approximation in terms of a variable such as $\tilde{x}_c = (\bar{x} - \bar{x}_c)/\mu(\varepsilon)$, where $\Omega_n(\bar{x}_c) = 0$. The details of such a treatment will be given in a forthcoming paper by Lesser. Formal completion of the solution for the problem posed requires meeting the condition $p = p_0 f(y/H)$ at $x = 0$ using the inner variable $\tilde{x} = \bar{x}/\varepsilon$, and finding the constants $\bar{C}$ by matching. The results for the simple problem posed are what one would expect, i.e., only traveling modes (Ω_n' real) appear in the outer region.

The extension of these ideas to more intricate situations, such as a sharp change in h or an open end wall, affects the gauge functions but not the general idea of the expansion.

We should note that Ahluwalia *et al.* (1974) have treated by quite similar methods the problem of propagation in a curved guide containing material with varying wave speeds. Unfortunately their treatment is marred by algebraic errors and is lacking in interpretation, and self-consistency. The ideas however are also related to work done on geometrical optics approximations to wave equations (Lewis and Keller, 1964) and WKB methods (Nayfeh, 1973).

IV. Nonlinear Acoustics

A. Introduction

Up to this point, we have concentrated our attention on problems of classical linear acoustics. While MAE methods provide an interesting and useful addition to more familiar techniques, as well as a unifying element for acousticians interested in linear problems, it is in nonlinear problems that MAE's real utility is to be seen. To demonstrate how MAE can be used to advantage in this huge and relatively new field, we will study a class of problems which has been discussed in a previous volume of this series (Volume IIB, 1965) concerned with nonlinear acoustics—the propagation in one dimension of waves of finite but small amplitude. As the field of nonlinear acoustics is extremely broad and our main concern is to show MAE at work on some relevant problems, we shall only treat limited aspects of this field.

The problem of propagation of finite amplitude waves is important not only in its own right, but also because it can help us better understand the limitations and strength of normal linear acoustic calculations carried out using the classical linear wave equation. As usual with the MAE approach, we shall see how the classical theory fits into the more complex model adopted for the class of problems under consideration. At the outset, we study the formation of a weak but finite shock wave, following the spirit of the treatment given by Moran and Shen (1966). These authors considered how a weak shock wave develops when a piston starts moving into a quiescent gas whose motions obey the Navier–Stokes equations. A result of the diffusion effects built into these equations is that the initial discontinuity, caused by the suddenly accelerated piston, tends to smooth out. At the same time, the fact that rear portions of the wave move at the local sound speed plus the finite (but small) velocity of the piston causes the well-known convective steepening effect. The balancing of these two opposed phenomena leads to a traveling wave of fixed structure, which in a fluid of sufficiently small viscosity presents the observer with the apparent discontinuity known as a shock wave. For sufficiently weak waves (small piston velocity), the internal shock structure has a characteristic length scale large enough in comparison with the mean free path of the gas to justify a treatment based on the continuum approximation implicit in the use of the Navier–Stokes equation. For a more detailed treatment of the physical aspects of weak shocks, the survey articles of Lighthill (1956) and of Hayes (1960) are highly recommended.

B. Development of Weak Shock Waves

In their treatment of the "piston" problem Moran and Shen (1966) took as a starting point the complete Navier–Stokes equation for a compressible perfect gas. In order to simplify the mathematics a little, we will employ a simplified version of the Navier–Stokes equation derived by Lighthill (1956), which to the accuracy considered in the present work gives the same answers as the full Navier–Stokes equation, as long as we are not concerned with heat transfer effects at the piston wall. In fact, Lighthill showed that these equations provide the same results as the Navier–Stokes equation to

$$O(U_0 \nu / \tau a^3) \qquad \text{and/or} \qquad O((\nu / a^2 \tau)^2) \tag{260}$$

where U_0 and τ are the characteristic velocity and time scales, a is the sound speed, and ν is the kinematic viscosity, $\nu = \mu/\rho$, ρ being the gas density. The reader can of course also consider the equations as a model that yields a parallel, but simplified treatment of the "piston problem," with many of the same results as obtained from the Navier–Stokes equations. Of course, there is a lot to be said for using equations that have minimal complications with regard to the intended calculation, an idea which is in fact implicit in the use of MAE methods.

Thus we assume the gas motion to be governed by the equations

$$\frac{\partial u}{\partial t} + u \frac{\partial u}{\partial x} + \frac{2}{\gamma - 1} a \frac{\partial a}{\partial x} = \delta \frac{\partial^2 u}{\partial x^2}, \tag{261}$$

$$\frac{\partial a}{\partial t} + u \frac{\partial a}{\partial x} + \frac{\gamma - 1}{2} a \frac{\partial u}{\partial x} = 0, \tag{262}$$

$$a/a_0 = (\rho/\rho_0)^{(\gamma - 1)/2}, \tag{263}$$

where γ is the ratio of specific heats, ρ_0 and a_0 are the quiescent density and sound speed, and using the Prandtl number $\sigma = \mu c_P/k$ (k = thermal conductivity) the parameter δ (called the diffusivity of sound by Lighthill, 1956) is given by

$$\delta = \nu \left\{ \frac{4}{3} + \frac{\mu'}{\mu} + \frac{\gamma - 1}{\sigma} \right\}, \tag{264}$$

μ and μ' being respectively the ordinary and the bulk viscosity. The reader should note that for a small perturbation, i.e. for $a = a_0 + a'$, $u = u' \ll a_0$, and small δ, dropping the nonlinear and diffusion terms leads to the familiar one-dimensional acoustic wave equation.

The piston problem is specified by the boundary and initial conditions

$$u(U_0 t, t) = H(t) U_0, \tag{265}$$

$$u(x \to \infty, t) \to 0, \tag{266}$$

$$a(x, t < 0) = a_0, \tag{267}$$

$$u(x, t < 0) = 0, \tag{268}$$

$H(t)$ being the Heaviside step function and U_0 the piston velocity. As our interest is in waves of small amplitude, a perturbation parameter is given by the piston Mach number $\varepsilon = U_0/a_0 \ll 1$.

We are now faced with the task of scaling variables in a manner consistent with the physics of the problem. If our understanding of the physical processes is both correct and correctly employed, this should at least lead to a consistent mathematical formulation. The initial discontinuity caused by the abrupt motion of the piston will be smoothed out by molecular transport effects in a time scale appropriate to the molecular collision frequency in the gas and on a length scale appropriate to the mean free path. Strict adherence to the physics would call for a kinetic theory (e.g., Boltzmann equation) treatment of the problem at this stage. However, in the spirit of Moran and Shen (1966) we assume that the continuum treatment of transport effects implicit in the diffusivity parameter δ provides sufficient accuracy for our purpose. This leads us to the initial length and time scales $l_1 = \delta/a_0$ and $\tau_1 = \delta/a_0{}^2$. The scaling for velocity and sound speed are arrived at from conditions (265) and (267). These scalings should lead to a perturbation formalism adequate for the initial behavior of the gas. Thus let

$$\varepsilon a_0 \, \hat{u} = u, \qquad a_0 \, (1 + \varepsilon \hat{a}) = a \tag{269}$$

and

$$\hat{x} l_1 = x, \qquad \hat{t} \, \tau_1 = t. \tag{270}$$

Substitution of Eqs. (269) and (270) into Eqs. (261) and (262) and the boundary conditions leads to the formal problem of approximating $\hat{u}$, $\hat{a}$ for small values of ε, these being implicitly defined by:

$$\frac{\partial \hat{u}}{\partial \hat{t}} + \varepsilon \hat{u} \, \frac{\partial \hat{u}}{\partial \hat{x}} + \frac{2}{\gamma - 1} \, (1 + \varepsilon \hat{a}) \, \frac{\partial \hat{a}}{\partial \hat{x}} = \frac{\partial^2 \hat{u}}{\partial \hat{x}^2}, \tag{271}$$

$$\frac{\partial \hat{a}}{\partial \hat{t}} + \varepsilon \hat{u} \, \frac{\partial \hat{a}}{\partial \hat{x}} + \frac{\gamma - 1}{2} \, (1 + \varepsilon \hat{a}) \, \frac{\partial \hat{u}}{\partial \hat{x}} = 0, \tag{272}$$

$$\hat{u}(\varepsilon \hat{t}, \, \hat{t}) = H(\hat{t}), \tag{273}$$

$$\hat{u}(\hat{x} \to \infty) \to 0, \tag{274}$$

$$\hat{a}(\hat{x}, \, \hat{t} < 0) = \hat{u}(\hat{x}, \, \hat{t} < 0) = 0. \tag{275}$$

Using the operator E_0 on Eqs. (271–275) gives the zeroth-order problem

$$\partial^2 \hat{u}^{(0)}/\partial \hat{t}^2 - \partial^2 \hat{u}^{(0)}/\partial \hat{x}^2 = \partial^3 \hat{u}^{(0)}/\partial \hat{x}^2 \partial \hat{t}, \tag{276}$$

$$\hat{u}^{(0)} \, (0, \, \hat{t}) = H(\hat{t})$$
$$\hat{u}^{(0)} \, (\hat{x} \to \infty, \, \hat{t}) = 0 \tag{277}$$
$$\hat{u}^{(0)} \, (\hat{x}, \, \hat{t} < 0) = 0.$$

Many readers will recognize Eq. (276) as the equation employed by Rayleigh (1945, p. 315) and others to study dissipative effects in acoustics. An integral representation of $\hat{u}^{(0)}$, readily obtained by Laplace transforms, is

$$\hat{u}^{(0)} = (1/2\pi i) \int_{-i\infty + C}^{i\infty + C} \frac{\exp\{s\hat{t} - [s/(1 + s)^{1/2}]\hat{x}\}}{s} \, ds, \tag{278}$$

and an expansion of this, valid for large $\hat{t}$, is (Lagerström *et al.*, 1949)

$$\hat{u}^{(0)}_{\hat{t} \to \infty} \sim \tfrac{1}{2} \, \mathrm{erfc}[(\hat{x} - \hat{t}/(2\hat{t})^{1/2}] + O \, (1/\sqrt{\hat{t}}). \tag{279}$$

One can proceed to seek an $O(\varepsilon)$ term using Eqs. (271) and (272). However, the basic problem that exists with the solution (278) would still be present, this problem being that as $\hat{t}$ increases, the length scale of the pulse shown by Eq. (279) increases as $\sqrt{\hat{t}}$, i.e., the initial diffusion of the square starting pulse continues. Either by using Eq. (279) to estimate the order of the neglected terms (Moran and Shen, 1966) or by physical reasoning, one notes that the expected convective steepening can not be found by an asymptotic series based on the variables $\hat{x}$ and $\hat{t}$. The conclusion is that the asymptotic expansion is not uniformly valid for large time—but how large?

Recall that the length scale and time scale were based on initial molecular transport effects, i.e. $\tau_1 = \delta/a_0^2$. After sufficient time has elapsed, the time scale of the process should be governed by the piston speed U_0, and we therefore expect $\tau_2 = \delta/U_0^2$ and $l_2 = \delta/U_0$ to be the appropriate time and length scales to govern the process (see Fig. 6a). It is also evident that Eq.

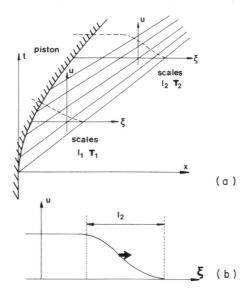

FIG. 6. Formation of shock wave by a moving piston and final steady state structure.

(279) shows some kind of traveling wave structure moving at the sound speed, and it is in a coordinate system fixed relative to the moving disturbance center that we expect τ_2 and l_2 to be the correct scales. To a certain extent, these reflections on scales may be the result of hindsight following some *ad hoc* cut and try procedures, it being, of course, somewhat a matter of taste how a problem is viewed. Even so, they provide insight and a reflection of how MAE connects formal procedures with physical reasoning. Thus we write

$$\bar{t} = t\ U_0^2/\delta = \varepsilon^2 \hat{t} \tag{280}$$

and

$$\bar{\xi} = U_0\ (x - a_0 t)/\delta = \varepsilon\ (\hat{x} - \hat{t}). \tag{281}$$

This leads to the transformed form of Eqs. (271) and (272) in what we shall call the *Burgers region*,

$$\frac{\partial}{\partial \bar{\xi}}\left(\frac{2}{\gamma - 1}\,\bar{a} - \bar{u}\right) = \varepsilon\left(\frac{\partial^2 \bar{u}}{\partial \bar{\xi}^2} - \frac{2}{\gamma - 1}\,\bar{a}\,\frac{\partial \bar{a}}{\partial \bar{\xi}} - \bar{u}\,\frac{\partial \bar{u}}{\partial \bar{\xi}} - \frac{\partial \bar{u}}{\partial \bar{t}}\right) \tag{282}$$

and

$$\frac{\partial}{\partial \bar{\xi}}\left(\frac{2}{\gamma - 1}\,\bar{a} - \bar{u}\right) = \varepsilon\left[\frac{2}{\gamma - 1}\left(\frac{\partial \bar{a}}{\partial \bar{t}} + \bar{u}\,\frac{\partial \bar{a}}{\partial \bar{\xi}}\right) + \bar{a}\,\frac{\partial \bar{u}}{\partial \bar{\xi}}\right]. \tag{283}$$

There is one small problem with these equations that has created a bit of mystery in the literature, in that in the limit $\varepsilon \to 0$, $\bar{\xi}$, $\bar{t}$ fixed, Eqs. (282) and (283) yield the *same* equation, i.e. only one equation results for the two unknowns. The problem is easily avoided if we choose as our two starting equations *either* (282) or (283) *and* the equation obtained by equating their right-hand sides, i.e.,

$$\frac{\partial^2 \bar{u}}{\partial \bar{\xi}^2} - \frac{\partial}{\partial \bar{t}}\left(\frac{2}{\gamma - 1}\,\bar{a} + \bar{u}\right) - \frac{\partial}{\partial \bar{\xi}}\left(\frac{1}{\gamma - 1}\,\bar{a}^2 + \frac{1}{2}\,\bar{u}^2\right)$$

$$- \frac{2}{\gamma - 1}\,\bar{u}\,\frac{\partial \bar{a}}{\partial \bar{\xi}} - \bar{a}\,\frac{\partial \bar{u}}{\partial \bar{\xi}} = 0. \tag{284}$$

Using the operator E_0 on Eq. (283) or (282) gives

$$\frac{\partial}{\partial \bar{\xi}}\left(\frac{2}{\gamma - 1}\,\bar{a}^{(0)} - \bar{u}^{(0)}\right) = 0. \tag{285}$$

This equation is readily integrated, and using the no disturbance condition at $\bar{\xi} \to \infty$, we find

$$\bar{a}^{(0)} = \tfrac{1}{2}(\gamma - 1)\,\bar{u}^{(0)}. \tag{286}$$

Equation (284) under E_0 just produces the changes $\bar{u} \to \bar{u}^{(0)}$, $\bar{a} \to \bar{a}^{(0)}$, so that using Eq. (286) we find

$$\frac{\partial \bar{u}^{(0)}}{\partial \bar{t}} + \Gamma\,\bar{u}^{(0)}\,\frac{\partial \bar{u}^{(0)}}{\partial \bar{\xi}} = \frac{1}{2}\,\frac{\partial^2 \bar{u}^{(0)}}{\partial \bar{\xi}^2}, \tag{287}$$

with $\Gamma = \tfrac{1}{2}(\gamma + 1)$. Equation (287) is the well-known Burgers equation, whose history can be found in the references. What is of prime interest here, of course, is how the derivation of the equation illustrates its connection with the full Navier–Stokes equation.

In the spirit of MAE, boundary or initial conditions for this equation are to be found by matching. Thus, let $\bar{E} = E\bar{T}$ and $\hat{E} = \bar{T}E\hat{T}$ define our expansion operators, and apply the rule

$$(\hat{E}_0\,\bar{E}_0 - \bar{E}_0\,\hat{E}_0)\,\bar{u} = 0. \tag{288}$$

Then

$$\hat{E}_0 \, \bar{E}_0 \, \bar{u}(\bar{\xi}, \bar{t}) = \hat{E}_0 \, \bar{u}^{(0)} \, (\bar{\xi}, \bar{t})$$
$$= \bar{T}E_0 \, \bar{u}^{(0)} \, (\varepsilon(\hat{x} - \hat{t}), \, \varepsilon^2\hat{t}). \tag{289}$$

Continuing with E_0 in Eq. (289) would appear to lead to $\bar{u}^{(0)}$ $(0, 0)$. However, before drawing this conclusion, consider the second member of Eq. (288),

$$\hat{E}_0 \, \bar{E}_0\hat{u}(\hat{x}, \hat{t}) = \bar{E}_0 \left\{\frac{1}{2} \operatorname{erfc} \left[\frac{\hat{x} - \hat{t}}{(2\hat{t})^{1/2}}\right] + O\left(\frac{1}{\hat{t}^{1/2}}\right)\right\}$$

$$= E_0 \left\{\frac{1}{2} \operatorname{erfc} \left[\frac{\bar{\xi}}{(2\bar{t})^{1/2}}\right] + O\left(\frac{\varepsilon}{\bar{t}^{1/2}}\right)\right\}$$

$$= \frac{1}{2} \operatorname{erfc} \left(\frac{\bar{\xi}}{(2\bar{t})^{1/2}}\right). \tag{290}$$

Therefore, for matching to be possible, the functional form of $\bar{u}^{(0)}$ must be

$$\bar{u}^{(0)} \, (\bar{\xi}, \bar{t}) = F(\bar{\xi}, \bar{t}, \bar{\xi}/\bar{t}^{1/2}). \tag{291}$$

If this is so, the final term in Eq. (289) is

$$\bar{T}\bar{E}_0 \, F(\varepsilon(\hat{x} - \hat{t}), \, \varepsilon^2 \, \hat{t}, \, (\hat{x} - \hat{t})/\hat{t}^{1/2}) = F(0, 0, \bar{\xi}/\bar{t}^{1/2}), \tag{292}$$

and hence, by matching,

$$\lim_{\substack{\bar{t} \to 0 \\ \bar{\xi}/\bar{t}^{1/2} \text{ fixed}}} \bar{u}^{(0)} \, (\bar{\xi}, \bar{t}) = \frac{1}{2} \operatorname{erfc} \left(\frac{\bar{\xi}}{(2\bar{t})^{1/2}}\right). \tag{293}$$

The problem of solving Eq. (287) so as to meet the condition (293) is considerably expedited by the existence of the *Hopf–Cole transformation*, which relates Eq. (287) to a simple diffusion equation. If we write

$$\bar{u}^{(0)} = - (1/\Gamma)(\partial/\partial\bar{\xi}) \ln \psi(\bar{\xi}, \bar{t}) \tag{294}$$

in Eq. (287), we find that the nonlinear term disappears, and that

$$\frac{\partial \psi}{\partial \bar{t}} = \frac{1}{2} \frac{\partial^2 \psi}{\partial \bar{\xi}^2}. \tag{295}$$

We can proceed by direct means to satisfy Eq. (293) or we can follow the physical reasoning employed by Moran and Shen (1966) and by Lighthill (1956), that in the Burgers region the problem is equivalent to the development of a step discontinuity, so that Eq. (293) may effectively be replaced by the simpler condition

$$\bar{u}^{(0)} \, (\bar{\xi}, 0) = H(-\bar{\xi}). \tag{296}$$

Equation (295) is then solved by standard means to obtain

$$\bar{u}^{(0)} = \left\{1 + \frac{\exp[\Gamma(\bar{\xi} - \frac{1}{2}\Gamma\bar{t})] \operatorname{erfc}[-\bar{\xi}/(2\bar{t})^{1/2}]}{\operatorname{erfc}[(\bar{\xi} - \Gamma\bar{t})/(2\bar{t})^{1/2}]}\right\}. \tag{297}$$

If we let $\bar{t} \to 0$ with $\bar{\xi}/\bar{t}^{1/2}$ fixed, it is apparent that Eq. (297) satisfies the matching condition (293). In addition, if we let $\bar{\xi}$ and $\bar{t}$ become large.

$$\bar{u}^{(0)} \to f_T(\bar{\xi} - \tfrac{1}{2}\Gamma\bar{t}) = 1/\{1 + \exp[\Gamma(\bar{\xi} - \tfrac{1}{2}\Gamma\bar{t})]\} \tag{298}$$

This form displays the structure of a weak shock wave and was first derived by Taylor (1910).

It is worthwhile to briefly examine Eq. (298) in physical coordinates, in terms of which we have, by Eqs. (280) and (281),

$$\frac{u}{U_0} = \frac{1}{1 + \exp[(\Gamma U_0/\delta)(x - (a_0 + \tfrac{1}{2}\Gamma U_0)\,t]}. \tag{299}$$

This represents a traveling wave of thickness δ/U_0 moving at speed $a_0 + \tfrac{1}{2}\Gamma U_0$, faster than the sound speed by the amount $\tfrac{1}{2}\Gamma\,\varepsilon\,a_0$ (see Fig. 6b). A uniformly valid solution to the piston boundary value problem could now be constructed by the composite expansion technique. By additive composition such a solution, at order zero, is

$$\bar{E}_0\,\bar{u} + \hat{E}_0\,\hat{u} - \bar{E}_0\,\hat{E}_0\,\bar{u}. \tag{300}$$

Before continuing, it is important to point out that Eq. (287) is as general a form of Burgers equation as found in the literature. Thus, although in the work of Moran and Shen (1966) and others one finds a parameter β multiplying the $\partial^2 \bar{u}^{(0)}/\partial\bar{\xi}^2$ term, that form of Burgers equation can always be put into the form

$$v_\tau + vv_\xi = v_{\xi\xi} \tag{301}$$

by the transformation

$$\tau = 2\Gamma^2\,\bar{t}/\beta, \qquad \xi = 2\Gamma\,\bar{\xi}/\beta. \tag{302}$$

C. SHOCK REFLECTION FROM A WALL

Consideration of Burgers equation and the above derivation indicates that it provides a valid approximation when only one direction of wave propagation is of principal importance, as in the above piston problem. The question naturally arises as to what modifications are needed to deal with problems where two propagation directions assume simultaneous importance. To obtain an answer to this question, Lesser and Seebass (1968) examined the problem of the time-dependent structure of a reflecting shock. Like Moran and Shen, they took the full Navier–Stokes equations as their starting point. Also, a major part of their investigation involved the examination of heat transfer effects during the reflection. Following the treatment of the previous section, we will use Lighthill's approximations Eqs. (261) and (262) to the Navier–Stokes equations to present the main ideas.

As our point of departure, we assume a weak shock, of the form given by Eq. (299), traveling toward a wall in such a way that its projected center would coincide with the wall position at $\bar{t} = 0$ if the wall were not present

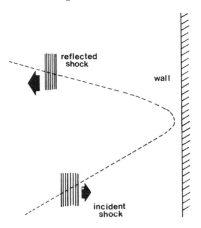

FIG. 7. Shock reflection from wall.

(see Fig. 7). Thus our starting point is now the Lighthill equations in Burgers variables, i.e. Eqs. (282) or (283) and (284), with the conditions that

$$\bar{u} \to f_T(\bar{\xi} - \tfrac{1}{2}\Gamma \bar{t}) + O(\varepsilon) \qquad \text{as} \quad \bar{\xi}, \bar{t} \to -\infty, \tag{303}$$

and

$$\bar{u} = 0 \qquad \text{on} \quad \bar{\xi} = -(1/\varepsilon)\bar{t}, \tag{304}$$

where the $O(\varepsilon)$ terms in Eq. (303) indicate higher order shock structure which is ignored in the present calculation. In the limit $\varepsilon \to 0$, $\bar{u} \to \bar{u}^{(0)}$, Eq. (303) satisfies (287) and the wall boundary condition Eq. (304) goes to $+\infty$ for $\bar{t} < 0$. As $\bar{t} \to 0$ [e.g., when $\bar{t} = O(\varepsilon)$], physical reasoning tells us that Eq. (287), the Burgers equation for waves traveling in the direction of increasing coordinate, cannot be correct. The reason for this is that the "shock" must reflect from the wall, and the structure of Eq. (287) does not permit this. Also, we expect in this nonlinear problem that the wave speed of the reflected shock will differ from $a_0 + \tfrac{1}{2}\Gamma U_0$, the wave speed of the incoming shock. Again, arguments based on the physical scales in the problem provide the solution. Recall [Eq. (280)] that the time scale for nonlinear events, i.e., the time scale of the *convection*, is δ/U_0^2. On the other hand, the *reflection* process takes place on a time scale equal to the shock thickness δ/U_0 divided by the sound speed a_0, i.e., $\delta/U_0 a_0$. This is the significant scale for the reflection process, so we define $\bar{t} = tU_0 a_0/\delta = tU_0^2(a_0/U_0)/\delta = \bar{t}/\varepsilon$ as the "acoustic" time scale. It is also convenient to move into the reference frame of a fixed wall; hence let

$$\tilde{x} - \bar{t} = \bar{\xi}. \tag{305}$$

Making this transformation $\tilde{T}(\bar{\xi}, \bar{t} \to \tilde{x} - \bar{t}, \varepsilon \bar{t})$ on Eqs. (282) and (283) we find

$$\frac{\partial \tilde{u}}{\partial \bar{t}} + \frac{2}{\gamma - 1} \frac{\partial \tilde{a}}{\partial \tilde{x}} = \varepsilon \left(\frac{\partial^2 \tilde{u}}{\partial \tilde{x}^2} - \tilde{u} \frac{\partial \tilde{u}}{\partial \tilde{x}} - \frac{2}{\gamma - 1} \tilde{a} \frac{\partial \tilde{a}}{\partial \tilde{x}} \right), \tag{306}$$

and

$$\frac{\partial \tilde{a}}{\partial \tilde{t}} + \frac{\gamma - 1}{2} \frac{\partial \tilde{u}}{\partial \tilde{x}} = \varepsilon \left(-\tilde{u} \frac{\partial \tilde{a}}{\partial \tilde{x}} - \frac{\gamma - 1}{2} \tilde{a} \frac{\partial \tilde{u}}{\partial \tilde{x}} \right). \tag{307}$$

The reader will note that the limit of these equations as $\varepsilon \to 0$ yields the one-dimensional wave equation,

$$\partial^2 \tilde{u}/\partial \tilde{x}^2 - \partial^2 \tilde{u}/\partial \tilde{t}^2 = 0. \tag{308}$$

This is expected on physical grounds as, for sufficiently short times, weak shocks behave as sound waves. What is new here is a formal means of producing quantitative results from this knowledge by matching procedures.

To arrive at initial conditions for Eq. (308) we apply the matching principle

$$(\bar{E}_0 \tilde{E}_0 - \tilde{E}_0 \bar{E}_0)\bar{u} = 0 \tag{309}$$

where

$$\bar{E} = \bar{T} E \tilde{T}.$$

Thus

$$\bar{E}_0 \tilde{E}_0 \bar{u} = \bar{E}_0 \tilde{u}^{(0)}(\bar{\xi} + \tilde{t}/\varepsilon, \tilde{t}/\varepsilon), \tag{310}$$

and

$$\begin{aligned} \bar{E}_0 \tilde{E}_0 \bar{u} &= \bar{E}_0 f_T (\bar{\xi} - \tfrac{1}{2}\Gamma \tilde{t}) \\ &= \bar{E}_0 f_T (\tilde{x} - \tilde{t} - \tfrac{1}{2}\Gamma \varepsilon \tilde{t}) \\ &= f_T (\bar{\xi}). \end{aligned} \tag{311}$$

Now, as $\tilde{u}^{(0)}$ satisfies the simple wave equation, we can write down the general form of the solution which satisfies the wall boundary condition $\tilde{u}^{(0)}(0, \tilde{t}) = 0$, namely

$$\tilde{u}^{(0)}(\tilde{x}, \tilde{t}) = g(\tilde{x} - \tilde{t}) - g(-\tilde{x} - \tilde{t}). \tag{312}$$

Substituting into Eq. (310), we have

$$\bar{E}_0 \tilde{u}^{(0)} = \bar{E}_0 [g(\bar{\xi}) - g(-\bar{\xi} - (2\tilde{t}/\varepsilon))] \tag{313}$$

and as $\tilde{t} < 0$ for this matching,

$$\bar{E}_0 \tilde{u}^{(0)} = g(\bar{\xi}) - g(+\infty). \tag{314}$$

Now $f_T(+\infty) = 0$, and therefore equating Eqs. (314) and (311) we see that

$$g(\bar{\xi}) = f_T(\bar{\xi}) \tag{315}$$

or, in acoustic variables,

$$\tilde{u}^{(0)} = f_T(\tilde{x} - \tilde{t}) - f_T(-\tilde{x} - \tilde{t}). \tag{316}$$

If we now attempt to improve this *acoustic* solution by using Eq. (316) to calculate terms of $O(\varepsilon)$, we would find that the result contained terms such as

$$\tilde{u}^{(1)} \sim \text{(bounded terms)} \, \tilde{t}$$

so that the acoustic expansion fails when $\tilde{t} = O(1/\varepsilon)$, as expected. To calculate the solution at large *positive* times, we must place ourselves in a coordinate system moving in the direction of decreasing $\tilde{x}$ at the acoustic speed and rescale $\tilde{t}$ to the "Burgers" time scale $\check{t} = \varepsilon \tilde{t}$. Thus let

$$\bar{\eta} = \tilde{x} + \tilde{t} \tag{317}$$

and apply the operator

$$\check{T}_+(\tilde{x}, \tilde{t} \to \bar{\eta}, \check{t})$$

to Eqs. (306) and (307). This results in an out-going wave form of Eq. (282) or (283), and in the limit $\varepsilon \to 0$, $(\check{t}, \bar{\eta})$ fixed, we obtain the appropriate Burgers equation,

$$(\partial \bar{u}_+^{(0)}/\partial \check{t}) + (\Gamma \bar{u}_+^{(0)} - \gamma + 1)\, \partial \bar{u}_+^{(0)}/\partial \bar{\eta} = \tfrac{1}{2}\, \partial^2 \bar{u}_+^{(0)}/\partial \bar{\eta}^2. \tag{318}$$

The traveling wave solution to Eq. (318) is

$$\bar{u}_+^{(0)}(\bar{\eta}, \check{t}) = f_T(\bar{\eta} - \tfrac{1}{4}(5 - 3\gamma)\check{t}), \tag{319}$$

which, as the reader can verify, can be matched to $\tilde{u}^{(0)}$ for $\check{t} > 0$ by the rule

$$(\bar{E}_{+0} \tilde{E}_{+0} - \tilde{E}_{+0} \bar{E}_{+0})\bar{u} = 0. \tag{320}$$

If the acoustic region solution is carried out and matched to $O(\varepsilon)$, one can form the composite expansion to $O(1)$ in the incoming and outgoing Burgers regions and to $O(\varepsilon)$ in the acoustic region. The details are quite cumbersome and can be found in Lesser and Seebass (1968). To $O(1)$, however, we have

$$\bar{u}_{\text{comp}} = f_T(\bar{\xi} - \tfrac{1}{2}\Gamma \check{t}) - f_T(-\bar{\eta}), \qquad \check{t} \leq 0, \tag{321}$$

$$= f_T(\bar{\eta} - \tfrac{1}{4}(5 - 3\gamma)\check{t}) + f_T(\bar{\xi}) - 1, \qquad \check{t} \geq 0. \tag{322}$$

Lesser and Seebass also studied the problem of reflection from an isothermal wall, which involves a *thermal boundary layer*—yet another asymptotic region obtained by rescaling $\tilde{x}$ by $\varepsilon^{-1/2}$. The calculations become somewhat lengthy and will not be presented here.

D. The Time-Harmonic Piston Problem

In the previous sections we have seen how, starting from the Navier–Stokes equations [or, in our presentation at any rate, the Lighthill approximation (261) and (262) to them] the MAE method leads to physically appropriate simplified equations for restricted domains of space and time. Thus, we have obtained the Burgers equations (287) and (318), describing right and left propagating waves, the linear inviscid wave equation (308), and the linear wave equation (276) with dissipative effects included. We have also shown how MAE leads to the requisite boundary and initial data for these simplified equations.

Now we want to extend the range of applicability of MAE by showing how Burgers equation itself can be solved with the aid of MAE for the classic problem of nonlinear wave excitation by a sinusoidally oscillating piston. We shall recover all the well-known results for this problem, but in a way which greatly clarifies their interrelationship. Further, we shall actually improve on some existing results, despite the fact that an *exact* solution to the problem is known from the Hopf–Cole transformation, Eqs. (294) and (295). The method can also be used to discuss problems in two- and three-dimensional nonlinear acoustics, where no transformation of the Hopf–Cole kind has yet been found.

We avoid irrelevant complications by assuming from the outset that the whole motion is described by the Burgers equation (287), which in physical coordinates reads

$$(\partial u/\partial t) + (a_0 + \Gamma u)(\partial u/\partial x) = \tfrac{1}{2}\delta \,\partial^2 u/\partial x^2. \tag{323}$$

This form of equation is suitable for initial value problems, in which $u(x, t = 0)$ is prescribed.

In the literature on nonlinear acoustics Burgers equation is introduced in a form suitable for the study of the evolution of a periodic wave form as it travels away from the source boundary (see, e.g., Blackstock, 1964a,b). Also the equation is viewed as a model or close analog to the full nonlinear acoustics equations. In Section IV, B we indicated a rigorous and logical relation between the equations of viscous gas dynamics and Eq. (323). A heuristic relation between this form of Burgers equation and the form commonly used in acoustic studies may be obtained by using the linear nondissipative relation $\partial u/\partial t = -a_0 \,\partial u/\partial x$ in the nonlinear and dissipative terms, and then taking retarded time $\tau = t - x/a_0$ and space x as independent variables. The error thus involved is formally of the same order as that involved in taking Eq. (323) as the governing equation.

A more careful study of this procedure is yet to be undertaken. However, for our present purpose we continue our analysis with the resulting equation

$$a_0^3(\partial u/\partial x) - \Gamma a_0 u(\partial u/\partial \tau) = \tfrac{1}{2}\delta \,\partial^2 u/\partial \tau^2. \tag{324}$$

This can be put in the form

$$(\partial \bar{u}/\partial \bar{x}) - \bar{u}(\partial \bar{u}/\partial \bar{\tau}) = \varepsilon \,\partial^2 \bar{u}/\partial \bar{\tau}^2 \tag{325}$$

with the definitions

$$\bar{\tau} = \omega\tau, \qquad \bar{x} = \Gamma(u_0/a_0)\,\omega x/a_0),$$
$$\bar{u} = u/u_0, \qquad \varepsilon = \omega\delta/2\Gamma u_0 a_0, \tag{326}$$

ω and u_0 denoting the piston frequency and velocity amplitude.

Consistent with the use of Eq. (325) to study the developing wave form we omit the consequences of *finite displacement* of the piston, which are in any case confined to the neighborhood of the piston.

This state of affairs is reflected by the parabolic structure of Eq. (325) where now $\bar{x}$ plays the role of what is normally the time variable. Thus we now study the "boundary value" problem (in mathematical terms actually an initial value problem) equivalent to the piston boundary value problem given by the condition

$$u(x=0,\,t) = \sin\,\omega t, \quad \text{or} \quad \bar{u}(\bar{x},\,\bar{\tau}) = \sin\,\bar{\tau}. \tag{327}$$

Our aim is to obtain an approximate solution of Eqs. (325) and (327) for small values of the inverse Reynolds number ε, such that the approximate solution is uniformly valid in the $(x,\,t)$ domain of interest. Observing that the differential equation preserves the parity and periodicity of the boundary value, we need to consider only $0 \le \bar{\tau} \le +\pi$, say, imposing the conditions

$$\bar{u}(\bar{x},\,\bar{\tau}=0) = \bar{u}(\bar{x},\,\bar{\tau}=\pi) = 0, \tag{328}$$

and defining $\bar{u}$ elsewhere as the odd periodic continuation of the function in $[0,\,\pi]$. The solution must be uniformly valid for $0 \le \bar{\tau} \le \pi$ and for all $\bar{x} \ge 0$.

For $(\bar{x},\,\bar{\tau}) = O(1)$ we assume that

$$\bar{u}(\bar{x},\,\bar{\tau}) = \bar{u}^{(0)}(\bar{x},\,\bar{\tau}) + o(1) \qquad \text{as} \quad \varepsilon \to 0,$$

and then

$$\frac{\partial \bar{u}^{(0)}}{\partial \bar{x}} - \bar{u}^{(0)}\frac{\partial \bar{u}^{(0)}}{\partial \bar{\tau}} = 0. \tag{329}$$

The general solution of this equation can be written as

$$\bar{u}^{(0)} = f(\sigma), \qquad \sigma = \bar{\tau} + \bar{x}f(\sigma), \tag{330}$$

where f denotes an arbitrary function and the variable σ can be interpreted as a distorted version of time $\bar{\tau}$, the distortion increasing with range $\bar{x}$ from the source. Choosing $f(\sigma) = \sin\,\sigma$ enables the conditions (327) and (328) to be met, so that

$$\bar{u}^{(0)} = \sin\,\sigma, \qquad \sigma = \bar{\tau} + \bar{x}\sin\,\sigma. \tag{331}$$

The next approximation can also be found, most easily by transforming the Burgers equation (329) to the variables $\bar{x}$ and σ as

$$(1 - \bar{x}\cos\sigma)\frac{\partial\bar{u}}{\partial\bar{x}} - (\bar{u} - \sin\sigma)\frac{\partial\bar{u}}{\partial\sigma} = \varepsilon\frac{\partial}{\partial\sigma}\left[\frac{\partial\bar{u}/\partial\sigma}{1 - \bar{x}\cos\sigma}\right].$$

Formal substitution of $\bar{u} = \bar{u}^{(0)} + \varepsilon\bar{u}^{(1)} + o(\varepsilon)$ into this equation gives

$$\frac{\partial}{\partial\bar{x}}(1 - \bar{x}\cos\sigma)\,\bar{u}^{(1)} = \frac{\partial}{\partial\sigma}\left(\frac{\cos\sigma}{1 - \bar{x}\cos\sigma}\right).$$

The solution satisfying

$$\bar{u}^{(1)}\,(\bar{x}=0,\,\tau)=0$$

is

$$\bar{u}^{(1)}\,(\bar{x},\,\sigma)=-\frac{\bar{x}\sin\sigma}{(1-\bar{x}\cos\sigma)^2},$$

and thus

$$\bar{u}(\bar{x},\,\sigma)=\sin\sigma\left\{1-\frac{\varepsilon\bar{x}}{(1-\bar{x}\cos\sigma)^2}+o(\varepsilon)\right\}. \tag{332}$$

These solutions can, of course, be written as Fourier sine series in $\bar{\tau}$. In particular, although $\bar{u}^{(0)}$ is given implicitly by Eq. (331), its Fourier coefficients can be found explicitly in terms of Bessel functions. This calculation yields the well-known *Fubini solution* (Fubini-Ghiron, 1935)

$$\bar{u}^{(0)}=2\sum_{n=1}^{\infty}\frac{J_n(n\bar{x})}{n\bar{x}}\sin n\bar{\tau}. \tag{333}$$

Now when $\bar{x}>1$ the relation $\sigma=\bar{\tau}+\bar{x}\sin\sigma$ no longer defines σ as a one-valued function of $\bar{\tau}$. This can easily be seen by noting that

$$d\bar{\tau}/d\sigma=1-\bar{x}\cos\sigma\qquad\text{and hence}\qquad d\sigma/d\bar{\tau}=0$$

when $\bar{x}=1/\cos\sigma$. In terms of $\bar{x}$ and $\bar{\tau}$, the transformation $\sigma=\sigma(\bar{x},\,\bar{\tau})$ fails along the "limit line"

$$\bar{\tau}=\cos^{-1}\,(1/\bar{x})-(\bar{x}^2-1)^{1/2}.$$

There is a range of values of $\bar{\tau}$ around $\bar{\tau}=0$ at which there are, according to Eq. (331), three different values of $\bar{u}^{(0)}$ (see Fig. 8). The utility of the solution (331) is not invalidated for all this range of $\bar{\tau}$, however, since the solution

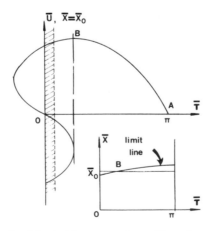

Fig. 8. Triple valued inviscid wave and resulting viscous adjustment zone.

corresponding to a given branch of $\sigma = \sigma(\bar{\tau})$ (A B in Fig. 8) is smooth right up to $\bar{\tau} = 0$, and would involve negligible values of the viscous terms omitted from Eq. (329). Things only go wrong, at $O(1)$ at any rate, very close to $\bar{\tau} = 0$, where this branch fails to satisfy the condition $\bar{u}^{(0)} = 0$ if $\bar{x} > 1$, and the value of $\bar{u}$ as $\bar{\tau} \to 0$ must there be rapidly brought down to zero by viscous force. This *shock wave* at $\bar{\tau} = 0$ is of dimensionless thickness ε, provided $\bar{x}$ is not too close to unity. The reason for this is that the processes competing in the shock wave are the nonlinear waveform steepening which, unchecked, would lead to the multivalued inviscid solution (331), and the waveform relaxing viscous forces, and therefore, if $\bar{u} = O(1)$ as $\bar{\tau} \to 0$, a balance between $u \, \partial u / \partial \tau$ and $\varepsilon \, \partial^2 u / \partial \tau^2$ can only be achieved if $\tau = O(\varepsilon)$.

The precise limit of $\bar{u}^{(0)} (\bar{x}, \bar{\tau})$ is obtained from

$$\bar{u}^{(0)} \sim \sin \sigma, \qquad \sigma \sim \bar{x} \sin \sigma \tag{334}$$

as $\bar{\tau} \to 0$ *provided $\bar{x}$ is bounded away from unity*. The second of these can be written in terms of spherical Bessel functions as

$$j_0(\sigma) = (\sin \sigma)/\sigma = \bar{x} - 1$$

so that $\sigma = j_0^{-1}(\bar{x}^{-1})$, where j_0^{-1} is the function inverse to j_0, and hence

$$\bar{u}^{(0)} (\bar{x}, \bar{\tau}) \sim \bar{x}^{-1} j_0^{-1}(\bar{x}^{-1}) = h(\bar{x}) \tag{335}$$

say, as $\bar{\tau} \to 0$.

We analyze the shock by writing $\hat{\tau} = \bar{\tau}/\varepsilon$ and assuming [since $\bar{u}^{(0)} = O(1)$ as $\bar{\tau} \to 0$ by Eq. (335)] that

$$\hat{u}(\bar{x}, \hat{\tau}) = \hat{u}^{(0)} (\bar{x}, \hat{\tau}) + o(1) \tag{336}$$

as $\varepsilon \to 0$, which leads to the anticipated shock equation

$$-\hat{u}^{(0)} \, \partial \hat{u}^{(0)} / \partial \hat{\tau} = \partial^2 \hat{u}^{(0)} / \partial \hat{\tau}^2. \tag{337}$$

This has the general solution

$$\hat{u}^{(0)} (\bar{x}, \hat{\tau}) = f(\bar{x}) \tanh \left[\tfrac{1}{2} f(\bar{x}) \, \hat{\tau} + g(\bar{x}) \right], \tag{338}$$

for arbitrary f and g, and enables the condition $\hat{u}^{(0)} = 0$ at $\hat{\tau} = 0$ to be satisfied for all $\bar{x}$ if $g \equiv 0$.

Now $f(\bar{x})$ must be found by matching. For simplicity we avoid our formal notation which is not needed to the order to which we carry out the expansions. Then

$$\hat{u}^{(0)}(\bar{x}, \hat{\tau} = \bar{\tau}/\varepsilon) = f(\bar{x}) \tanh[\tfrac{1}{2} f(\bar{x})\bar{\tau}/\varepsilon]$$
$$\sim |f(\bar{x})|$$

as $\varepsilon \to 0$ with $\bar{\tau} > 0$, and this matches Eq. (335) if

$$f(\bar{x}) = h(\bar{x}) = \bar{x}^{-1} j_0^{-1} (\bar{x}^{-1}). \tag{339}$$

Thus the shock structure is defined by

$$\hat{u}(\bar{x}, \hat{\tau}) = h(\bar{x}) \tanh(\tfrac{1}{2} h(\bar{x})\hat{\tau}) + o(1). \tag{340}$$

This formula describes the shock at all ranges $\bar{x} = O(1)$ from the piston *except in the "initial shock region"* around $\bar{x} = 1$ (see Fig. 9), the range at which the first shock appears. If $\bar{x}$ is sufficiently close to unity, the inner limit of the inviscid solution is not Eq. (335) but, rather, a function of $\bar{x}$, $\bar{\tau}$, and ε. This leads to an initial shock in which all three terms in Eq. (325) are of comparable magnitude, so that the description of the initial shock region is complicated and involves the solution of the full Burgers equation (325). The difference between the "initial shocks" around $\bar{x} = 1$ and the shocks at ranges $\bar{x}$ bounded away from unity does not seem to have been noticed before. The details of the initial shock region will be reported elsewhere (Crighton, forthcoming article), as that region does not play an essential part in the evolution of the wave.

The shock, Eq. (340), and the inviscid flow, Eq. (331), around it, can be combined to give a composite approximation (good for all $\bar{x} = O(1)$ bounded away from unity, and for all τ from 0 to π)

$$u_c^{(0)}(\bar{x}, \bar{\tau}, \hat{\tau}) = \sin \sigma(\bar{x}, \bar{\tau}) + h(\bar{x}) \tanh(\tfrac{1}{2}h(\bar{x})\hat{\tau}) - h(\bar{x}). \qquad (341)$$

This takes a simple form for $\bar{x} \gg 1$, for then $\sigma = \bar{\tau} + \bar{x} \sin \sigma$ can be satisfied only if σ is close to π. In fact, one easily finds that

$$\sigma(\bar{x}, \bar{\tau}) \sim \pi - \left(\frac{\pi - \bar{\tau}}{\bar{x} + 1}\right) \qquad (342)$$

and that

$$h(\bar{x}) \sim \pi/(\bar{x} + 1) \qquad (343)$$

for $\bar{x} \gg 1$. Thus

$$u_c^{(0)}(\bar{x}, \bar{\tau}, \hat{\tau}) = \left(\frac{\pi - \bar{\tau}}{\bar{x} + 1}\right) + \left(\frac{\pi}{\bar{x} + 1}\right) \tanh \left[\frac{\pi\hat{\tau}}{2(\bar{x} + 1)}\right] - \left(\frac{\pi}{\bar{x} + 1}\right) \qquad (344)$$

$$= \left(\frac{\pi}{\bar{x} + 1}\right)\left[\tanh \frac{\pi\hat{\tau}}{2(\bar{x} + 1)} - \bar{\tau}\right], \qquad (345)$$

in which the inviscid solution has now developed from the Fubini form (333) to the *Sawtooth form*

$$\bar{u}^{(0)}(\bar{x}, \bar{\tau}) \sim \left(\frac{\pi - \bar{\tau}}{\bar{x} + 1}\right) = \sum_{n=1}^{\infty} \frac{\sin n\bar{\tau}}{n}\left(\frac{2}{\bar{x} + 1}\right), \qquad (346)$$

while the thickness (with respect to $\bar{\tau}$) of the shock is not just $O(\varepsilon)$, but in fact $O(\varepsilon\bar{x})$. This shows that the shock thickness becomes comparable with the wavelength when $\bar{x} = O(\varepsilon^{-1})$, and there Eq. (345) cannot necessarily be expected to hold any longer.

A well-known result is found by expressing the whole of Eq. (345) in Fourier sine series form. If

$$u^{(0)}\left(\bar{x}, \bar{\tau}, \hat{\tau} = \frac{\bar{\tau}}{\varepsilon}\right) = \sum_{n=1}^{\infty} A_n(\bar{x}, \varepsilon) \sin n\bar{\tau}, \qquad (347)$$

the contribution from the tanh in Eq. (345) can be found as an infinite series, leading to

$$A_n(\bar{x}, \, \varepsilon) = \frac{4\varepsilon}{\pi\delta} \left(\frac{1}{n} + 2n \sum_{p=1}^{\infty} \frac{(-1)^p(1 - e^{-2p\pi/\delta} \cos n\pi)}{(2p/\delta)^2 + n^2} \right)$$

where the thickness $\delta = 2\varepsilon(\bar{x} + 1)/\pi$. If we now let $\varepsilon \to 0$ with $\bar{x} = O(1)$ we have

$$A_n(\bar{x}, \, \varepsilon) \sim \frac{4\varepsilon}{\pi\delta} \left(\frac{1}{n} + 2n \sum_{p=1}^{\infty} \frac{(-1)^p}{(2p/\delta)^2 + n^2} \right)$$

$$= 2\varepsilon \, \mathrm{cosech} \, n\varepsilon \, (\bar{x} + 1),$$

so that

$$u_c^{(0)}(\bar{x}, \, \bar{\tau}, \, \varepsilon) = 2\varepsilon \sum_{n=1}^{\infty} \frac{\sin n\bar{\tau}}{\sinh n\varepsilon(\bar{x} + 1)}. \tag{348}$$

This is the well-known *Fay solution* (Fay, 1931), which is now seen to be the Fourier series version of a composite solution made up from the saw-tooth solution (346) and the shock solution (340), in which viscous forces cancel the n^{-1} fall-off of the Sawtooth Fourier coefficients and replace them with the decay $\exp[-n\varepsilon(\bar{x} + 1)]$ for large n.

The Fay solution cannot be expected to hold at ranges $\bar{x}$ so large that the shock thickness $\varepsilon\bar{x}$ has become comparable with the scale, $O(1)$, of the inviscid sawtooth waves between the shocks (though in fact it does, as we shall see in a moment). To follow the motion into its *old age*, in which viscous forces become significant everywhere and not just in narrow shock regions, we need to rescale $\bar{x}$, defining $x^* = \varepsilon\bar{x}$ and finding another asymptotic solution which can be matched to Eq. (348). Note that we are now matching in $\bar{x}$, rather than in $\bar{\tau}$ as previously, *the need for two regions in $\bar{\tau}$ having been removed by forming the composite equation* (348), which is uniformly valid in $\bar{\tau}$. Now $\bar{x} = \varepsilon^{-1}$ marks the range at which a superficial balancing of $\partial u/\partial\bar{x}$ against $\varepsilon\partial^2 u/\partial\bar{\tau}^2$ would lead us to expect the dominance of viscous forces over inertia, so that for fixed x^* we might expect to have to match Eq. (348) to a solution of the diffusion equation. That, however, is not the case, and all three terms in the Burgers equation are in a delicate balance around $\bar{x} = \varepsilon^{-1}$, as we shall now see.

We have

$$u^{(0)}\left(\bar{x} = \frac{x^*}{\varepsilon}, \, \bar{\tau}, \, \varepsilon\right) \sim 2\varepsilon \sum_{n=1}^{\infty} \frac{\sin n\bar{\tau}}{\sinh nx^*} \tag{349}$$

as $\varepsilon \to 0$, and therefore assume that

$$u^*(x^*, \, \bar{\tau}) = \varepsilon u_0^*(x^*, \, \bar{\tau}) + o(\varepsilon),$$

and so find that

$$\frac{\partial u_0^*}{\partial x^*} - u_0^* \frac{\partial u_0^*}{\partial\bar{\tau}} = \frac{\partial^2 u_0^*}{\partial\bar{\tau}^2}. \tag{350}$$

The general solution can be written, with the aid of the Hopf–Cole transformation, as

$$u_0{}^* = 2(\partial/\partial\bar\tau)\ln F(x^*, \bar\tau) \tag{351}$$

where

$$\partial F/\partial x^* = \partial^2 F/\partial\bar\tau^2.$$

It is remarkable that this $u_0{}^*$ is in fact just the function in Eq. (348). For according to the definition of the theta function (Abramowitz and Stegun, 1964)

$$(\partial/\partial u)\ln\vartheta_4(u, q = e^{-\alpha}) = 2\sum_{n=1}^{\infty}\sin 2nu/\sinh n\alpha \tag{352}$$

where

$$\vartheta_4(u, q) = 1 + 2\sum_{n=1}^{\infty}(-1)^n q^{n^2}\cos nu, \tag{353}$$

and therefore the function

$$F(x^*, \bar\tau) = \vartheta_4(\bar\tau/2, q = e^{-x^*}) \tag{354}$$

evidently satisfies the diffusion equation and matches Eq. (349) uniformly in $\bar\tau$. Thus

$$u^*(\bar x^*, \bar\tau) = 2\sum_{n=1}^{\infty}\sin n\bar\tau/\sinh nx^* \tag{355}$$

is the required solution, in which the various Fourier coefficients do not simply satisfy the diffusion equation. If, however, $x^* \gg 1$, then

$$u_0{}^* \sim 4e^{-x^*}\sin\bar\tau \tag{356}$$

which is the *old age solution*, showing that at ranges much greater than ε^{-1} the wave contains only a single Fourier component, which decays under viscous action alone. This is also a well-known result, and it shows that Eq. (348) is uniformly valid for arbitrarily large ranges, and that no further asymptotically distinct regions are required.

Thus we have recovered all the essential results for this problem, and have unified them in a way which had not previously been possible—even with the aid of the exact though rather intractable solution to Eqs. (325) and (327),

$$u(\bar x, \bar\tau) = 2\varepsilon(\partial/\partial\bar\tau)\{\ln[1 + 2\sum_{n=0}^{\infty}(-1)\,I_n(1/2\varepsilon)e^{-n^2\varepsilon\bar x}\cos n\bar\tau]\} \tag{357}$$

which can be obtained from the Hopf–Cole transformation. Further details of this problem, and other similar applications are given in a forthcoming article by Crighton. The various regions and solutions are depicted in Fig. 9.

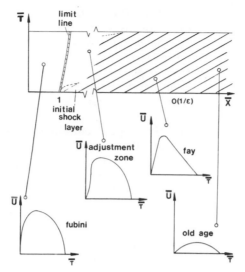

F ɪ ɢ. 9. Evolution of a periodic signal in terms of distance from source.

It was stated in the article on nonlinear acoustics in this series (Beyer, 1965, p. 247) that the relations between Eq. (357) and forms such as Fay's solution (348) have not been worked out. Although these relationships have now been partially clarified by methods other than MAE (see, for example, Blackstock, 1964a,b), it is evident that MAE provides both new insight and new results for this problem.

Limitations of space preclude us from dealing with other aspects of nonlinear acoustics. Fortunately, however, three other areas have already been discussed in the literature from the MAE viewpoint. First, Crow (1970) gives a particularly penetrating analysis of the problems interest in the formulation of theories of aerodynamically generated sound, a subject also treated by Lauvstad (1968), while the particular (linear) applications of MAE to aerodynamic noise are given by Amiet and Sears (1970), Crighton (1972, 1975, and references given there), and Crighton and Leppington (1971). Second, numerous authors have used MAE in problems of "acoustic streaming" induced by nonlinear inertia forces in unsteady viscous flows [see, for example, Wang (1968), and the article by Riley (1967)]. Third, the interaction of underwater sound with linear and nonlinear free surface motions has been treated by Harper and Simpkins (1975).

V. Conclusions

At this point the reader has had ample opportunity to appreciate the constructive aspects of the theory of matched asymptotic expansions, and perhaps has seen some applications to problems of interest to him. In the remaining pages we wish to sum up the major points, indicate some side branches of interest, and highlight some of the research toward making the theory more useful and mathematically sound. Therefore, in this last section we shall first discuss some of the current trends in the development of the theory. This will be followed by a very brief discussion of other "modern" perturbation methods with enough references to enable the interested reader to enter into a study of such techniques. Finally, we shall close with some personal opinions and prejudices regarding the present status of singular perturbation theory in acoustics.

Perhaps one of the most active areas of current research *vis a vis* M.A.E. is the technique of "matching." To convey some of the feelings among those working on the subject, consider the following quotations, all from investigators who have made weighty contributions to the development of M.A.E.; first Lagerström and Casten (1972) in a discussion of the Van Dyke principle:

"Another important point, which has been made earlier, is that whatever techniques are used, *matching is always based on overlap*. It obviously does not make sense to compare two approximations which do not have a common domain of validity."[1]

while on the other hand Fraenkel (1969) tells us:

"The asymptotic matching principle correctly determines unknown constants, and leads to uniformly valid composite series, even though it is applied to inner and outer expansions which contain too few terms to overlap to the order of the terms being matched.... Although it is not astonishing, the point is emphasized here because it seems to be widely believed that the success of the asymptotic matching principle depends on and implies overlapping of the series used to the order of the terms being matched."[2]

Finally Eckhaus (1973), after a discussion of several matching techniques, says:

"In applications one usually has no *a priori* information on the structure of the uniform expansion, nor does one possess any *a priori* estimates of the overlap domain between different approximations. It is for this reason that the choice of the matching rules to be used is usually dictated by the authors taste, habits and ingenuity."[3]

Perhaps the fairest thing that can be said at present is as follows. The matching principle, Eq. (22), has been rigorously proved (in various forms) by Fraenkel (1969), Crighton and Leppington (1973, Appendix), and Eckhaus

[1] Reprinted with permission from *SIAM Rev.* (1972). Copyright 1972 by Society for Industrial and Applied Mathematics.

[2] Reprinted from *Proceedings of the Cambridge Philosophical Society*, with permission of Cambridge University Press.

[3] Reprinted from "Matched Asymptotic Expansions and Singular Perturbations," with permission of North-Holland Publishing Company.

(1973), subject to constraints like Eq. (52) on the gauge functions, and subject to constraints on the asymptotic validity and domains of validity of the expansions being matched. The proofs show that "overlap," in some properly defined sense, is *sufficient* for Eq. (22) to hold. Overlap is not necessary, however, as evidenced by Fraenkel's statement above, and the details of his papers. On the other hand, it is easy to construct examples (see Crighton and Leppington, 1973) in which two expansions do not overlap and cannot be matched by Eq. (22), indicating that both a generalization of the matching principle and a wider notion than overlap are probably needed in order to find *necessary* conditions.

But in any case the apparent conflict over matching rules, and the problem of knowing whether an expansion produced by formal manipulations guided by physical and mathematical experience actually constitutes any kind of asymptotic series, is no reason for despair. Many of the methods in common use in physics fall into the same category and many others have only recently been removed from this status. What is clear is that more research is needed in the foundations of the subject. For example, even the term "singular perturbation" is not always clear. The most widely accepted current meaning of the phrase appears to be that there does not exist one single expansion of Poincaré form valid everywhere in the domain of interest of a singular perturbation problem.

Another area of research in the foundations of the method concerns the very important practical question of how to find the scalings and regions appropriate to a given problem. At this stage this seems to be very much an art, although some first steps towards an answer to this question can be found in the monograph of Eckhaus (1973). For the present, the main approach to this problem lies in physical intuition, and experience in applying the method. Examples of this were given above; for example the classical ideas about problems in which the wavelength is long with respect to geometric scales were formalized so as to continue expansions beyond the obvious first terms.

In the present exposition we have concentrated on only one of several newly developed techniques for dealing with so-called singular perturbation problems. Our reason for so doing is that we believe it to be both the most widely applicable method and the one most closely coupled to physical models. This certainly does not mean that it is the easiest to apply, or that all problems will yield to it. Another method, at present most applicable to nonlinear wave problems, involves variations of the multiple scales and PLK (Poincaré–Lighthill–Kuo) techniques. A modern survey, in which the initiators of many of the new and most exciting developments along these lines speak for themselves, can be found in Leibovich and Seebass (1974). Without going into details of these techniques, we can simply state that they seek to transform the independent variables simultaneously with the perturbation procedure, in such a manner as to eliminate nonuniformities such as singular behavior and secular growth. A collection of such problems can be found in the book by Nayfeh (1973), who also gives an exhaustive bibliography to which the interested reader is referred. Several papers of particular note are those by Lighthill (1949), Lin (1954), Fox (1955), Morrison (1966), and Lick (1967, 1970).

Finally, we wish to say a few words about the place of MAE in the future of acoustics. There are clearly several roles which can be played, as illustrated above. In the discussion of linear problems, typically involving waveguides and diffraction effects, the method shows how to combine solutions of "simple" problems to obtain results for complex geometrical situations. In a way this can be considered a refinement of the traditional concepts of impedance, transmission line analog, etc., which are used to decompose problems into simpler elements. The utility of such a decomposition was discussed in the work of Schwinger and Saxon (1968). There it was pointed out that for the technological application of a theory it is useful to have a buffer theory which employs constructs that are more easily manipulated than the concepts of the parent discipline, the latter being used to calculate the parameters of the buffer theory. A familiar example of this is the relation between solutions of the wave equation for simple geometries and the circuit type theories in which circuit element parameters like radiation impedance are obtained from such solutions. MAE provides not only a very attractive way of dealing with such calculations but also an excellent means of assessing their validity and of forming new buffer theories. It is thus of central interest in dealing with such a classical subject as diffraction theory to carry out the approximate calculations to relatively high order so that fruitful results can be obtained for cases in which, for example, the wavenumber is not very small. To this end the recent work on Padé approximants (Graves-Morris, 1973) and computer formulation of analytical problem solving procedures (Schwartz, 1974) is of great interest. A number of cases now exist where restructuring of a perturbation series into rational fractions or Padé approximants has greatly extended the usefulness of the results [see, for example, the discussion by Shanks (1955) and Van Dyke (1970) for a more recent application].

Another contribution from MAE and other singular perturbation methods is directed toward a better and more systematic treatment of nonlinear and more extensive physical effects in acoustics. The illustration of how Burgers' equation arises naturally from a MAE treatment of the Navier–Stokes equations is but one of many examples. The current critical examination of aerodynamic noise theory (Crow, 1970) and the examination of acoustic streaming effects are two others (Riley, 1967; Wang, 1968). At the very least, as a supplement to extensive and perhaps all-too-often performed numerical calculations, singular perturbation methods provide a powerful tool for expanding our understanding in these difficult areas. The same is likely to be true of solid mechanics, another field in which singular perturbation ideas have yet to be adequately recognized and employed—somewhat surprisingly, in view of the inevitably greater complexity which arises in linear boundary value problems in elasticity compared with their acoustic counterparts, not to mention nonlinear elastodynamics. The analogs in elasticity theory of the types of diffraction problem discussed in Section II above have, however, recently been solved using MAE by Datta (1974). Another paper dealing with the so-called "contact problem" (Schwartz and Harper, 1971), though treated in a nondynamic context, is nevertheless of interest for potential applications of MAE in the area; while surface waves in solids, a topic of great current interest, may perhaps be dealt with by extensions of the techniques used by Leppington (1972) to treat high frequency surface water wave scattering.

We close with the hope that we have conveyed some of the excitement generated by these new methods and that many of the readers of this small, acoustically oriented survey of a very wide field will find MAE and related singular perturbation methods a powerful new tool in attacking problems of interest.

ACKNOWLEDGMENTS

In closing we must express our thanks to our respective institutions and staffs for support and interest. In particular we wish to thank Drs. Ryhming and Lundberg of Institut CERAC for careful reading of several versions of this work, and Mme. Ruth Labastrou for her patience and excellence as our typist, while D. G. C. would like to thank Dr. Bo Lemcke, Director of Institut CERAC, for hospitality in several visits during which this article was written.

REFERENCES

Abramowitz, M., and Stegun, L. A. (1964). "Handbook of Mathematical Functions." National Bureau of Standards, Washington, D.C.

Ahluwalia, P. S., Keller, J. B., and Matkowsky, B. J. (1974). *J. Acoust. Soc. Amer.* **55**, 7.

Amiet, R., and Sears, W. R. (1970). *J. Fluid Mech.* **44**, 227.

Beyer, R. T. (1965). *In* "Physical Acoustics" (W. P. Mason, ed.) Vol. 2B, p. 231. Academic Press, New York.

Birkhoff, G. (1960). "Hydrodynamics." Princeton Univ. Press. Princeton, New Jersey.

Blackstock, D. T. (1964a). *J. Acoust. Soc. Amer.* **36**, 217.

Blackstock, D. T. (1964b). *J. Acoust. Soc. Amer.* **36**, 534.

Bowman, J. J., Senior, T. B. A., and Uslenghi, P. L. E. (1969). "Electromagnetic and Acoustic Scattering by Simple Shapes." North-Holland Publ., Amsterdam.

Cole, J. D. (1968). "Perturbation Methods in Applied Mathematics." Ginn (Blaisdell), Waltham, Massachusetts.

Crighton, D. G. (1972). *J. Fluid Mech.* **51**, 357.

Crighton, D. G. (1975). *Progr. Aerosp. Sci.* **16/1**, 31.

Crighton, D. G., and Leppington, F. G. (1971). *J. Fluid Mech.* **46**, 577.

Crighton, D. G., and Leppington, F. G. (1973). *Proc. Roy. Soc., Ser. A* **335**, 313.

Crow, S. C. (1970). *Stud. Appl. Math.* **49**, 21.

Datta, S. K. (1974). *Int. J. Solids Struct.* **10**, 123.

Eckhaus, W. (1973). "Matched Asymptotic Expansions and Singular Perturbations." North-Holland Publ., Amsterdam.

Eisner, E. (1964). *In* "Physical Acoustics" (W. P. Mason, ed.) Vol. 1, Part B, p. 353. Academic Press, New York.

Fay, R. D. (1931). *J. Acoust. Soc. Amer.* **3**, 222.

Fox, P. A. (1955). *J. Math. Phys.* **34**, 133.

Fraenkel, L. E. (1969). *Proc. Cambridge Phil. Soc.* **65**, 209.

Friedrichs, K. O. (1955). *Bull. Amer. Math. Soc.* **61**, 484.

Fubini-Ghiron, S. (1935). *Alta Freq.* **4**, 530.

Graves-Morris, P. R., ed. (1973). "Padé Approximants and their Applications." Academic Press, New York.

Harper, E. Y. (1969). *J. Math. Phys.* **10**, 1975.

Harper, E. Y., and Simpkins, P. G. (1975). *J. Sound Vib.* (to appear).

Harper, E. Y., Chang, I-D., and Grube, G. W. (1971). *J. Math. Phys.* **12**, 1955.

Hayes, W. D. (1960). "Gasdynamic Discontinuities." Princeton Univ. Press. Princeton, New Jersey.

Jones, D. S. (1953). *Proc. Roy. Soc., Ser. A* **217**, 153.

Jones, D. S. (1955). *Phil. Trans. Roy. Soc. London, Ser. A* **247**, 499.

Jones, D. S. (1964). "The Theory of Electromagnetism." Pergamon, Oxford.

Kanwal, R. P. (1967). *J. Math. Phys.* **8**, 821.

Kaplun, S. (1957). *J. Math. Mech.* **6**, 595.

Kaplun, S. (1967). *In* "Fluid Mechanics and Singular Perturbations: A Collection of Papers by Saul Kaplan" (P. A. Lagerström, L. N. Howard, and C.-S. Liu, eds.), pp. 1–369. Academic Press, New York.

Lagerström, P. A., and Casten, R. G. (1972). *SIAM (Soc. Ind. Appl. Math.) Rev.* **14**, 63.

Lagerström, P. A., Cole, J. D., and Trilling, L. (1949). "GALCIT Report." Calif Inst. Technol , Pasadena.

Landahl, M., Ryhming, I., Sorenson, H., and Drougge, G. (1971). *In* "Third Conference on Sonic Boom Research" (I. R. Schwartz, ed.), US Gov. Printing Office, Washington, D.C.

Lauvstad, V. R. (1968). *J. Sound Vib.* **7**, 90.

Leibovich, S., and Seebass, A. R., eds. (1974). "Nonlinear Waves." Cornell Univ. Press, Ithaca, New York.

Leppington, F. G. (1972). *J. Fluid Mech.* **56**, 101.

Lesser, M. B., and Lewis, J. A. (1972a). *J. Acoust. Soc. Amer.* **51**, 1664.

Lesser, M. B., and Lewis, J. A. (1972b). *J. Acoust. Soc. Amer.* **52**, 1406.

Lesser, M. B., and Lewis, J. A. (1974). *J. Sound Vib.* **33**, 13.

Lesser, M. B., and Seebass, R. (1968). *J. Fluid Mech.* **31**, 501.

Lewis, R. M., and Keller, J. B. (1964). "Asymptotic Methods for Partial Differential Equations: The Reduced Wave Equation and Maxwells Equation." Courant Inst. Rep. EM-194. New York University, New York.

Lick, W. (1967). *Advan. Appl. Mech.* **10**, 1.

Lick, W. (1970). *Ann. Rev. Fluid Mech.* **2**, 13.

Lighthill, M. J. (1949). *Phil. Mag.* (7) **40**, 1179.

Lighthill, M. J. (1956). *In* "Surveys in Mechanics" (G. K. Batchelor and R. M. Davies, eds.), p. 250. Cambridge Univ. Press, London and New York.

Lin, C. C. (1954). *J. Math. Phys.* **33**, 117.

Moran, J. P., and Shen, S. F. (1966). *J. Fluid Mech.* **25**, 705.

Morrison, J. A. (1966). *SIAM (Soc. Ind. Appl. Math.) Rev.* **8**, 66.

Morse, P. M., and Ingard, K. U. (1968). "Theoretical Acoustics." McGraw-Hill, New York.

Murray, J. D. (1974). "Asymptotic Analysis." Oxford Univ. Press (Clarendon), London and New York.

Nayfeh, A. H. (1973). "Perturbation Methods." Wiley (Interscience), New York.

O'Malley, R. E. (1974). "Introduction to Singular Perturbations." Academic Press, New York.

Oseen, C. W. (1910). *Ark. Mat., Astron, Fys.* **6**, Art. 29.

Oseen, C. W. (1913). *Ark. Mat., Astron. Fys.* **9**, Art. 16.

Proudman, I., and Pearson, J. R. A. (1957). *J. Fluid Mech.* **2**, 237.

Rayleigh, Lord. (1945). "Theory of Sound," Vol. 2. Dover, New York.

Riley, N. (1967). *J. Inst. Math. Appl.* **3**, 419.

Schneider, W. (1973). *J. Fluid Mech.* **50**, 785.

Schwartz, J., and Harper, E. (1971). *Int. J. Solids Struct.* **7**, 1613.

Schwartz, L. W. (1974). *J. Fluid Mech.* **62**, 553.

Schwinger, J., and Saxon, D. S. (1968). "Discontinuities in Waveguides." Gordon & Breach, New York.

Shanks, D. (1955). *J. Math. Phys.* **34**, 1–42.

Stakgold, I. (1967). "Boundary Value Problems of Mathematical Physics," Vol. 1, p. 85. Macmillan, New York.

Stokes, G. G. (1851). *Trans. Cambridge Phil. Soc.* **9**, No. 2, 8.

Taylor, G. I. (1910). *Proc. Roy. Soc. Ser. A* **84**, 371.

Van Dyke, M. D. (1964). "Perturbation Methods in Fluid Mechanics." Academic Press, New York.

Van Dyke, M. (1970). *J. Fluid Mech.* **44**, 365.

Wang, C.-Y. (1968). *J. Fluid Mech.* **32**, 55.

Whitehead, A. N. (1889). *Quart. J. Math.* **23**, 143.

—3—

Ultrasonic Diffraction from Single Apertures with Application to Pulse Measurements and Crystal Physics

EMMANUEL P. PAPADAKIS

*Ford Motor Company, Manufacturing Development Center,
Detroit, Michigan*

I.	Introduction	152
II.	Theory	153
	A. Formulation for Pressure and Phase	153
	B. Formulation for Spatial Phase Dot Product	157
	C. Expressions for Anisotropy Parameter b for Longitudinal Wave Propagation	158
	D. Physical Limits on the Parameter b	159
III.	Computations	160
IV.	Experiments	165
	Amplitude and Loss	165
V.	Diffraction Corrections	173
	A. Diffraction Corrections for Attenuation	173
	B. Diffraction Corrections for Velocity	174
	C. Dispersion Introduced by Diffraction	174
	D. Use of Buffer Rods	175
	E. Corrections for Attenuation: Experiments with Bonded Transducer Plates	177
	F. Phase and Time Delay Experiments	177
	G. Experiments with Buffer Rods	178
VI.	Input Amplitude Profile	186
	A. Background	186
	B. Calculations	187
	C. Discussion	190
	D. Conclusions	191
VII.	Broadband Pulses	191
	A. Introduction	191
	B. Theory and Calculations	192
VIII.	Specimens of Finite Width	205

IX. Surface Waves 206
 X. Summary 208
    References 208
    Appendix 210

I. Introduction

All ultrasonic diffraction can be divided into two parts: diffraction from apertures several wavelengths across, and diffraction from arrays of spacing of the order of a wavelength. The two areas of ultrasonic diffraction are of unlike character, and present different problems and varied opportunities to the experimentalist, the theoretician, and the device designer. In the present chapter we will treat only diffraction from single apertures several wavelengths across.

Diffraction from large apertures became important with the advent of pulse-echo ultrasonic investigations utilizing piezoelectric plates as transceivers. Early calculations applicable to narrow band bursts of rf waves soon made their way into the practical literature (Mason, 1958; McMasters, 1959).

Qualitatively, the effects of diffraction from large apertures are as follows:

1. As in optics, there is a Fresnel region and a Fraunhoffer region separated by an ill-defined intermediate region.

2. In the Fresnel region for monochromatic waves, there are zeroes of pressure along the axis of the transducer, and local minima and maxima of pressure and phase across the diameter of the beam.

3. In the Faunhoffer region for monochromatic waves, the beam is divided into lobes, a central lobe and symmetrically positioned higher-order lobes.

4. The net effect of the Fresnel zone pressure and phase variations and the Fraunhoffer zone lobe divergence is commonly called "beam spreading," but is more complicated than simply considering the part of the central lobe and others which are not within the cylindrical beam defined by the perimeter of the radiating aperture (transducer).

5. When the same transducer which generates the monochromatic beam (as a piston source in an infinite baffle) is used as a receiver in pulse-echo measurements, the integrated effect of the pressure and phase variation in the field is to add a nonmonotonic loss to the echoes, and to add a monotonic but nonlinear increment to the phase as a function of distance. As distance increases without limit, the added loss approaches 6 dB per doubling of distance, and the added phase approaches $\pi/2$. Similar results are obtained, of course, when two transducers are used on opposite ends of a specimen. If the transducers are of different size or are not coaxial, the results vary accordingly, but are qualitatively similar.

6. When a transducer is used to examine a specimen for small flaws, the apparent flaw size as measured by an echo amplitude varies with the position

of the flaw in the field of the transducer because of the pressure variations as a function of position in the monochromatic field.

7. When a transducer producing monochromatic waves is placed on a specimen of regular geometry and finite lateral extent (e.g. on the end of a cylinder), the "beam spreading" (as lobes interact with the sidewalls) causes constructive and destructive interference versus distance from the transducer. Incremental loss and gyrations in phase are noted.

8. "Beam spreading," i.e. the effect of diffraction, is a function of the anisotropy of the specimen. "Beam spreading" can be accelerated, retarded, or "smoothed out," depending on the symmetry involved.

9. Broadband pulses (i.e. generated by spiked or rectangular voltage pulses applied to heavily damped transducers) tend to average out the effects of diffraction because they contain a spectrum of frequencies, each of which experiences diffraction effects at a different distance from the transducer.

The diffraction problem is generally solved in two steps: (1) finding the pressure and phase at all field points of interest in the field of the radiating aperture (transducer), and (2) finding the response of the receiving transducer which spans those field points of interest.

For pulse-echo work in a specimen of length L, the field points of interest are on planes at distances $2nL$ in the propagation direction and are within a cylinder defined by the perimeter of the transducer as directrix, and the normal direction to the transducer as generatrix. (Surface waves and plate waves would be two-dimensional problems.)

For through-transmission work, the single plane of interest is at the distance L, and the set of points is over the area of the receiving transducer which may be of a different size from the transmitting transducer. In any specific instance, one may solve for the pressure and phase experienced by the receiving transducer as a function of distance from the transmitter.

II. Theory

A. Formulation for Pressure and Phase

The pressure p in the field of an aperture in a rigid baffle in a fluid when the baffle is irradiated by a plane pressure wave (Rayleigh, 1945) is given by

$$p = \frac{j\omega\rho_0 V}{2\pi} \int_{\sigma} \frac{\exp[j(\omega t - \beta r)]\, d\sigma}{r}, \tag{1}$$

where ω is the angular frequency, ρ_0 the material density, V the particle velocity amplitude at the aperture, β the magnitude of the propagation vector, r the distance to the field point from the element of area $d\sigma$ on the aperture of area σ, and t is the time.

Several authors have applied this formula to solve the beam-spreading problem for monochromatic waves in practical cases (Seki *et al.*, 1956;

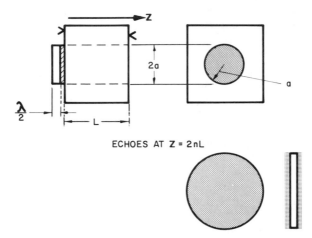

FIG. 1. All-plated transducer bonded to a sample for pulse-echo measurements. Stippled region is plating, while crosshatched layer is bond. From Papadakis, (1966) by permission of the American Institute of Physics.

Tjadens, 1961; Papadakis, 1959, 1963a, 1964a, 1966; Lord, 1966a,b; Benson and Kiyohara, 1974). Gitis and Khimunin (1969) have reviewed the field recently. The solution for fluids applies adequately to isotropic solids as long as the transducer is bonded adequately to one of a pair of plane parallel faces of a slab considerably larger in lateral extent than the transducer diameter (see Fig. 1). Then the result (Seki *et al.*, 1956) for loss due to diffraction from circular apertures is shown in Fig. 2. This curve applies to isotropic solids and to fluids. Its abscissa $S = z\lambda/a^2$ is propagation distance z normalized by the

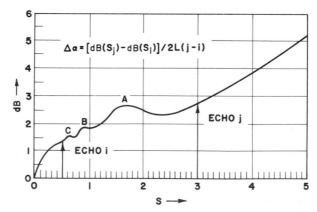

FIG. 2. Loss for a circular piston radiating longitudinal waves into isotropic media according to Seki *et al.* (1956). From Papadakis (1966), by permission of the American Institute of Physics.

Fresnel length a^2/λ, making a universal curve. (Other losses may be associated with reflections at the transducer or at a reflector in a fluid.) Seki *et al.* (1956) and Benson and Kiyohara (1974) used mathematical expansions of the field in terms of cylindrical functions to solve for the pressure in the field of the transmitter and the response of the receiver. Papadakis (1964a, 1966) used a brute force approach involving numerical integration by computer. The latter approach is constructive to study because it illustrates the procedure to be followed in any method. The essence of the theory was presented by Seki *et al.* (1956).

The formulation due to the author (Papadakis, 1963a, 1964a, 1966) is applicable to anisotropic solids. As will be seen, longitudinal waves in certain pure-mode directions yield the only simple solutions.

In general, the pressure at a field point is

$$p = \frac{j\omega\rho_0}{2\pi} \int_\sigma \frac{V(\rho, \psi)\,\exp j[\omega t - \boldsymbol{\beta}\cdot\mathbf{r} - \zeta(\rho, \psi)]\,d\sigma}{|\mathbf{r}|}. \tag{2}$$

The integration is preformed over the area σ of the transmitting transducer. See Fig. 3 for the geometry of the situation in the case of propagation along

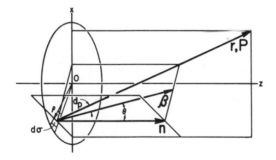

FIG. 3. Coordinates and relevant vectors for the integration to find the pressure in front of a circular piston radiator. For anisotropic media, the propagation vector $\boldsymbol{\beta}$ deviates from the Poynting vector **P**, but these and the surface normal **n** are coplanar for 3-, 4-, and 6-fold axes along **n**. From Papadakis (1966), by permission of the American Institute of Physics.

axes of 3-, 4-, or 6-fold symmetry for cylindrical geometry. The equations could be written and solved in Cartesian coordinates, as well, for rectangular transducers. The vector **r** goes from the element of area $d\sigma$ on the transducer to the field point (x, z). The Poynting vector **P** lies along **r**. The coordinates of the element $d\sigma$ are (ρ, ψ). The segment of a plane wave carrying energy from $d\sigma$ to the field point (x, z) along **r** has a propagation vector $\boldsymbol{\beta}$ that deviates from **r** by an angle d_p (Waterman, 1959). The vector $\boldsymbol{\beta}$ makes an angle θ with the transducer normal **n**. The angular frequency of the elastic wave is ω and the density of the medium is ρ_0. $V(\rho, \psi)$ is the particle velocity input amplitude at $z = 0$. In the case of a piston source, $V(\rho, \psi)$ would be a constant

V_0 over the transducer and zero elsewhere. The incremental phase $\zeta(\rho, \psi)$ refers to the particle velocity input, and is zero for a piston source. Thus, for the piston source considered in this section, the pressure is

$$p = \frac{j\omega\rho_0 V_0}{2\pi} \int_\sigma \frac{\exp j[\omega t - \boldsymbol{\beta} \cdot \mathbf{r}]\, d\sigma}{|\mathbf{r}|}. \tag{3}$$

Dropping the coefficient and using trigonometric identities, one finds

$$p \propto \sin\omega t \int_\sigma \frac{\cos\boldsymbol{\beta} \cdot \mathbf{r}\, d\sigma}{|\mathbf{r}|} - \cos\omega t \int_\sigma \frac{\sin\boldsymbol{\beta} \cdot \mathbf{r}\, d\sigma}{|\mathbf{r}|}. \tag{4}$$

This is written as $\tilde{p}$, and is given by

$$\tilde{p} = A_1 \sin\omega t - A_2 \cos\omega t,$$
$$= C \cos(\omega t - \delta), \tag{5}$$

where A_1 is the first integral and A_2 is the second integral of Eq. (4). Then we have

$$C^2 = A_1{}^2 + A_2{}^2$$

and   (6)

$$\tan\delta = -A_1/A_2.$$

The response of the receiving transducer p_M is proportional to the maximum over a period of the integral of $\tilde{p}$ over its area σ'.

$$p_\mathrm{M} \propto \int_{\sigma'} \tilde{p}\, d\sigma' \bigg|_{\text{max over time}} = \int_{\sigma'} C \cos(\omega t - \delta)\, d\sigma' \bigg|_{\text{max}} \tag{7}$$

To find the maximum, a differentiation with respect to time is performed as follows:

$$\partial p_\mathrm{M}/\partial t \propto -\int_{\sigma'} C \sin(\omega t - \delta)\, d\sigma'$$

$$= \sin\omega t \int_{\sigma'} C \cos\delta\, d\sigma' - \cos\omega t \int_{\sigma'} C \sin\delta\, d\sigma'$$

$$= H_2 \sin\omega t - H_1 \cos\omega t. \tag{8}$$

Then, this is set equal to zero, and ωt_M for the maximum is found from

$$\tan\omega t_\mathrm{M} = H_1/H_2. \tag{9}$$

The value of ωt_M found from H_1 and H_2 is substituted in Eq. (6) to give p_M in the form

$$p_\mathrm{M} \propto \int_{\sigma'} C \cos(\omega t_\mathrm{M} - \delta)\, d\sigma'. \tag{10}$$

The maximization procedure might determine the minimum instead; but since the minimum in this case is just the negative of the maximum, it is sufficient to take the absolute value of p_M in finding the decibel level of the pressure. Thus,

$$dB = 20 \log_{10}\{|\,p_M\,|_1/|\,p_M\,|\}. \tag{11}$$

The subscript 1 refers to the reference level at a certain distance along the z axis. It corresponds to the first echo in a pulse-echo measurement.

The quantity ωt_M defines the excess phase ϕ experienced by the propagating wave as sensed by the transducer. It turns out that ωt_M is not linear in propagation distance, but that $\omega t_M - \beta_0 z$ increases from zero at the origin to $\pi/2$ as $z \to \infty$. Here, β_0 is the magnitude of $\boldsymbol{\beta}$ for the direction of $\mathbf{r}$ pointing along the z-axis, the propagation direction. In any computer calculation, it is convenient to choose $\beta_0 z$ in multiples of 2π to eliminate its perturbing effect.

There is another method, the "angular spectrum of plane waves" representation, which arrives at the same results by a different system of integration (Kharusi and Farnell, 1970). It should be studied for use in those cases in which it proves advantageous. One is propagation not along pure-mode axes. The ASPW method will not be treated in this paper.

B. FORMULATION FOR SPATIAL PHASE DOT PRODUCT

When plane elastic waves propagate in crystals, pure modes can travel along certain axes called pure-mode axes. One longitudinal mode and two shear modes, either distinct or degenerate, may be propagated along each axis. Surface waves may propagate on free surfaces, also.

Axes of 3-, 4-, and 6-fold symmetry offer certain simplifications in the longitudinal bulk wave propagation problem since they are directions of either maximum or minimum longitudinal velocity, not saddlepoints. This is important when one considers the implications of Fig. 3. There, the propagation vector $\boldsymbol{\beta}$ deviates from the puremode axis ($\mathbf{n}$ and the z-axis) by an angle θ. However, $\mathbf{n}$, $\boldsymbol{\beta}$, and $\mathbf{P}$ are coplanar. Thus the velocity v along $\boldsymbol{\beta}$ differs from the velocity v_0 along the pure-mode axis where the magnitude of $\boldsymbol{\beta}$ is β_0. Waterman (1959) has given v in the form

$$v_3 = v_3(1 + \Delta v_3/v_3), \tag{12}$$

where the subscript 3 refers to the third mode that he treated, the longitudinal one. The increment Δv_3 is proportional to θ^2, so one may write

$$v = v_0(1 - b\theta^2), \tag{13}$$

where

$$b = -\Delta v_3/v_3\,\theta^2 \tag{14}$$

of Waterman's paper. Since the propagation constant is $|\,\boldsymbol{\beta}\,| = 2\pi f/v$, one may write

$$|\,\boldsymbol{\beta}\,| = \beta_0(1 + b\theta^2) \tag{15}$$

when $b\theta^2 \ll 1$. Also, since d_{P} is proportional to θ, one may write

$$d_{\mathrm{P}} = (2B)^{1/2}\theta, \tag{16}$$

thus defining B. The dot product $\boldsymbol{\beta} \cdot \mathbf{r}$ becomes

$$\begin{aligned}
\boldsymbol{\beta} \cdot \mathbf{r} &= |\boldsymbol{\beta}|\,|\mathbf{r}|\,(1 - B\theta^2) \\
&= \beta_0|\mathbf{r}|[1 + (b - B)\theta^2].
\end{aligned} \tag{17}$$

It was shown (Papadakis, 1964) that $B = 2b^2$, so

$$(b - B) = b(1 - 2b). \tag{18}$$

Thus the expression in Eq. (3) for pressure becomes

$$p = \frac{j\omega\rho_0 V_0}{2\pi} \int_\sigma \frac{\exp j\{\omega t - \beta_0|\mathbf{r}|[1 + b(1 - 2b)\theta^2]\}\,d\sigma}{|\mathbf{r}|}. \tag{19}$$

The introduction of anisotropy into the spatial part of the phase in the pressure integral is the only change from the theory of fluids to the present theory of diffraction in solids. The validity of this change is established by the comparison of experiment with calculations using Eq. (19) in Eqs. (4)–(11) to find the diffraction loss and phase change.

Expressions similar to Eq. (13) hold for the extremal velocity directions of surface waves. Hence, expressions for the parameter b can be found on crystal surfaces. For bulk longitudinal waves in other than extremal directions, and for bulk shear waves in all directions, the complication in the functional dependence of b upon direction proliferates (Waterman, 1959; Papadakis, 1966). Even the seemingly simple case of [110] propagation of longitudinal waves in a cubic crystal is overwhelming because that direction yields a saddle point in phase velocity.

C. Expressions for Anisotropy Parameter b for Longitudinal Wave Propagation

The expressions for $\Delta v_3/v_3$ (Waterman, 1959) are converted to expressions for the parameter b by means of Eq. (14) and are presented here. Only those for 3-, 4-, and 6-fold axes are shown.

1. Cubic system:
   (a) [100] propagation

$$b = (c_{11} - c_{12} - 2c_{44})(c_{11} + c_{12})/2c_{11}(c_{11} - c_{44}). \tag{20}$$

   (b) [111] propagation

$$b = 2(2c_{44} + c_{12} - c_{11})(c_{11} + 2c_{12} + c_{44})/3(c_{12} + c_{44})(c_{11} + 2c_{12} + 4c_{44}). \tag{21}$$

2. Hexagonal system: c-axis propagation

$$b = (c_{33} - c_{13} - 2c_{44})(c_{33} + c_{13})/2c_{33}(c_{33} - c_{44}). \tag{22}$$

3. Tetragonal system: c-axis propagation

$$b = (c_{33} - c_{13} - 2c_{44})(c_{33} + c_{13})/2c_{33}(c_{33} - c_{44}). \tag{23}$$

4. Trigonal system: propagation along 3-fold axis

$$b = (c_{33} - c_{13} - 2c_{44})(c_{33} + c_{13})/2c_{33}(c_{33} - c_{44}). \tag{24}$$

Cases 2, 3, and 4 are identical; the other moduli, c_{12}, c_{25}, and c_{66}, are not present even if the crystal system has such moduli. Indeed, case 1(a) reduces to the form of 2, 3, and 4 if one uses the equalities $c_{11} = c_{33}$ and $c_{12} = c_{13}$ valid in the cubic system. Only case 1(b) is different. Equations (20)–(24) are needed to evaluate b in the diffraction problem.

D. PHYSICAL LIMITS ON THE PARAMETER b

Considering the propagation of elastic waves in crystals, one finds 0.5 to be an upper limit on b. By algebraic manipulation, Eq. (20) for b in the [100] direction in a cubic crystal may be rewritten as follows:

$$b = 0.5\left[1 + \frac{c_{12}}{c_{11}} - \frac{c_{12} + c_{44}}{c_{11} - c_{44}} - \frac{c_{12}}{c_{11}}\left(\frac{c_{12} + c_{44}}{c_{11} - c_{44}}\right)\right], \tag{25}$$

or

$$b = 0.5(1 + \delta), \tag{26}$$

where

$$\delta = \frac{c_{12}}{c_{11}} - \frac{c_{12} + c_{44}}{c_{11} - c_{44}} - \frac{c_{12}}{c_{11}}\left(\frac{c_{12} + c_{44}}{c_{11} - c_{44}}\right). \tag{27}$$

It can be shown in the following manner that δ is always negative. Call $c_{12}/c_{11} \equiv \varepsilon$, and factor the sums $(c_{12} + c_{44})$ into $c_{12}(1 + c_{44}/c_{12})$, etc. Then,

$$\delta = \varepsilon - \varepsilon \frac{(1 + c_{44}/c_{12})}{(1 - c_{44}/c_{11})} - \varepsilon^2 \frac{(1 + c_{44}/c_{12})}{(1 - c_{44}/c_{11})}. \tag{28}$$

We must have $c_{44} < c_{11}$ and all the c_{ij} positive to ensure that the shear wave velocities are smaller than the longitudinal wave velocities in all directions (Mason, 1958, pp. 370–372). Otherwise, there would be directions in which the quasilongitudinal and quasi-transverse waves would mix and perturb each other. New modes would form in a way shown diagrammatically in Fig. 4. With modes like these, the quasi-longitudinal and quasi-transverse character of particle motion would lose all meaning in the transition region. There, the particle motion of both modes would be at nearly 45° to the propagation direction, not near it or near 90° to it. In Eq. (28), the conditions $c_{44} < c_{11}$ and c_{ij} positive imply that the ratio $(1 + c_{44}/c_{12})/(1 - c_{44}/c_{11})$ is greater than unity. Call

$$(1 + c_{44}/c_{12})/(1 - c_{44}/c_{11}) \equiv 1 + \eta. \tag{29}$$

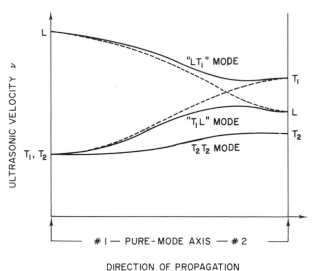

FIG. 4. The mixing of elastic modes necessary if a crystal is to have an anisotropy parameter $b > 0.5$. At least one shear mode must have a higher velocity than the longitudinal mode along one pure-mode axis. From Papadakis (1964a), by permission of the American Institute of Physics.

Then Eq. (28) yields

$$\delta = \varepsilon(-\varepsilon - \eta - \varepsilon\eta) \tag{30}$$

which is identically negative. With δ negative, Eq. (25) reveals that the anisotropy parameter b is less than 0.5. Thus, any normal cubic crystal is characterized by $b < 0.5$. A completely analogous analysis holds for the hexagonal crystals. The result is $b < 0.5$ if $c_{44} < c_{33}$ and $c_{ij} > 0$.

III. Computations

The calculations indicated in Eqs. (4) through (11) were carried out (Papadakis, 1966) with the aid of an electronic computer for values of b from -5.0 to $+0.4$ for circular transducers many wavelengths in diameter ($\beta_0 a \simeq 80$). The transmitting and receiving transducers were made the same size and coaxial at a distance z apart. Both dB and ϕ were calculated from $S = 0.1$ to $S = 5.0$ in steps of $S = 0.1$. ($S = z\lambda/a^2$ is the Seki parameter, propagation distance normalized by the Fresnel length.) To assure zero contribution to ϕ from the spatial part of the phase in the integral of Eq. (19), the value of a was chosen to make $\beta_0 z$ an integer multiple of 2π at each value of S. Thus ωt_M gave ϕ directly. Elements of area $d\sigma$ and $d\sigma'$ were placed close enough together so that the arguments of the sinusoidal functions in the integrals changed by no more than 0.1 rad from element to element.

The dB loss is plotted as a function of S with the anisotropy b as a parameter in Figs. 5 and 6. The first contains graphs with b between -5 and

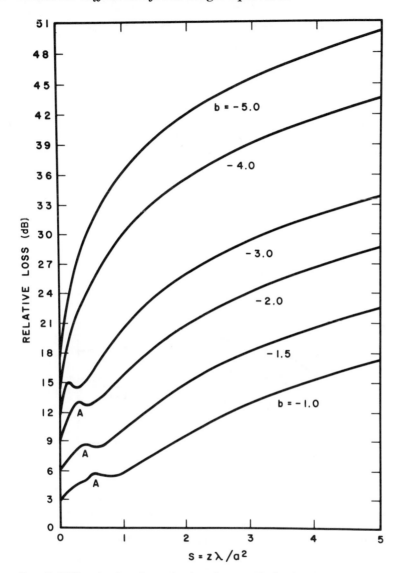

FIG. 5. Diffraction loss for a circular piston radiating longitudinal waves into an anisotropic medium along direction of 3-, 4-, or 6-fold symmetry. Values of the anisotropy parameter b from -5 to -1 are shown. This and the following figure are to be used in computing attenuation corrections. From Papadakis (1966), by permission of the American Institute of Physics.

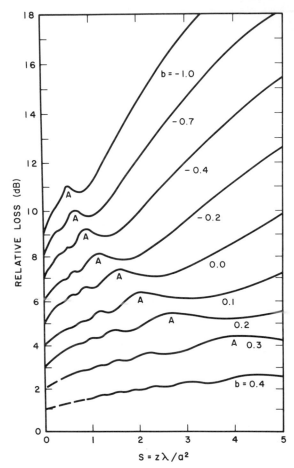

FIG. 6. Diffraction loss for a circular piston radiating longitudinal waves into an anisotropic medium along a direction of 3-, 4-, or 6-fold symmetry. Values of the anisotropy parameter b from -1 to $+0.4$ are shown. The upper limit of b is 0.5. The position of peak "A" is given by $S_A = 0.8/(0.5 - b)$. This and the preceding figure are to be used in computing attenuation corrections. From Papadakis (1966), by permission of the American Institute of Physics.

-1, while the second covers b from -1 to $+0.4$. (It was shown above that b cannot be greater than 0.5.) The most striking feature of the curves is their similarity in shape. It should be noted that

    1. The locations of the loss peaks move toward the origin as b decreases.

    2. The location of peak A shown in Figs. 1, 5, and 6 is given by the equation

$$S_A = 0.8/(0.5 - b), \tag{31}$$

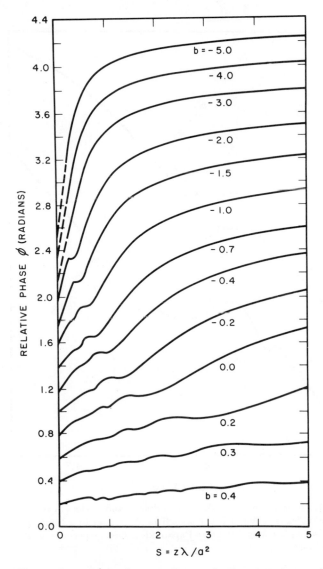

FIG. 7. Phase advance ϕ for the wave from a circular piston source. This advance occurs as the secondary lobes leave the region of the main beam. The total advance from 0 to infinity is $\pi/2$ rad. The phase advance can be used in correcting both phase velocity and group velocity measurements. For group velocity, it would be advantageous to place the echoes where $\partial\phi/\partial S = 0$. From Papadakis (1966), by permission of the American Institute of Physics.

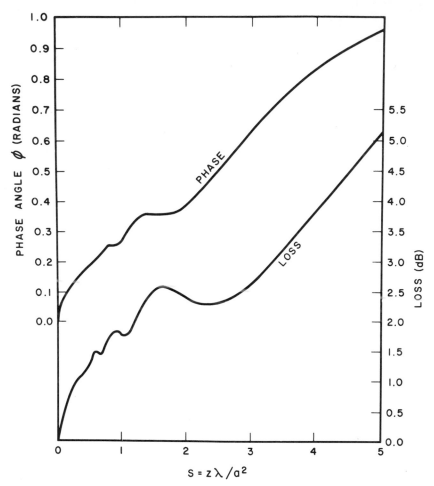

FIG. 8. Relationship between the phase advance and the loss in the field of a piston source. Phase plateaus occur at loss peaks. The phase goes to a limit of $\pi/2$ rad while the loss increases logarithmically. From Papadakis (1966), by permission of the American Institute of Physics.

a new universal equation for pulse-echo experiments relating anisotropy to Fresnel length.

3. The magnitude of a given peak, e.g. "peak A," remains almost constant, increasing only slightly with increasing b.

4. The diffraction loss per unit S grows larger as b becomes smaller algebraically.

5. Beyond the peaks the loss is monotonic increasing without limit but with a monotonic decreasing slope.

6. The curve for $b = 0$, the isotropic case, duplicates the series expansion calculations (Seki *et al.*, 1956; Tjadens, 1961; Benson and Kiyohara, 1974). *Figures 5 and 6 are to be used for diffraction corrections to the attenuation.*

The phase change ϕ is plotted as a function of S with anisotropy b as a parameter in Fig. 7 for the range $-5.0 \leqslant b \leqslant 0.4$. These curves are very similar to those for loss with two exceptions: (i) the phase has plateaus instead of peaks, and (ii) it goes to a limit $\pi/2$ as S increases. The slope $d\phi/dS$ can be large. *The curves in Fig. 7 are to be used for diffraction corrections to the velocity.*

The relationship between the phase plateaus and the loss peaks is shown in Fig. 8 in which the curves for $b = 0$ are repeated. The plateaus in phase coincide with the peaks in loss, and are about as wide in S as the distance between the points of inflection next to the loss peaks.

Computations of Benson and Kiyohara (1974) extend the range of S to 50, and actually show that the phase is asymptotic to $\pi/2$, and that the loss goes to 6 dB per doubling of distance, the point-source limit, at large S.

Calculated values of the pressure amplitude and phase in the field of the radiating transducers are given in Appendix A for certain values of S, the normalized propagation distance, and x/a, the fraction of the transducer radius from the centerline (Papadakis and Fowler, 1971).

IV. Experiments

AMPLITUDE AND LOSS

Many pulse-echo experiments have been performed in which echo amplitudes have been recorded (Roderick and Truell, 1952; Seki *et al.*, 1956; Papadakis, 1963a, 1964a, 1966). These experiments yield copious evidence as to the correctness of Eq. (31), and thus verify the diffraction theory for anisotropic solids enunciated by the author (Papadakis, 1963a) in Eq. (3). From the echo amplitudes, the loss was calculated and plotted versus S for each crystal studied. A representative group of these graphs is reproduced in Figs. 9 through 16. Represented are propagation along the 6-fold axis in a hexagonal crystal, the 3-fold axis in a trigonal crystal, and the 2-fold, 3-fold,

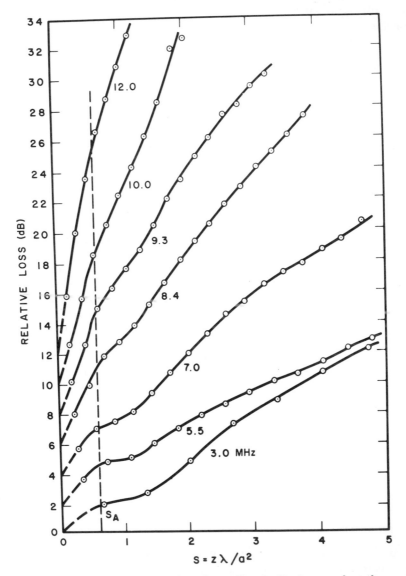

FIG. 9. Loss versus S for successive echoes of longitudinal waves along the c-axis of cadmium, an hexagonal crystal. There is evidence for the existence of peak "A" in the diffraction loss at a position 50% higher than the correct values of S_A. The curves are separated by 2 dB at $S = 0$ for clarity. From Papadakis (1966), by permission of the American Institute of Physics.

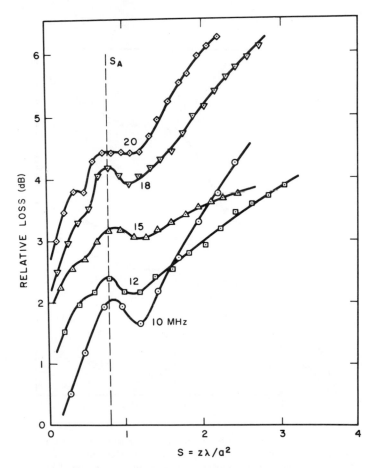

FIG. 10. Loss versus S for successive echoes of longitudinal waves along the 3-fold axis of calcite—$CaCO_3$, a trigonal crystal. The loss peaks are in the proper places according to Eq. (16), and are between 60% and 100% of the theoretical peak amplitudes. The curves are separated an arbitrary amount at $S = 0$ for clarity. From Papadakis (1966), by permission of the American Institute of Physics.

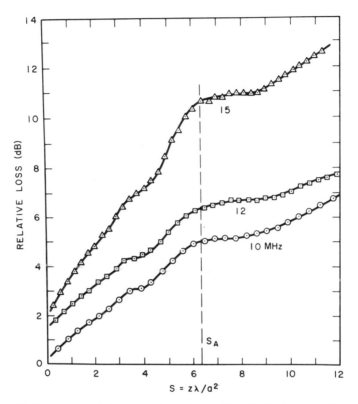

FIG. 11. Loss versus S for successive echoes of longitudinal waves along a 4-fold axis, [100], of KBr, a cubic crystal with large positive b. Evidence for two loss peaks, "A" and "B," can be seen. These are at the locations along S predicted by theory. The intrinsic loss is about twice the diffraction loss at 12 MHz, and three times at 15 MHz. From Papadakis (1966), by permission of the American Institute of Physics.

and 4-fold axes in cubic crystals. Values of b from -1.4 to $+0.4$ are represented. The position S_A of the loss peak A is indicated in each graph. These S_A values are tabulated for all the available data (Papadakis, 1966) in Table I.

There is excellent agreement between theory and experiment concerning the peak positions. The magnitude of the loss measured at peak A in the graphs agrees approximately with theory in the crystals with low intrinsic loss.

TABLE I[a]

EXPERIMENTS ON PEAK POSITIONS WITH CIRCULAR ALL-PLATED
LONGITUDINAL TRANSDUCERS

Material	b	S_A Theory	S_A Experiment
Zn c axis	-5.23	0.137	Not seen
Cd c axis	-1.408	0.410	0.6
Ge [100]	-0.581	0.74	0.7
$CaCO_3$ 3-fold	-0.567	0.75	0.8
Si [100]	-0.461	0.83	0.9
Quartz 3-fold	-0.250	1.07	1.1
NaCl [111]	-0.212	1.12	1.3
Steel	0.000	1.60	1.8
Si [111]	0.162	2.37	2.4
NaCl [100]	0.196	2.63	2.7
KBr [100]	0.373	6.30	6.3
KI [100]	0.380	6.67	5.7

[a] From Papadakis (1966), by permission of the American
Institute of Physics.

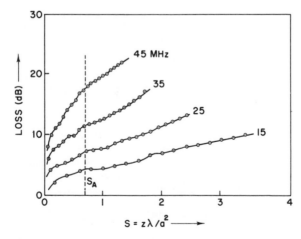

FIG. 12. Ultrasonic losses for longitudinal waves in [100] germanium showing the diffraction-loss peak A as occurring at $S_A = 0.74$. The curves are separated by 2 dB at the first echo for clarity. The true loss is zero at $S = 0$. (Data of Seki, Granato, and Truell). From Papadakis (1964a), by permission of the American Institute of Physics.

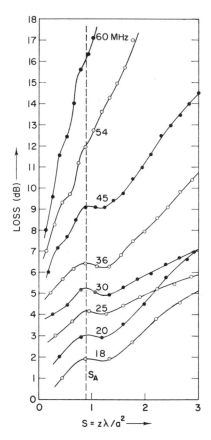

Fig. 13. Ultrasonic losses in silicon for longitudinal waves propagating in the [100] direction. The piston source was an all-plated X-cut quartz disk. The experimental peak position S_A of 0.9 compares favorably with the value of 0.83 computed from Eq. (22). The curves are separated by 1 dB at the first echo for clarity. The true loss is zero at $S = 0$. From Papadakis (1964a), by permission of the American Institute of Physics.

Seki *et al.* (1956), without knowledge of the theory for anisotropy, misinterpreted their data on [100] germanium. They supposed that the loss peak at $S = 1.7$ (Fig. 12 and Table I) was the B-peak, while actually it was the A-peak at $S_A = 0.74$, according to Eq. (31). Unfortunately, this misinterpretation has been developed in review literature (Truell *et al.*, 1969) despite the intervening publications clarifying the matter (Papadakis, 1964a, 1966). In

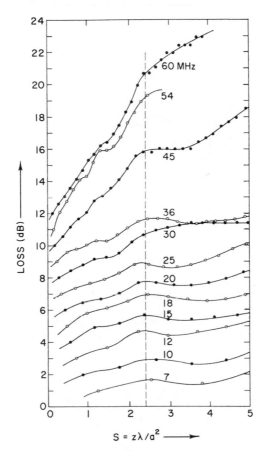

FIG. 14. Ultrasonic losses in silicon for longitudinal waves propagating in the [111] direction. The piston source was an all-plated X-cut quartz disk. The experimental peak position S_A of 2.4 agrees with the predicted value of 2.37 from Eq. (22). The curves are offset 1 dB at the first echo for clarity. The true loss is zero at $S = 0$. From Papadakis (1964a), by permission of the American Institute of Physics.

reading the literature, one must be careful to distinguish also between the *loss peaks* referred to here, e.g. Figs. 1, 5, 6, 8–16, and the *echo amplitude-peaks* referred to elsewhere (Seki *et al.*, 1956; Truell *et al.*, 1969).

The nth echo amplitude peak follows the nth loss peak (which, of course, is observed as a valley in the echo pattern). In the notation of echo amplitude peaks, Seki *et al.* mistook the third echo amplitude peak in [100] germanium for the second.

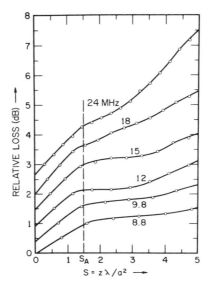

FIG. 15. Diffraction loss for longitudinal waves in the [110] direction in silicon. The peaks show up only as a change of slope near S_A. S_A is actually variable since b is a function of the azimuth of the vector from an element of area on the transducer to a field point. The azimuthal dependence of the anisotropy apparently averages out the sharp peaks in the diffraction loss. From Papadakis (1966), by permission of the American Institute of Physics.

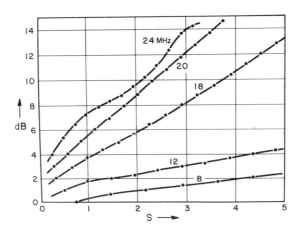

FIG. 16. Diffraction loss for longitudinal waves propagating along the [110] direction, a 2-fold axis, in NaCl, a cubic crystal. The peaks are much weaker than those in the 3-fold and 4-fold directions in NaCl (Papadakis, 1963) or in other crystals reported here and elsewhere. From Papadakis (1964a, 1966), by permission of the American Institute of Physics.

V. Diffraction Corrections

A. Diffraction Corrections for Attenuation

The author (Papadakis, 1959) outlined the method for computing the corrections for diffraction in ultrasonic attenuation experiments in isotropic bodies with plane parallel faces. The method is applicable to anisotropic media as long as the diffraction loss-distance characteristic is known; the procedure is recapitulated here. The diffraction loss–distance characteristic is a plot, as in Figs. 5 and 6, of decibel loss from beam spreading versus S, the normalized distance $z\lambda/a^2$. In pulse-echo work, the echoes are at specific points along S. Writing $\lambda = v/f$ and $z = 2Ln$ for the nth echo where L is the length of the specimen, we have

$$S_n = 2Lnv/a^2f. \tag{32}$$

Between echoes numbered m and n, the incremental loss is

$$\Delta dB = dB(S_n) - dB(S_m), \tag{33}$$

where $n > m$. For loss per unit length, ΔdB is divided by $2L(n - m)$, the incremental path, so

$$\alpha_{\mathrm{Ld}} = \frac{dB(S_n) - dB(S_m)}{2L(n - m)}. \tag{34}$$

Here the subscript "Ld" stands for "per unit *length* due to *diffraction*." Similarly, for loss per unit time, ΔdB is divided by $2L(n - m)/v$, the incremental travel time (or by the time if it is measured directly), so

$$\alpha_{\mathrm{Td}} = \frac{v[dB(S_n) - dB(S_m)]}{2L(n - m)}. \tag{35}$$

"Td" stands for "per unit *time* due to *diffraction*." These corrections, α_{Ld} and α_{Td}, are to be subtracted from the measured attenuation, α_{L}' or α_{T}' per length or time, to find the true attenuation in the medium under study. Thus

$$\alpha_{\mathrm{L}} = \alpha_{\mathrm{L}}' - \alpha_{\mathrm{Ld}},$$

and

$$\alpha_{\mathrm{T}} = \alpha_{\mathrm{T}}' - \alpha_{\mathrm{Td}}. \tag{36}$$

In most work, the majority of the measurements will be made in the Fresnel region where the diffraction loss is *not* monotonic in S. From Fig. 2 one can see that the diffraction correction between certain pairs of echoes might even be negative. Thus it is very important that the corrections be computed accurately from the proper curves of dB versus S.

In general, for a given size of transducer, the diffraction loss between a given pair of echoes will be higher at the lower frequencies. When the frequency dependence of the attenuation is important (i.e. most of the time), the dependence on the raw data may be masked by the diffraction loss.

B. Diffraction Corrections for Velocity

The phase increment mentioned in Fig. 7 represents an error in the travel time in phase velocity measurements. The expression for the increment in time is

$$t_D = [\phi(S_n) - \phi(S_m)]/2\pi f \tag{37}$$

with ϕ in radians and $n > m$, these integers being the echo numbers. The increment t_D is to be added to the measured value t_M of the travel time to get the true travel time $t_M^{(T)}$. Thus,

$$t_M^{(T)} = t_M + t_D. \tag{38}$$

Other things being equal, t_D decreases as f increases because f is in the denominator of t_D and in the denominators of the S_j's also. This finding is in agreement with the work of McSkimin (1960), and of Barshauskas *et al.* (1964) who found excess velocity at the lower frequencies.

It should be noted at this point that the pulse-superposition method of McSkimin (1961) partially masks the effect of diffraction by overlapping several echoes at different S values simultaneously. Some kind of weighted automatic averaging is performed. The pulse-echo-overlap method (Papadakis, 1967) uses only pairs of echoes, and permits exact diffraction corrections.

C. Dispersion Introduced by Diffraction

As already mentioned, the phase velocity of waves is a function of S through diffraction. Integration over the phase profile makes this so. Since S is a function of f and v, one can write the propagation constant β_0 as a function of the phase and get the group velocity as follows. By definition,

$$v_g = d\omega/d\beta_0 = 2\pi/(d\beta_0/df). \tag{39}$$

This is equivalent to

$$v_g = V_0/[1 - (\omega_0/v_0)(dv/d\omega)_0], \tag{40}$$

where the subscript "0" refers to the center frequency of a pulse spectrum. The time of flight for the apparent phase velocity is

$$t_M' = t_M + t(S_m) - t(S_n), \tag{41}$$

where $t(S_j)$ is a small increment due to diffraction. Using Eq. (37), one obtains

$$t_M' = t_M - (1/\omega)[\phi(S_n) - \phi(S_m)], \tag{42}$$

so the apparent phase velocity is

$$v = \frac{2L(n-m)}{\{t_M - (1/\omega)[\phi(S_n) - \phi(S_m)]\}} \tag{43}$$

or

$$v \simeq [2L(n-m)/t_M]\{1 + [\phi(S_n) - \phi(S_m)]/\omega t_M\}. \tag{44}$$

The partial to insert for $(dv/d\omega)_0$ is

$$\left(\frac{\partial v}{\partial \omega}\right)_0 = -\frac{v_0}{t_M \omega_0{}^2}\left\{[\phi(S_n) - \phi(S_m)] + \left[S_n \frac{\partial \phi}{\partial S}\bigg|_{S_n} - S_m \frac{\partial \phi}{\partial S}\bigg|_{S_m}\right]\right\}. \quad (45)$$

This makes the group velocity

$$v_g = v_0\left(1 + \frac{1}{\omega_0 t_M}\left\{[\phi(S_n) - \phi(S_m)] + \left[S_n \frac{\partial \phi}{\partial S}\bigg|_{S_n} - S_m \frac{\partial \phi}{\partial S}\bigg|_{S_m}\right]\right\}_0\right)^{-1}$$

$$\quad (46)$$

As shown in Fig. 7, $\partial\phi/\partial S$ fluctuates between zero and fairly large slopes, so the second term in the denominator of Eq. (46) can be large. Thus it can be seen that the group velocity can be considerably lower than the phase velocity in the Fresnel region. This dispersion is caused by diffraction alone. Experiments should be performed on video pulse transmission in the Fresnel region in intrinsically nondispersive media to test this finding.

D. USE OF BUFFER RODS

The same curves, shown in Figs. 5 and 6, of loss versus normalized distance (dB versus $S = z\lambda/a^2$) can be used in the case of buffer rods or liquid columns which convey the ultrasonic waves from the transducer to the specimen and back. See Fig. 17. The theory for diffraction corrections in buffer/specimen systems is presented here (Papadakis *et al.*, 1973).

In a buffer/specimen system, one needs the amplitudes of three separate echoes, A, B, and C (or A', A, and B where echo A' is echo A before the specimen is attached in Fig. 17) for the calculation of the attenuation and

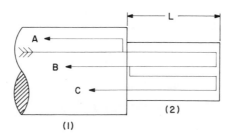

$$A = R$$

$$B = (1 - R^2)e^{-2\alpha L}$$

$$C = -R(1 - R^2)e^{-4\alpha L}$$

FIG. 17. Definition of echoes A, B, and C from a specimen at the end of a buffer. From Papadakis *et al.* (1973), by permission of the American Institute of Physics.

reflection coefficients. In bulk specimens, these three echoes are affected to different degrees by diffraction (beam spreading). To be specific, each echo is smaller than it would have been in the absence of diffraction. *The principle invoked in diffraction corrections in this case is the correction of each echo amplitude to its undiffracted value, and the subsequent calculation of R and α from the corrected echo amplitudes.*

First, S is calculated for each echo.

$$S = z\lambda/a^2 = zv/a^2f \tag{47}$$

is the normalized distance, where z is the propagation distance, v is the velocity, a is the transducer radius, and f is the frequency. Use capital letters Z and V for distance and velocity in the buffer, and small z and v for corresponding quantities in the specimen. $Z = 2L$ where L is the length of the buffer, and $z = 2l$ where l is the length of the specimen. The S values for echoes A, B, and C (or A', A, and B) are

$$S_A' = S_A = 2LV/a^2f,$$
$$S_B = 2LV/a^2f + 2lv/a^2f, \tag{48}$$

and

$$S_C = 2LV/a^2f + 4lv/a^2f.$$

Then, from the curve of dB versus S, one finds

$$dB_A' = dB_A = dB(S_A),$$
$$dB_B = dB(S_B), \tag{49}$$

and

$$dB_C = dB(S_C).$$

From the definition $dB = 20 \log_{10}(X_0/X)$, and the measured amplitudes A, B, and C (or A', A, and B), one can find the corrected values A_0', A_0, B_0, and C_0 by inverting the equations

$$dB_A' = 20 \log_{10}(A_0'/A'),$$
$$dB_A = 20 \log_{10}(A_0/A), \tag{50}$$
$$dB_B = 20 \log_{10}(B_0/B),$$

and

$$dB_C = 20 \log_{10}(C_0/C).$$

The values of A_0, B_0, and C_0 (or A_0', A_0, and B_0) are used in the calculation of R and $α$ by the method of separate echoes (Papadakis, 1968a, 1971a; Lynnworth, 1974). The formulas are recapitulated here for completeness.

Echoes A, B, and C

$$R = [\tilde{A}_0 \tilde{C}_0/(\tilde{A}_0 \tilde{C}_0 - 1)]^{1/2}, \tag{51}$$

and

$$\alpha = [\ln(-R/\tilde{C}_0)]/2l, \tag{52}$$

where

$$\tilde{A}_0 = A_0/B_0 \quad \text{and} \quad \tilde{C}_0 = C_0/B_0.$$

Echoes A', A, and B

$$R = A_0/A_0', \tag{53}$$

and

$$\alpha = \{\ln[A_0'(1-R^2)/B_0]\}/2l. \tag{54}$$

In the present formulation, the relative signs of A', A, B, and C must be used. The experimenter *must* note whether A', B, or C are inverted with respect to A.

Similar calculations use S_A, S_B, and S_C to find the phase ϕ to correct travel time measurements.

E. CORRECTIONS FOR ATTENUATION: EXPERIMENTS WITH BONDED TRANSDUCER PLATES

Diffraction corrections for attenuation (Section V,A above) have been used to determine the functional relationship between the intrinsic attenuation in materials and the ultrasonic frequency used for measurement (Papadakis, 1960, 1961, 1963b, 1964b,c, 1965, 1968b, 1970; Papadakis *et al.*, 1973). The major result has been the detection of attenuation in polycrystalline media dependent upon powers of frequency higher than 2.0 for data which did not show such a high power before diffraction corrections were applied. One expects attenuation to depend on powers as high as 4.0 in the Rayleigh scattering region ($\lambda \gg$ grain diameter) which was the condition in the experiments cited. The discovery of Rayleigh scattering attenuation is evidence for the correctness of the diffraction corrections for attenuation. The experiments referred to as well as others are completely documented in an earlier volume of this set (Papadakis, 1968b).

F. PHASE AND TIME DELAY EXPERIMENTS

The author (Papadakis, 1967) used the pulse-echo-overlap method to study the phase shift in an ultrasonic beam due to diffraction versus distance. The travel times t_M in several specimens were measured between the first echo and several other echoes with X-cut quartz transducers resonant at 15 and 20 MHz. In the process of measuring and comparing the McSkimin Δt's (McSkimin, 1961), it was found that the value of bond thickness was equivalent to 8° of phase, resulting in a reflection phase angle γ_R of 3°. The round-trip travel times t were computed from the values of t_M and γ_R by McSkimin's formulas; so a set of values of t was found, one for echoes 1 and 2, another for 1 and 3, etc. up to 1 and 15. A monotonic trend was found in the data. These values of t were averaged, and a standard deviation found. Then the diffraction phase correction t_D (Papadakis, 1966) was calculated and added to t_M to get $t_M^{(T)}$ (for true). See Section V,B above. Again, the round-trip travel times $t^{(T)}$ were found. The corrected data showed a much weaker trend, had an average differing from the original by about 60 ppm, and had a *lower* standard deviation. The lower standard deviation resulted from the correctness of the diffraction correction for phase based upon Fig. 7 which took the diffraction trend out of the data and made all the values of round-trip travel time more equal.

TABLE II[a]

STANDARD DEVIATIONS IN TRAVEL TIME BEFORE AND
AFTER DIFFRACTION CORRECTIONS FOR PHASE

	Standard deviation (μsec)	
Specimen	Uncorrected	Corrected
1 in. fused quartz	0.00016	0.00013
2 in. fused quartz	0.00015	0.00010
[100] Silicon	0.00021	0.00017
[110] Silicon	0.00014	0.00008
[111] Silicon	0.00021	0.00012

[a] From Papadakis (1967), by permission of the American Institute of Physics.

Table II shows the resulting standard deviations, before and after. The resultant error in absolute time is of the order of ± 0.1 nsec. This accuracy would be impossible without the corrections for the diffraction phase effect as calculated by the methods outlined in this chapter. *All published data on elastic moduli found by ultrasonics should be reviewed in the light of the diffraction phase effect.*

G. EXPERIMENTS WITH BUFFER RODS

1. *Outline*

The most important uses of diffraction corrections in buffer rods have been in conjunction with broadband pulses and spectrum analysis (Papadakis *et al.*, 1973). Two basic propositions have been proved: (a) that ultrasonic attenuation can be measured accurately with the buffer method using spectrum analysis of broadband pulses to retrieve amplitude data as functions of frequency, and (b) that broadband ultrasonic transducers radiate considerable energy down to zero frequency, and act as almost critically damped oscillators. Both these proofs were effected with broadband pulses which will be treated further in Section VII below.

2. *Spectrum/Buffer Method*

The method of spectrum analysis is outlined in the block diagram in Fig. 18. The system is organized around a pulser–receiver–gate unit which performs several functions. First, it delivers a very short electrical pulse to energize the transducer which generates a broadband ultrasonic pulse and

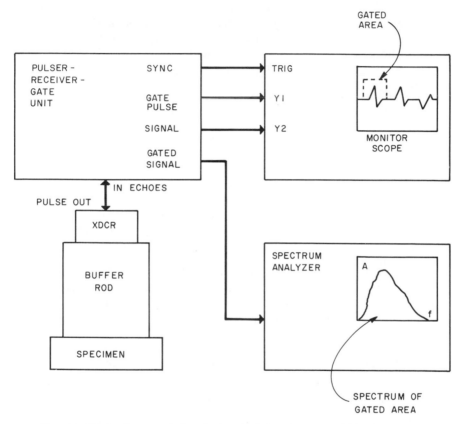

Fig. 18. Block diagram of electrical and ultrasonic system for spectrum analysis of echoes in a buffer/specimen system. From Papadakis *et al.* (1973), by permission of the American Institute of Physics.

receives broadband ultrasonic echoes. The pulser–receiver–gate unit amplifies the echoes and sends this signal into two channels: (1) directly to a monitoring oscilloscope, and (2) through a stepless gate to a spectrum analyzer. Thus, a single echo can be selected by the movable, variable-width gating pulse. The gate pulse is displayed along with the amplified signal on the two-channel oscilloscope. The spectrum of the selected echo is displayed on the spectrum analyzer.

The broadband signal in the buffer rod is partially reflected at the buffer–specimen interface, and is partially transmitted into the specimen. There, it reverberates, and part of the energy returns to the buffer at each reverberation. See Fig. 17. The amplitudes A, B, and C of the first three echoes are given as an inset in Fig. 17 in terms of the particle amplitude reflection coefficient R and the attenuation α. With the formulas of Eqs. (51) and (52), one can calculate R and α from A, B, and C, provided diffraction corrections be made to A, B, and C before the calculation of R and α.

In the present system, echo A is gated into the spectrum analyzer and its spectrum photographed; then echoes B and C are treated likewise. (If the echoes are very different in amplitudes, calibrated attenuation may be introduced to achieve comparable amplitudes. Correction for this must be made in subsequent calculations.) The output of the spectrum analyzer is $A\,(f)$, $B\,(f)$, and $C\,(f)$, the amplitudes as functions of frequency.

3. Spectrum Analysis Compared with rf Bursts in Bonded Transducers

The standard method for pulse-echo ultrasonic attenuation measurements (Roderick and Truell, 1952) utilizes direct bonding of quartz crystals to plane-parallel specimens. It is pertinent to ask whether the spectrum/buffer method presented in Section V,D agrees with the standard method. To this end, an experiment was performed (Papadakis *et al.*, 1973) on a specimen of grade A nickel which had been measured previously (Papadakis, 1965) in the standard manner. Diffraction corrections had been applied to these data in the recognized way [using Eqs. (35) and (36)].

The nickel specimen was mounted on a water buffer column and interrogated by a broadband 5 MHz longitudinal wave transducer (Panametrics, Inc., Waltham, Massachusetts) mounted in the other end of the buffer column. Data on the specimen, buffer column, and transducer appear in Table III.

The spectra of echoes A, B, and C were photographed and measured for $A(f)$, $B(f)$, and $C(f)$. These amplitudes were corrected for diffraction by the

TABLE III[a]

DATA ON WATER BUFFER, NICKEL SPECIMEN, AND TRANSDUCER

Datum	Transducer	Buffer	Specimen
Mode	Longitudinal	—	—
Crystal	5 MHz	—	—
Diameter	1.27 cm	1.53 cm	3.78 cm
Length	—	1.335 cm	1.775 cm
Velocity	—	0.150 cm/μsec	0.573 cm/μsec

[a] From Papadakis *et al.* (1973), by permission of the American Institute of Physics.

method in Section V,D above, to yield $A_0(f)$, $B_0(f)$, and $C_0(f)$. Then $R(f)$ and $\alpha(f)$ were calculated. $R(f)$ held fairly constant around $R = 0.9435 \pm 0.0045$ in agreement with the theoretical value $R = (Z_1 - Z_2)/(Z_1 + Z_2) = 0.945$ where the Z_i are the specific acoustic impedances. The attenuation α is plotted in Fig. 19 as a function of frequency. Superimposed are the values reported earlier on the same specimen measured by the standard methods of bonded quartz crystals and pulse-echo diffraction corrections. The agreement is exact within the standard error of either measurement method. We can conclude that the spectrum analysis method is as accurate as the bonded-transducer method. As with other measurements, one is limited by the need

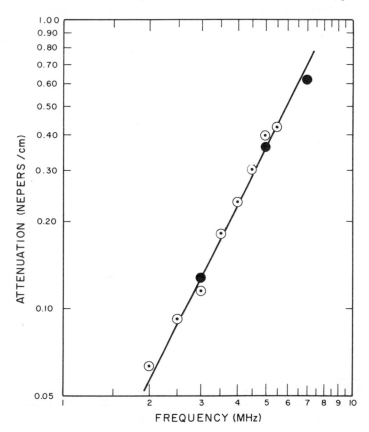

FIG. 19. Attenuation in a nickel specimen tested by broadband pulses and spectrum analysis utilizing the buffer method, and by rf bursts with bonded quartz transducers. Open circles are data with broadband pulses, buffer rods, and spectrum analysis. Solid circles are data (Papadakis, 1965) taken with rf bursts and quartz transducers bonded to the nickel. Agreement is essentially perfect, substantiating the buffer/specimen method with diffraction corrections as given in Section V,D. From Papadakis *et al.* (1973), by permission of the American Institute of Physics.

for sufficient amplitude and by the requirement that the sidewalls of the buffer and specimen do not interfere with the measurement. Because of beam spreading in the relatively long buffer, the use of a buffer may sometimes introduce sidewall interferences not found in pulse-echo experiments with bonded transducers. Under the proper conditions, the buffer method for attenuation by spectrum analysis yields accurate, quantitative data.

4. *Spectrum Analysis With and Without Buffers Compared*

Broadband NDT transducers adversely affect the second and subsequent echoes in a pulse-echo pattern if they are coupled directly to the specimen. The adverse effect comes about because: (a) the efficient piezoelectric element extracts energy from the beam; (b) the matched backing behind the piezoelectric element absorbs the part of the wave passing through the transducer; and (c) the wear plate in front of the piezoelectric element distorts the pulse. The net effect should be twofold: to distort the frequency spectrum of the second and subsequent echoes, and to lower the amplitude of the second and subsequent echoes.

An experiment was performed (Papadakis *et al.*, 1973) to test these hypotheses and to demonstrate that the buffer method should be used in preference to the direct contact method when NDT transducers (search units) are used to measure attenuation by spectrum analysis.

A block of ATJ graphite was tested with a 1 MHz longitudinal wave transducer (Panametrics, Inc.) by spectrum analysis in two ways: by direct contact with a thin fluid couplant layer, and, by the A′ AB buffer method with a rubber buffer, also coupled with thin fluid layers. Data on the transducer, buffer, and specimen are given in Table IV.

TABLE IV[a]

DATA ON RUBBER BUFFER, GRAPHITE SPECIMEN, AND TRANSDUCER

Datum	Transducer	Buffer	Specimen
Mode	Longitudinal	—	—
Crystal	1 MHz	—	—
Diameter	2.54 cm	5.0 cm	4.0 cm
Length	—	2.6 cm	4.0 cm
Velocity	—	0.116 cm/μsec	0.225 cm/μsec

[a] From Papadakis *et al.* (1973), by permission of the American Institute of Physics.

The attenuation data by contact were corrected for diffraction in the standard way, while the attenuation data by buffer rod were corrected for diffraction by the method outlined in Section V,D, above. The two resulting curves for attenuation versus frequency are drawn on the same graph in Fig. 20. The attenuation shown by direct contact spectrum analysis is distinctly

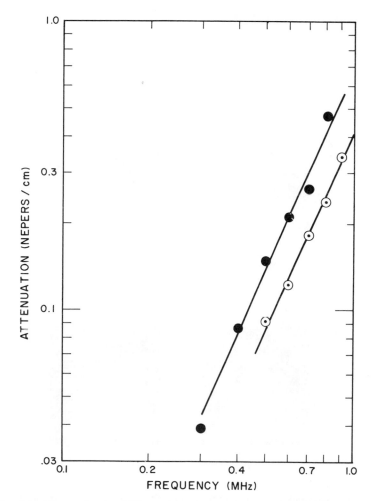

Fig. 20. Attenuation in ATJ graphite by spectrum analysis. Lower curve (open circles) is with a buffer rod. Upper curve is with a damped broadband transducer coupled directly to the specimen. Direct coupling leads to higher attenuation readings because of absorption in the transducer. The buffer method is preferred. From Papadakis *et al.* (1973), by permission of the American Institute of Physics.

higher than the attenuation measured with the interposed buffer using spectrum analysis. This result bears out the contention that the NDT transducer used in direct contact extracts energy from the beam at each echo, attenuating the echoes, and yielding erroneous results. In this case the excess attenuation is as much as 0.2 dB/cm at 0.9 MHz, or 1.5 dB/echo at the transducer. Thus it is concluded that the buffer method is to be preferred over direct contact for attenuation measurements by spectrum analysis.

5. *Spectrum Analysis with Diffraction Corrections to Yield True Transducer Output*

The spectrum of echo A′ at the end of a plane, parallel, lossless buffer with no specimen attached is representative of the transducer efficiency upon two transductions—transmission and reception. The spectra from the A′ echoes of a typical 10 MHz transducer (Panametrics, Inc.) radiating into various low-loss buffer materials are shown in Fig. 21 (Papadakis *et al.*,

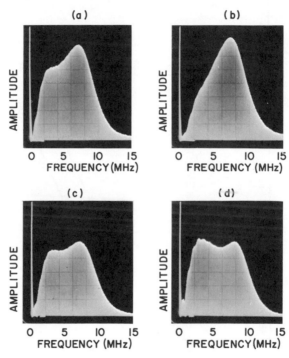

FIG. 21. The spectra of the first echo in bare buffer rods. A 10 MHz transducer was used for both transmission and reception. (a) 2.54 cm fused quartz. (b) 6.22 cm fused quartz. (c) 2.54 cm aluminum. (d) 1.89 cm hardened steel. From Papadakis *et al.* (1973), by permission of the American Institute of Physics.

1973). Information on the transducer and buffers are given in Table V. The spectra are seen to peak well below 10 MHz, and to contain appreciable energy down to 2 MHz. The spectra appear to depend upon buffer length and material at this juncture.

As discussed above, beam spreading is greater at low than at high frequencies. This fact implies that the low frequency output of the transducer is actually greater than in Fig. 21. If the correction of Eq. (50) is applied to the echo A′, then the undiffracted plane wave amplitude emitted aud received by the transducer should be represented by $A_0′$ as in Eq. (50). This calculation

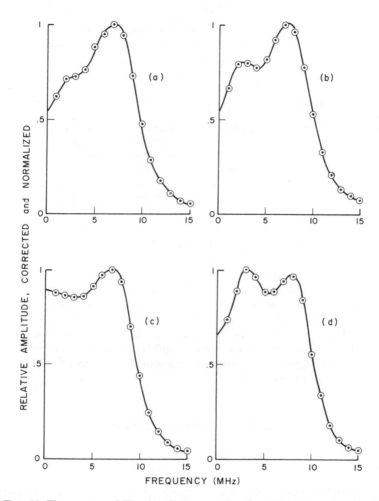

Fɪɢ. 22 The spectra of Fig. 21 after correction for diffraction and normalization for the droop in the input spectrum. The transducer appears close to critically damped. Buffers: (a) 2.54 cm fused quartz; (b) 6.22 cm fused quartz; (c) 2.54 cm aluminum; (d) 1.89 cm hardened steel. From Papadakis *et al.* (1973), by permission of the American Institute of Physics.

was performed on the observed spectra of Fig. 21, and the results were then normalized to account for the input spectrum of the pulser used. The resulting corrected amplitudes are shown in Fig. 22. The four graphs are not identical, but do show the following common characteristics: a peak in the vicinity of 7 MHz, rapid fall-off above this peak, and amplitude less but not much less at 1 MHz (and probably at 0) than at the peak.

TABLE V[a]

DATA ON TRANSDUCER AND VARIOUS BUFFERS

Datum	Transducer	Fused quartz No. 1	Fused quartz No. 2	Aluminum	Steel
Mode	Longitudinal	—	—	—	—
Crystal	10 MHz	—	—	—	—
Diameter	1.27 cm	5.08 cm	4.45 cm	5.08 cm	3.45 cm
Length	—	2.54 cm	6.22 cm	2.54 cm	1.89 cm
Velocity	—	0.596 cm/ μsec	0.593 cm/ μsec	0.638 cm/ μsec	0.590 cm/ μsec

[a] From Papadakis *et al.* (1973), by permission of the American Institute of Physics.

The shapes of the curves differ between 0 and 7 MHz, but have a general rise around 3 MHz. The cause of the variation in the curves for the same transducer radiating into different low-loss materials is not explained at present, but may relate to (a) radiation pattern of a transducer with a wear plate—any deviation from piston profile could change diffraction, (b) quality of specimen surfaces, and (c) impedance mismatch between transducer and specimen. However, the general conclusion is that the transducers act like very highly damped oscillators with amplitude peaking somewhat at a frequency significantly lower than the fundamental resonance of the piezo-electric plate sandwiched inside.

VI. Input Amplitude Profile

A. BACKGROUND

Some evidence exists that the diffraction loss changes if the transducer radiates as other than a piston source. Roderick and Truell (1952) pointed out that sidelobes would be suppressed if an unplated crystal transducer were activated by a convex metal electrode placed above it. The author (Papadakis, 1963a) found that the diffraction loss became more variable in magnitude as a function of S, and was displaced along S, if fringing fields near the rim of the transducer were important in doughnut-plated or wrap-around-plated transducers. In addition, it was found that the diffraction loss could be made more nearly monotonic in S if wave energy were absorbed from the outer portion of the beam impinging on the side of the specimen not bonded to the transducer. Martin and Breazeale (1971) have achieved narrow beams without sidelobes by utilizing fringing fields to lower the electric field strength at the periphery of a radiator.

These pieces of evidence lead to the hypothesis that the diffraction loss can be affected by the particle displacement profile across the face of the input transducer. In particular, lower rim amplitudes should lead to smoother diffraction-loss curves. The pressure in the field of a transducer was given by Eq. (2) including amplitude and phase profiles $V(\rho, \psi)$ and $\zeta(\rho, \psi)$ as functions of radius ρ and azimuth ψ. The calculations between Eq. (2) and Eq. (31) dealt with the case $V = 1$, $\zeta = 0$, but included anisotropy in $\boldsymbol{\beta}$ along 3-, 4-, and 6-fold axes in crystals. This section deals with the case $|\boldsymbol{\beta}| = \text{const}$, $\zeta = 0$, and $V = V(\rho)$. The results can be extended to anisotropic conditions by the formula (Papadakis, 1966)

$$S_A = S_A^{(0)}/2(0.5 - b), \tag{55}$$

where S_A is the location along S of the last diffraction-loss peak, $S_A^{(0)}$ is the value of S_A for isotropy calculated in this section, and b is the anisotropy parameter defined in Eq. (14).

In the present section (Papadakis, 1971b) the normal derivative $V(\rho, \psi)$ of the velocity potential is taken to be circularly symmetrical, not dependent upon azimuth ψ. Hence $V = V(\rho)$. Various plausible functions are investigated to simulate transducers driven in such a way as to make $V(\rho)$ maximum at $\rho = 0$ and minimum at $\rho = a$, the rim. Round transducers plated in a circularly symmetrical manner from $\rho = 0$ to $\rho = a_p$ with $a_p < a$ should respond in the way $V(\rho)$ is defined. The actual response of a transducer may not correspond exactly to any of the functions $V(\rho)$ chosen, however.

B. CALCULATIONS

Integration of Eq. (2) was carried out by numerical methods with a digital computer. The program written for Section III was modified in two respects for the new calculations:

1. $V(\rho)$ was employed as a weighting function for the input radiation from each element of area σ of the transducer acting as a transmitter.

2. Reception was limited to an area with a radius $a_p < a$ corresponding to a partially plated top surface, and the weighting function over this area was made unity for reception.

The second point is justified on the basis that the transducer elements receive amplitude and phase information independently and contribute it to an integrated output. This same assumption was made by Seki *et al.* (1956) and has been used ever since with good results.

The functions used as $V(\rho)$ are listed in Table VI. In all cases, the value of a_p, the radius of the plated area, was taken to be $0.8a$ where a is the radius of the transducer disk.

The functions $V(\rho)$ are drawn in Fig. 23. For these functions, the diffraction-loss curves are plotted in Fig. 24 and the diffraction phase shifts are plotted in Fig. 25. The abscissa is $S^{(p)}$, defined by the plated radius a_p as

TABLE VI[a]

FUNCTIONS $V(\rho)$ FOR VELOCITY POTENTIAL
DERIVATIVE

Number	Type	Function
I	Sinusoid	$\cos(\pi\rho/4a)$
II	Sinusoid	$\cos(\pi\rho/2a)$
III	Gaussian	$\exp(-\rho^2/1.28a^2)$
IV	Gaussian	$\exp(-\rho^2/2.56a^2)$
V	Fermi	$1/[1 + \exp(5\rho/a - 4)]$
VI	Fermi	$1/[1 + \exp(25\rho/a - 20)]$

[a]From Papadakis (1971b), by permission of the
American Institute of Physics.

$S^{(p)} = z\lambda/a_p{}^2$. It seemed expedient to compare the diffraction-loss and phase
curves in terms of the assumed receiving area. In experiments, it is frequently
assumed that a_p is the effective radius for both transmitting and receiving.
The curves for the all-plated piston source are included for comparison.

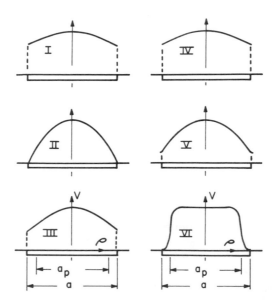

FIG. 23. Profiles of the normal derivative $V(\rho)$ of the velocity potential across the
transmitting transducers. The functions are listed in Table VI. From Papadakis (1971b),
by permission of the American Institute of Physics.

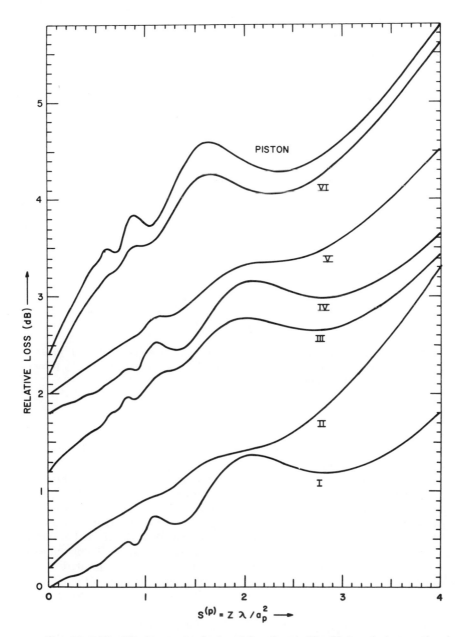

FIG. 24. Diffraction-loss curves for input functions in Fig. 23. Loss is decreased and smoothed by monotonic decreasing functions $V(\rho)$. From Papadakis (1971b), by permission of the American Institute of Physics.

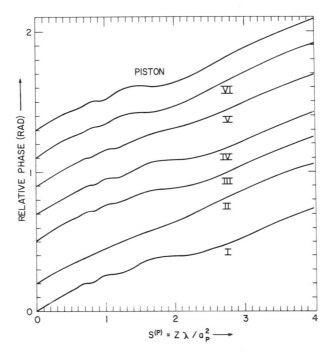

FIG. 25. Diffraction-phase-change curves for input functions in Fig. 23. Phase is smoothed by monotonic decreasing functions $V(\rho)$. From Papadakis (1971b), by permission of the American Institute of Physics.

C. DISCUSSION

The principal effect of reducing the amplitude of motion of the outer part of the radiating transducer is to lower the diffraction loss and its fluctuation with distance (see Fig. 24). This result confirms the hypothesis stated in Section VI,A. In addition, the position of the loss peaks along $S^{(p)}$ depends upon the amount of radiation generated outside of a_p. This is to be expected, because the receiver of radius a_p is sensing the field of a larger transmitter with an effective radius a_t in the range $a_p < a_t < a$. Phase is affected in such a way as to smooth the curves (see Fig. 25). Plateaus found for the piston case are changed into regions of low slope. These regions are still located at the diffraction-loss peaks if present, or at diffraction-loss plateaus if the effect is strong enough to smooth out the diffraction-loss curves to monotonic functions.

The proper function to be used for $V(\rho)$ for partially plated transducers and for other nonpiston configurations will have to be determined. Experimental methods might include multiple-beam interferometry, laser Bragg scattering, and Lang X-ray topography.

D. Conclusions

If a transducer deviates from a piston source, the ultrasonic radiation pattern will be affected in such a way as to modify the diffraction-loss and diffraction-phase-change curves for pulse-echo experiments. When the normal derivative $V(\rho)$ of the velocity potential on the face of the transmitter is monotonic decreasing with radius ρ from the center to the rim at $\rho = a$, the diffraction loss is decreased and smoothed in the near field (up to $S = z\lambda/a^2 \simeq 3$), and the diffraction phase change is smoothed. Since transducers are plated over only a fraction of their area frequently, such gradients in the velocity potential are to be expected. The exact form of $V(\rho)$ has not been determined, but plausible functions were tried. When diffraction curves are plotted against $S^{(\mathrm{p})} = z\lambda/a_\mathrm{p}^2$, where a_p is the plating radius, the positions along $S^{(\mathrm{p})}$ of characteristic features of the curves (such as loss peaks) depend upon the fraction of the radiation generated in the rim area. Thus, both the magnitudes of the diffraction loss and phase change, and their functional dependences upon S, depend upon the form of $V(\rho)$. Further work should be done on the definition of $V(\rho)$ if partially plated transducers are to be used in experiments.

VII. Broadband Pulses

A. Introduction

Although narrowband ultrasonic measurements have been widely used for many applications, such as the determination of ultrasonic attenuation and velocity as functions of frequency and for ultrasonic flaw detection, there are also many applications which either require or are enhanced by broadband pulsed operation of ultrasonic transducers. These include digital storage delay lines (Eveleth, 1965), high resolution nondestructive flaw detection (Lees and Barber, 1968; Papadakis and Fowler, 1971), and pulse-echo spectrum analysis (Gericke, 1965, 1966; Papadakis and Fowler, 1971: Papadakis et al., 1973).

In all cases, elastic pulses short in space and time must be generated, propagated, and received. Bandwidths upward of 40% are required to make the pulse duration less than two wavelengths of the center frequency of the pulse spectrum. Much work has already been published (Ivanov et al., 1962; Redwood, 1963; Kossoff, 1966; Sittig, 1967, 1969) on the ultrasonic-pulse problem. Plane wave analysis has been used in this previous work. The present section follows Papadakis and Fowler (1971) and Papadakis (1972), and deals with the theory of pulse propagation as a single-aperture diffraction problem, utilizing a weighted superposition of the fields of single-frequency piston sources at various frequencies within the band. New pressure and phase profiles are calculated for broadband operation. The diffraction loss and phase change are calculated as functions of S for pulse-echo operation.

The profiles could be utilized in the future for the calculation of scattering factors for objects of various shapes and sizes such as flaws.

B. Theory and Calculations

1. *Radiation Field*

The method of obtaining the pressure and phase in the broadband pulse follows. Let the pressure of a monochromatic transducer can be expressed as

$$p = C \exp[j(\omega t - \beta_0 z - \delta)], \tag{56}$$

where the phase $\omega t - \beta_0 z$ for a plane wave cancels, leaving the phase δ relative to the plane wave. Then p, C, and δ are functions of S, the normalized distance along the axis of the transducer, and of x, the radial distance from the axis. They are implicitly functions of frequency f, since $S = z\lambda/a^2 = zv/a^2 f$ is a function of f. The equation for p reduces to

$$p(f) = C(f)[\cos\delta(f) - j \sin\delta(f)]. \tag{57}$$

Introducing a spectral density function $B(f)$ for the broadband pulse gives

$$dp(f) = B(f)C(f)[\cos\delta(f) - j \sin\delta(f)] \, df. \tag{58}$$

Integrating, one obtains

$$p = \int_0^\infty BC(\cos\delta - j \sin\delta) \, df$$

$$= \int_0^\infty BC \cos\delta \, df - j \int_0^\infty BC \sin\delta \, df. \tag{59}$$

This may be equated to

$$p = A(\cos\varepsilon - j \sin\varepsilon), \tag{60}$$

with

$$A \cos\varepsilon = \int_0^\infty BC \cos\delta \, df = A_1$$

and $\tag{61}$

$$A \sin\varepsilon = \int_0^\infty BC \sin\delta \, df = A_2 .$$

This introduces a new amplitude A and an effective phase ε for the broadband pulse, where

$$A = (A_1{}^2 + A_2{}^2)^{1/2}$$

and $\tag{62}$

$$\varepsilon = \tan^{-1}(A_2/A_1).$$

For computational purposes, the integrals were turned into summations

$$A_1 = \sum_{n=1}^{9} B_n C_n \cos\delta_n$$

and (63)

$$A_2 = \sum_{n=1}^{9} B_n C_n \sin\delta_n .$$

For the calculation of the profiles, the quantity S_c was defined to be the value of S at the center frequency f_c of the spectral distribution,

$$S_c = zv/a^2 f_c . \tag{64}$$

For any other frequency f, the equivalent value of S was

$$S = S_c f_c/f. \tag{65}$$

Values of S as a function of f for $f_c = 5$ MHz and S_c specified between 0.2 and 2.5 are given in Table VII.

TABLE VII

VALUES OF S AT VARIOUS FREQUENCIES WITH f_c AND S_c SPECIFIED[a]

f_c \ S_c	0.200	0.250	0.312	0.500	0.750	1.000	1.500	2.500
1	1.000	1.250	1.560	2.500	3.750	5.000	7.500	12.500
2	0.500	0.625	0.780	1.250	1.870	2.500	3.750	6.250
3	0.333	0.417	0.520	0.833	1.250	1.670	2.500	4.170
4	0.250	0.312	0.390	0.625	0.940	1.250	1.870	3.120
5	0.200	0.250	0.312	0.500	0.750	1.000	1.500	2.500
6	0.167	0.208	0.260	0.417	0.625	0.833	1.250	2.080
7	0.143	0.179	0.223	0.357	0.536	0.714	1.070	1.790
8	0.125	0.156	0.195	0.312	0.469	0.625	0.940	1.560
9	0.111	0.139	0.173	0.278	0.417	0.555	0.833	1.390

[a] From Papadakis and Fowler (1971), by permission of the American Institute of Physics.

Since C and δ were known as functions of S and x from previous computations (Papadakis, 1966; see Sections II and III), it was necessary only to tabulate C and δ in the appropriate order for the summations. (See Appendix A for C and δ at the values of S in Table VII.)

The spectral density was represented by an idealized symmetrical function $B(f)$. The function is given in Table VIII for several percentage bandwidths. Since the real bandwidth of a transducer represents two transductions, B^2 is representative of the amplitude seen in the spectrum of an echo.

TABLE VIII[a]

TABULATION OF SPECTRAL DENSITY $B(f)$ FOR VARIOUS BANDWIDTHS

	Bandwidth (%)				
f_c	10	20	40	80	120
1	0.001	0.002	0.010	0.100	0.562
2	0.002	0.010	0.030	0.450	0.840
3	0.015	0.100	0.180	0.840	0.943
4	0.100	0.320	0.840	0.970	0.983
5	1.000	1.000	1.000	1.000	1.000
6	0.100	0.320	0.840	0.970	0.983
7	0.015	0.100	0.180	0.840	0.943
8	0.002	0.010	0.030	0.450	0.840
9	0.001	0.002	0.010	0.100	0.562

[a] From Papadakis and Fowler (1971), by permission of the American Institute of Physics.

A program was written to perform the summations and calculate A and ε, the pressure and phase. In Figs. 26a–26h, the results for the pressure profiles are presented at various values of S_c for various pulse bandwidths. The cw monochromatic case shows zeros on the centerline at $S_c = 0.25$ and 0.50 as predicted by theory. (See Figs. 26b and 26d.) $S_c = 0.50$ is the $Y_1^{(-)}$ point at which the last zero in pressure occurs along the centerline of the transducer for monochromatic waves (McMasters, 1959). The finite bandwidths do not have zeros. Indeed, above 80% bandwidth the field is fairly flat out to three-fourths the radius of the transducer. Phase profiles are given in Figs. 27a–27h. They are smoother for higher bandwidths.

Figure 28 contains a plot of the pressure profiles at various distances from a transducer of 120% bandwidth for double transduction. The pressure in the near field is relatively smooth, and no strong minima appear along the axis. This behavior is distinctly different from the rf burst case (Seki *et al.*, 1956; McMasters, 1959).

In Fig. 29, the relative pressure along the centerline is plotted with bandwidth as a parameter. These pressure amplitudes plotted here represent the relative amplitudes as a function of path and do not represent the degree of sensitivity of the transducers versus bandwidth. The ratio of the heights of the last minimum to the last maximum is seen to increase with bandwidth. This ratio, plotted in Fig. 30, is suggested as one figure of merit for broadband nondestructive testing transducers. It represents the difference in amplitude of the echoes from equal flaws at different depths, not counting attenuation effects. Note that not much more field uniformity is gained by a bandwidth above 80%.

Frequently, transducers are used for immersed testing. Then part of the path is in water and part in solid material. Figure 31 contains plots of the amplitude along the centerline in the solid for two water paths. One might want to work in the region of increasing amplitude with distance to cancel attenuation effects in the solid.

Broadband pulses are advantageous for flaw detection because the acoustic pressure does not show any serious minima, because they are narrow in space and time (permitting high depth resolution), and because the spectrum contains information about the material interrogated or the flaw observed. Increased bandwidth is achieved at the expense of sensitivity. Provided the signal-to-noise ratio is great enough, the loss in sensitivity can be compensated for by electronic amplification.

Concerning resolution, the graph in Fig. 32 shows the predicted pulse shape of an echo from a broadband transducer with a 10 MHz piezoelectric plate. Sittig's (1967, 1969) analysis was used to account for the mechanical properties of the backing and face-plate materials and for the electrical terminations. The inset in this figure is an oscillogram of an echo from a transducer built to the same specifications. The details of pulse width and relative amplitude agree exactly. The width of the first half-cycle, about 60 nsec, represents a round trip through about 0.1 mm of a metal.

2. Diffraction Loss and Phase

Further, the amplitude and phase profiles $A(r)$ and $\varepsilon(r)$ calculated above for broadband pulses were used to study the effect of bandwidth on the diffraction loss and phase change in pulse-echo work (Papadakis, 1972). The response of the receiving transducer was calculated as follows:

The phase ϕ for diffraction phase corrections is the weighted average over receiver area and wave amplitude of the effective phase profile ε,

$$\phi = \int_0^a A(r)\varepsilon(r)2\pi r\,dr \Big/ \int_0^a A(r)2\pi r\,dr. \tag{66}$$

The pressure sensed by the transducer is proportional to the integral of the amplitude weighted by the cosine of the deviation from the average phase,

$$p = \int_0^a A(r)\cos[\varepsilon(r) - \phi]\pi r\,dr. \tag{67}$$

The loss in decibels relative to the pressure at some arbitrary point along the axis of the transducer is

$$dB = 20\log_{10}(p_0/p). \tag{68}$$

The phase ϕ and loss in decibels are functions of the normalized distance

$$S_c = z\lambda_c/a^2 = zv/a^2 f_c, \tag{69}$$

where z is the propagation distance, a the transducer radius, v the velocity, f_c the center frequency of the broadband pulse, and λ_c is the wavelength at the center frequency. Curves of decibels versus S_c and ϕ versus S_c result.

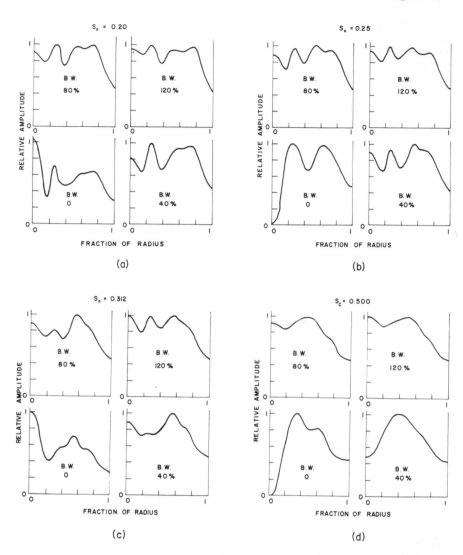

FIG. 26 (a)–(h). Pressure profiles for various bandwidths at several values of S_c. B. W.—bandwidth. From Papadakis and Fowler (1971), by permission of the American Institute of Physics.

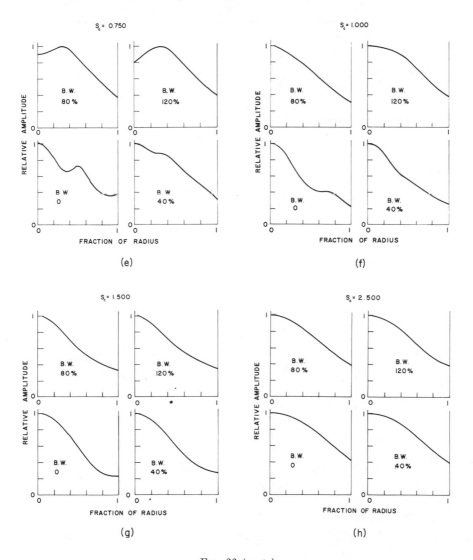

FIG. 26 (cont.)

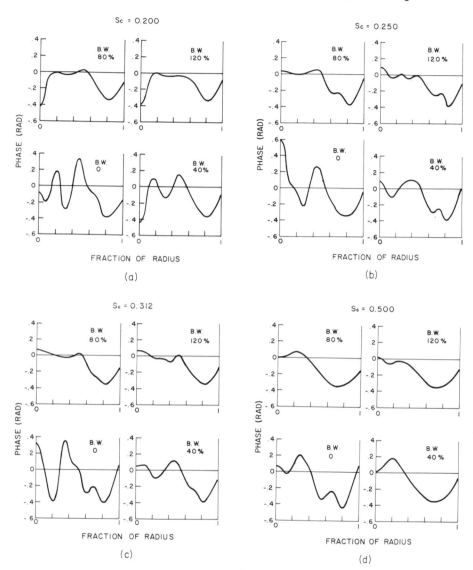

F<small>IG</small>. 27. (a)–(h). Phase profiles for various bandwidths at several values of S_c. B. W.— bandwidth. From Papadakis and Fowler (1971), by permission of the American Institute of Physics.

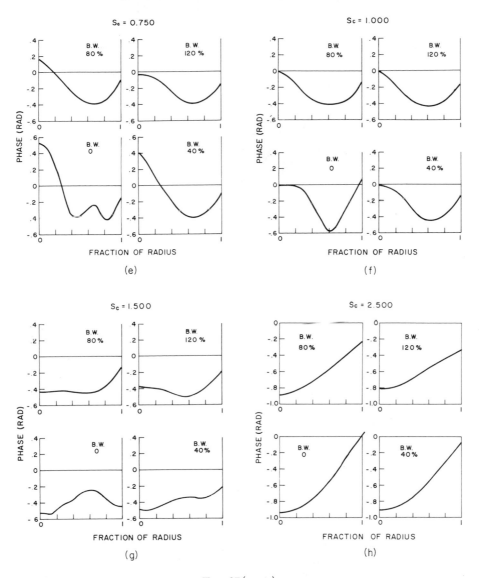

FIG. 27 (cont.)

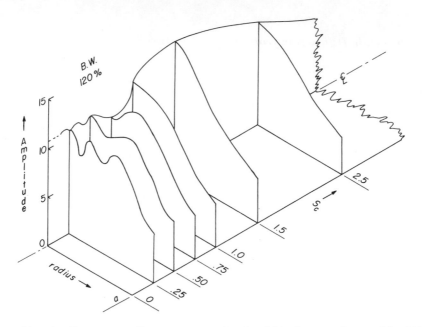

FIG. 28. Pressure profile contour map for the field of a transducer with 120% bandwidth. B. W.—bandwidth. From Papadakis and Fowler (1971), by permission of the American Institute of Physics.

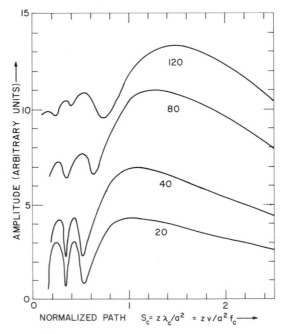

FIG. 29. Centerline pressure of the transducer field for various bandwidths. From Papadakis and Fowler (1971), by permission of the American Institute of Physics.

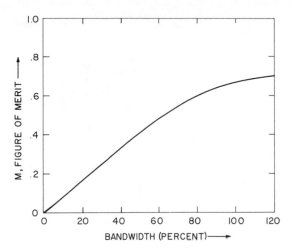

Fig. 30. Ratio of the depth of the $Y_1^{(-)}$ minimum to the $Y_0^{(+)}$ maximum center-line pressure as a function of bandwidth. This ratio is proposed as a figure of merit for NDT broadband transducers. From Papadakis and Fowler (1971), by permission of the American Institute of Physics.

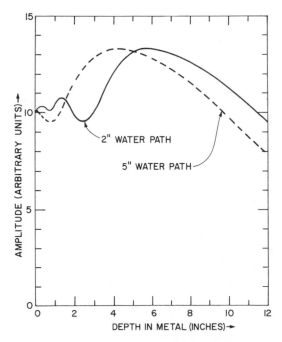

Fig. 31. Pressure along the transducer centerline in immersed testing. The pressure in the solid under test is plotted for two water paths. 120% Bandwidth, 5 MHz center frequency, $\frac{1}{2}$ in. diam. transducer. From Papadakis and Fowler (1971), by permission of the American Institute of Physics.

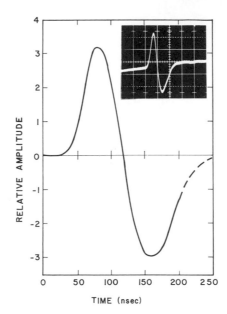

Fig. 32. Pulse shape and duration for the echo from a broadband transducer with a 10 MHz piezoelectric plate. Experiment and theory agree quantitatively (Sittig, 1967, 1969). From Papadakis and Fowler (1971), by permission of the American Institute of Physics.

For computational purposes, the integrals were turned into summations,

$$\phi = \sum_n A_n \, \varepsilon_n \, 2\pi r_n \, \Delta r \big/ \sum_n A_n \, 2\pi r_n \, \Delta r, \tag{70}$$

and

$$p = \sum_n A_n \cos[\varepsilon_n - \phi] 2\pi r_n \, \Delta r. \tag{71}$$

For each bandwidth (10%, 20%, 40%, 80%, 120%) the quantities ϕ and p were computed at several values of S_c. Then the decibel loss was calculated relative to p at $S_c = 0.20$. The curves were extrapolated back to $S_c = 0$ in a way consistent with the cw case.

Curves of loss in decibels and phase in radians versus S_c appear in Figs. 33 and 34. The curves are displaced vertically for clarity. It can be seen that both the loss and phase smooth out as the bandwidth increases. For moderate and large percentage bandwidths, one would not expect to see the loss peaks or the phase plateaus present in monochromatic experiments. In anisotropic materials, one would redraw the curves with an abscissa scale factor as suggested in Eq. (55).

The curves for broadband pulses in Figs. 33 and 34 differ from the curves for monochromatic pulses (also shown) in that they are smoother. It is

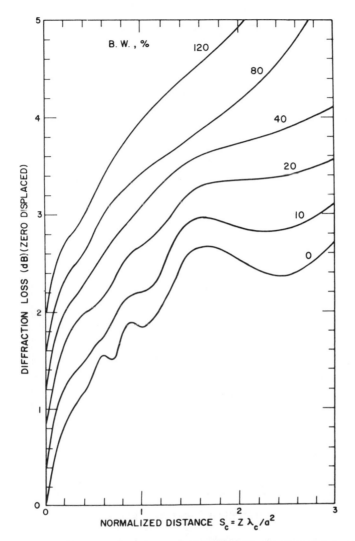

Fig. 33. Diffraction loss versus distance for broadband pulses with bandwidth as the parameter. Loss becomes monotonic as the bandwidth increases. B. W.—bandwidth. From Papadakis (1972), by permission of the American Institute of Physics.

suggested that these curves are appropriate for making diffraction corrections for broadband pulses using Eqs. (36) and (38). Care must be taken in cases in which the attenuation is a strong function of frequency because the pulse spectrum would change with distance, invalidating the calculations. The maximum error one could encounter due to diffraction could be calculated on the basis of the center frequency of the most attenuated echo used in the

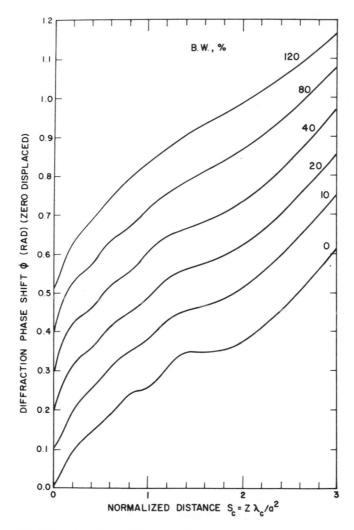

FIG. 34. Diffraction phase shift versus distance for broad-band pulses with band-width as the parameter. B. W.—bandwidth. From Papadakis (1972), by permission of the American Institute of Physics.

measurement. Comparison could be made with the diffraction effects upon less attenuated echoes, arriving at an average probable correction. This correction, while itself in error to some degree, would be much better than no correction at all. The total difference between t and t' can never be greater than $\tau_c^{(\alpha)}/4$ where $\tau_c^{(\alpha)}$ is the period of the center frequency of the most attenuated echo, because the phase curve ϕ is asymptotic to $\pi/2$ rad. The loss curve, on the other hand, rises as the logarithm of S_c, and the limits must be computed in each individual case. It is suggested that a preferable method would be to use spectrum analysis with a buffer rod system to obtain monochromatic information from broadband pulses (Papadakis *et al.*, 1973). Diffraction corrections for the buffer rod system were explained in Section V,D, and the spectrum analysis system is shown in Section V,G.

VIII. Specimens of Finite Width

Problems arise when one encounters specimens of dimensions not much larger than the transducer normal to the propagation direction. This frequently happens in solids where the specimen is a rod, a section of a plate, or a valuable crystal. Fluids contained in tubes also exhibit the problems. One encounters what is called qualitatively "sidewall effects" and technically "multimode guided-wave propagation." This means that although the cylinder or plate are several or even many wavelengths in lateral extent, they are not large enough to support free-field propagation. Rather, they support all the rod modes or plate modes excited by the transducer and not beyond cutoff due to the geometry. These modes, being dispersive (Meeker and Meitzler, 1964), interfere with each other as they propagate down the specimen. It is possible under certain circumstances for destructive interference to be almost complete, canceling the pressure wave over the face of a receiving transducer (Carome and Witting, 1961; Carome *et al.*, 1961). The interference phenomena are expressible in terms of the normalized distance S—parameter $S = z\lambda/a^2$. Universal curves obtain, with destructive interference at sequential positions along S in the approximate ratio $1:3:5:7 \cdots$. Obviously, attenuation measurements in this regime must be made judiciously because of the apparent loss caused by the interferences. In a pulse-echo experiment it is best to measure loss using echoes at antinodes of the interference pattern.

It was also pointed out (Papadakis, 1969) that there is a phase shift associated with the destructive interference. This phase shift is large enough to interfere with the correct overlap in the pulse-echo-overlap method (Papadakis, 1967) and in the pulse-superposition method (McSkimin, 1961) when McSkimin's Δt Criterion is used with rf bursts. Sidewall effects can introduce several rf cycles of error into a travel time measurement if McSkimin's Δt Criterion is used to determine the overlap in specimens supporting multimode guided-wave propagation. A solution was suggested (Papadakis, 1969) employing broadband pulses with center frequency equal to the desired rf burst frequency to determine the proper overlap. This procedure is possible although the multimode interference distorts the broadband echoes somewhat. There is a small difference between the broadband pulse measurement and the rf burst measurement set at the same cycle-for-cycle overlap because the rf burst echo is affected more severely by the adjacent node in the echo pattern. A system for making pulse-echo-overlap measurements in the presence of multimode guided-wave propagation is shown in Fig. 35. The theory

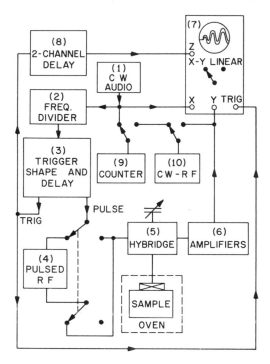

Fɪɢ. 35. Apparatus for pulse-echo overlap measurements employing video pulses to establish the proper cyclic matching. For video matching, the square pulse goes to the transducer; for rf matching, the square pulse triggers the rf generator which activates the transducer. From Papadakis (1969), by permission of the American Institute of Physics.

for the phase shift at points of destructive interference in liquid columns in pipes has been worked out (Del Grosso, 1968), and shows a phase increment and then a decrement about the nodal S-value as S increases. These changes are superimposed on a curve quite similar to Fig. 7. These phase shifts can affect resonant measurements (interferometers) as well as pulse measurements. In either case, the travel time of the ultrasonic wave can be determined to better than $\pm \frac{1}{2}$ period of the rf frequency if the proper matching or overlap is carried out, even if the exact diffraction correction cannot be computed.

IX. Surface Waves

Ultrasonic surface waves are a special two-dimensional case of the general diffraction problem. Their use in the study of ultrasonic diffraction is of particular importance because the surface can be probed acoustically, electrically, and optically to study the beam (Richardson and Kino, 1970; Slobodnik, 1970; Weglein *et al.*, 1970).

Earlier experiments in three dimensions have probed the interiors of transparent substances with laser Bragg scattering (Cohen, 1967) and the ends of specimens with absorbers (Papadakis, 1963a). With surface wave probing, one can study directly both the beam-spreading angle, defined either as the angle between the center of the beam and the null between the main beam and the first sidelobe or as the half-power angle away from the sidelobe, and the beam-steering or "walk-off" angle, i.e. the deviation angle between the energy flux (Poynting) vector and the normal to the wavefronts (Weglein *et al.*, 1970).

In all physical cases for straight interdigital transducers, the beam-spreading angle is positive. The limit of zero spreading is approached as the anisotropy parameter b approaches 0.5 from below. As stated earlier, b is the quadratic coefficient in the expansion of the velocity surface in the propagation direction. For pure mode directions, the cubic coefficient is zero, and the beam does not deviate from the wave normal. There is an exact analogy for surface waves. For various cuts of crystals, there are various propagation directions for which the wave surface is parabolic, and the cubic coefficient is zero. In these directions, the beam-steering or walk-off angle is zero. Examples are Y-cut Li NbO$_3$ with Z-axis propagation and with propagation at $\pm 21.9°$ from Z (Weglein *et al.*, 1970), certain directions on other cuts of LiNbO$_3$ (Slobodnik and Conway, 1970; Slobodnik and Szabo, 1971), and certain directions on X-cut and Z-cut quartz (Coquin and Tiersten, 1967), etc.

One can choose to propagate a wave with propagation vector along a pure-mode direction and calculate the anisotropy parameter b from measurements of the beam-spreading angle. One can also choose to propagate a wave with propagation vector at a small angle θ to the pure-mode direction by misorienting the interdigital transducer, and then calculate b from measurements of the beam-steering angle. The beam-steering angle (walk-off angle) is the same as d_P, the deviation angle between the propagation vector and the Poynting vector, and was given previously in Eq. (16) as $d_P = (2B)^{1/2}\theta$. The relationship $B = 2b^2$ (Papadakis, 1964a) yields $d_P = 2b\theta$. This is analytically true for bulk waves where b can be written in terms of the elastic moduli as in Eqs. (20)–(24), and is also true for surface waves although it is not possible to write an analytical expression for b for surface waves in terms of the elastic moduli (Weglein, 1973).

Measurements of the half-power beam-spreading angle in Y-cut LiNbO$_3$ showed b approaching 0.5 in the Z direction (Weglein *et al.*, 1970). Measurements of the deviation angle d_P on a similar sample yielded $b = 0.455$ (de-Klerk, 1970). These measurements show Y-cut Z-propagating LiNbO$_3$ to give a highly collimated beam in agreement with diffraction theory which predicts perfect collimation for $b = 0.5$. A general treatment of surface waves and their Poynting vectors has been given in an earlier volume of this series (Farnell, 1970). The diffraction loss for surface waves is given with fair accuracy by the two-dimensional calculation (Papadakis, 1964a) if the interdigital transducers for transmission and reception are of equal width and placed for propagation along an equivalent pure-mode direction, and if the equivalent b is known. It has been shown that the length of the interdigital structure can have a small effect upon the diffraction loss and phase (Kharusi and Farnell, 1971). Diffraction must be taken into consideration when interdigital transducers are apodized to tailor the bandpass characteristics of dispersive delay lines. A narrow pattern at low frequencies will have more beam spreading than a narrow pattern at high frequencies.

X. Summary

Early and recent work on ultrasonic diffraction from single apertures has been reviewed and recapitulated. Topics covered have included bulk waves and surface waves; monochromatic bursts and broadband pulses; piston sources and shaded sources; diffraction corrections for attenuation and velocity; theory of the diffraction process; and critical experiments on diffraction and its effects. It has been shown that the echo pattern fluctuations seen in pulse-echo work are predictable, and that accurate measurements can be made in the presence of diffraction. The effects of anisotropy upon pulse-echo work with bulk waves and on through-transmission work with surface waves have been explained. The stage is set for future work on unsolved problems in bulk shear wave diffraction and in engineering applications involving pulses, echoes, and spectra.

ACKNOWLEDGMENTS

This work was supported in part by the Office of Naval Research, Acoustics Branch. The author is indebted to Mr. J. W. Orner, Prof. A. R. von Hippel, Dr. J. C. King, Dr. J. E. May, Jr., and Dr. E. H. Carnevale who have supported his acoustics studies including this work on diffraction.

REFERENCES

Barshauskas, K., Ilgunas, V., and Kubilyunene, O. (1964). *Sov. Phys.—Acoust.* **10**, 21.
Benson, G. C., and Kiyohara, O. (1974). *J. Acoust. Soc. Amer.* **55**, 184.
Carome, E. F., and Witting, J. M. (1961). *J. Acoust. Soc. Amer.* **33**, 187.
Carome, E. F., Witting, J. M., and Fleury, P. A. (1961). *J. Acoust. Soc. Amer.* **33**, 1417.
Cohen, M. G. (1967). *J. Appl. Phys.* **38**, 3821.
Coquin, G. A., and Tiersten, H. F. (1967). *J. Acoust. Soc. Amer.* **41**, 922.
deKlerk, J. (1970). Int. Symp. Acoust. Surface Waves, IBM, Yorktown Heights, New York.
Del Grosso, V. A. (1968). NRL Rep. No. 6852. Nav. Res. Lab., Washington, D.C.
Eveleth, J. H. (1965). *Proc. IEEE* **53**, 1406.
Farnell, G. W. (1970). *In* "Physical Acoustics: Principles and Methods" (W. P. Mason and R. N. Thurston, eds.), Vol. 6, pp. 109–166. Academic Press, New York.
Gericke, O. R. (1965). *Mater. Res. Stand.* **5**, 23.
Gericke, O. R. (1966). *J. Metals* **18**, 932.
Gitis, M. B., and Khimunin, A. S. (1969). *Soviet Phys.—Acoust.* **14**, 413.
Ivanov, V. E., Merkulov, L. G., and Yakovlev, L. A. (1962). *Zavodsk. Lab.* **28**, 1459.
Kharusi, M. S., and Farnell, G. W. (1970). *J. Acoust. Soc. Amer.* **48**, 665.
Kharusi, M. S., and Farnell, G. W. (1971). *IEEE Trans. Sonics Ultrason.* **18**, 35.
Kossoff, G. (1966). *IEEE Trans. Sonics Ultrason.* **13**, 20.

Lees, S., and Barber, F. E. (1968). *Science* **161**, 477.

Lord, A. E. (1966a). *J. Acoust. Soc. Amer.* **39**, 650.

Lord, A. E. (1966b). *J. Acoust. Soc. Amer.* **40**, 163.

Lynnworth, L. C. (1974). *Ultrasonics* **12**, 72.

McMasters, R. C. (1959). "Nondestructive Testing Handbook," Vol. 2, Sect. 44, pp. 12–19. Ronald Press, New York.

McSkimin, H. J. (1960). *J. Acoust. Soc. Amer.* **32**, 1401.

McSkimin, H. J. (1961). *J. Acoust. Soc. Amer.* **33**, 12.

Martin, F. D., and Breazeale, M. A. (1971). *J. Acoust. Soc. Amer.* **49**, 1668.

Mason, W. P. (1958). "Physical Acoustics and the Properties of Solids," pp. 96–97. Van Nostrand, Princeton, New Jersey.

Meeker, T. R., and Meitzler, A. H. (1964). *In* "Physical Acoustics: Principles and Methods," (W. P. Mason, ed.), Vol. 1, Part A, pp. 111–167. Academic Press, New York.

Papadakis, E. P. (1959). *J. Acoust. Soc. Amer.* **31**, 150.

Papadakis, E. P. (1960). *J. Acoust. Soc. Amer.* **32**, 1628.

Papadakis, E. P. (1961). *J. Appl. Phys.* **32**, 682.

Papadakis, E. P. (1963a). *J. Acoust. Soc. Amer.* **35**, 490.

Papadakis. E. P. (1963b). *J. Appl. Phys.* **34**, 265.

Papadakis, E. P. (1964a). *J. Acoust. Soc. Amer.* **36**, 414.

Papadakis, E. P. (1964b). *J. Appl. Phys.* **35**, 1474.

Papadakis, E. P. (1964c). *J. Appl. Phys.* **35**, 1586.

Papadakis, E. P. (1965). *J. Acoust. Soc. Amer.* **37**, 711.

Papadakis, E. P. (1966). *J. Acoust. Soc. Amer.* **40**, 863.

Papadakis, E. P. (1967). *J. Acoust. Soc. Amer.* **42**, 1045.

Papadakis, E. P. (1968a). *J. Acoust. Soc. Amer.* **44**, 1437.

Papadakis, E. P. (1968b). *In* "Physical Acoustics: Principles and Methods," (W. P. Mason, ed.), Vol. 4, Part B, pp. 269–328. Academic Press, New York.

Papadakis, E. P. (1969). *J. Acoust. Soc. Amer.* **45**, 1547.

Papadakis, E. P. (1971a). *J. Appl. Phys.* **42**, 2990.

Papadakis, E. P. (1971b). *J. Acoust. Soc. Amer.* **49**, 166.

Papadakis, E. P. (1972). *J. Acoust. Soc. Amer.* **52**, 847.

Papadakis, E. P., and Fowler, K. A. (1971). *J. Acoust. Soc. Amer.* **50**, 729.

Papadakis, E. P., Fowler, K. A., and Lynnworth, L. C. (1973). *J. Acoust. Soc. Amer.* **53**, 1336.

Rayleigh, Lord, Ed. (1945). "The Theory of Sound," 2nd Ed., Vol. 2, p. 105. Dover, New York.

Redwood, M. (1963). *Appl. Mater. Res.* **2**, 76.

Richardson, B. A., and Kino, G. S. (1970). *Appl. Phys. Lett.* **16**, 82.

Roderick, R. L., and Truell, R. (1952). *J. Appl. Phys.* **23**, 267.

Seki, H., Granato, A., and Truell, R. (1956). *J. Acoust. Soc. Amer.* **28**, 230

Sittig, E. K. (1967). *IEEE Trans. Sonics Ultrason.* **14**, 167.

Sittig, E. K. (1969). *IEEE Trans. Sonics Ultrason.* **16**, 2.

Slobodnik, A. J., Jr. (1970). *Proc. IEEE* **58**, 488.

Slobodnik, A. J., Jr., and Conway, E. D. (1970). *Electron. Lett.* **6**, 171.

Slobodnik, A. J., Jr., and Szabo, T. L. (1971). *Electron. Lett.* **7**, 257.

Tjadens, K. (1961). *Acoustica* **11**, 127.

Truell, R., Elbaum, C., and Chick, B. B. (1969). "Ultrasonic Methods in Solid State Physics," pp. 100–103. Academic Press, New York.

Waterman, P. C. (1959). *Phys. Rev.* **113**, 1240.

Weglein, R. D. (1973). Personal communication.

Weglein, R. D., Pedinoff, M. E., and Winston, H. (1970). *Electron. Lett.* **6**, 654.

| x/a | 0.025 | | 0.125 | | 0.225 | | 0.325 | | 0.425 | |
S	C	δ	C	δ	C	δ	C	δ	C	δ
0.111	2.35	0.00	1.60	-0.30	1.10	0.18	1.05	-0.02	1.25	0.02
0.125	2.35	-1.30	1.62	-0.31	1.22	-0.06	0.91	-0.11	1.29	-0.13
0.139	1.85	0.30	1.50	0.02	1.33	-0.30	0.77	0.10	1.30	-0.21
0.143	1.15	0.00	1.45	0.11	1.37	-0.22	0.76	0.09	1.30	-0.23
0.156	0.85	-0.84	1.20	0.19	1.45	-0.03	0.81	-0.04	1.27	-0.26
0.167	1.50	-1.40	0.97	0.06	1.50	0.11	0.87	0.01	1.25	-0.19
0.173	2.05	1.15	0.82	-0.04	1.52	0.15	0.92	0.04	1.23	-0.14
0.179	2.05	0.86	0.65	-0.16	1.55	0.18	0.98	0.00	1.20	-0.05
0.195	2.30	0.10	0.70	-0.36	1.60	0.24	1.07	-0.24	1.14	0.08
0.200	2.26	-0.10	0.75	-0.16	1.60	0.18	1.10	-0.28	1.10	0.09
0.208	2.15	-0.38	0.80	0.42	1.60	-0.02	1.15	-0.23	1.05	0.12
0.223	1.68	-0.86	0.90	0.40	1.58	-0.17	1.23	-0.05	0.96	0.14
0.250	0.02	0.60	1.05	0.07	1.49	-0.12	1.28	-0.14	0.98	0.26
0.260	0.10	1.20	1.10	-0.04	1.45	-0.07	1.30	-0.14	1.06	0.30
0.278	1.75	0.88	1.18	-0.14	1.31	-0.14	1.26	0.08	1.21	0.19
0.312	2.30	0.31	1.28	-0.11	1.00	-0.37	0.98	0.32	1.39	0.12
0.333	2.40	0.00	1.30	-0.06	0.85	-0.22	0.85	0.22	1.48	0.18
0.357	2.30	-0.32	1.27	-0.10	0.93	0.14	0.96	0.12	1.57	0.18
0.390	1.90	-0.65	1.27	-0.23	1.10	0.42	1.19	0.16	1.63	0.02
0.417	1.40	-0.92	1.13	-0.31	1.21	0.39	1.36	0.25	1.65	-0.11
0.469	0.45	-1.32	0.85	-0.23	1.47	0.21	1.63	0.24	1.55	-0.12
0.500	0.05	0.07	0.64	-0.02	1.48	0.12	1.75	0.16	1.47	-0.01
0.520	0.35	1.16	0.77	0.26	1.49	0.09	1.77	0.11	1.40	0.02
0.536	0.60	1.16	0.75	0.42	1.48	0.07	1.79	0.06	1.40	0.02
0.555	0.90	1.10	0.90	0.56	1.40	0.06	1.80	-0.02	1.45	-0.02
0.625	1.50	0.93	1.30	0.64	1.30	0.10	1.66	-0.17	1.80	-0.20
0.714	2.00	0.62	1.75	0.50	1.45	0.18	1.47	-0.20	1.58	-0.34
0.750	2.15	0.52	1.90	0.43	1.55	0.19	1.40	-0.18	1.48	-0.38
0.780	2.25	0.43	2.00	0.37	1.60	0.18	1.37	-0.16	1.40	-0.40
0.833	2.35	0.30	2.12	0.26	1.70	0.14	1.39	-0.12	1.33	-0.40
0.940	2.50	0.10	2.28	0.09	1.90	0.04	1.45	-0.08	1.18	-0.31
1.000	2.50	0.00	2.32	0.00	1.96	-0.02	1.53	-0.10	1.19	-0.27
1.070	2.48	-0.10	2.15	-0.11	1.95	-0.10	1.43	-0.12	1.15	-0.24
1.250	2.37	-0.31	2.26	-0.30	1.97	-0.27	1.67	-0.24	1.31	-0.23
1.390	2.25	-0.44	2.17	-0.42	1.98	-0.38	1.71	-0.33	1.40	-0.28
1.500	2.16	-0.52	2.09	-0.54	1.94	-0.46	1.71	-0.40	1.44	-0.34
1.560	2.11	-0.56	2.05	-0.55	1.90	-0.53	1.69	-0.47	1.44	-0.38
1.670	2.01	-0.63	1.95	-0.61	1.84	-0.56	1.67	-0.50	1.45	-0.42
1.790	1.92	-0.69	1.88	-0.67	1.78	-0.63	1.63	-0.56	1.44	-0.47
1.870	1.85	-0.72	1.82	-0.71	1.73	-0.67	1.60	-0.60	1.43	-0.51
2.080	1.71	-0.81	1.69	-0.79	1.62	-0.75	1.52	-0.69	1.39	-0.60
2.500	1.47	-0.94	1.45	-0.92	1.41	-0.89	1.35	-0.83	1.27	-0.75
3.120	1.20	-1.06	1.19	-1.05	1.17	-1.02	1.14	-0.97	1.09	-0.90
3.750	1.02	-1.15	1.01	-1.13	1.00	-1.11	0.98	-1.06	0.95	-1.00
4.170	0.92	-1.19	0.91	-1.18	0.91	-1.15	0.89	-1.11	0.87	-1.06
5.000	0.77	-1.24	0.77	-1.24	0.76	-1.21	0.76	-1.18	0.75	-1.13
6.250	0.64	-1.32	0.64	-1.31	0.63	-1.29	0.62	-1.26	0.61	-1.23
7.500	0.57	-1.40	0.54	-1.35	0.50	-1.33	0.50	-1.31	0.50	-1.28
12.500	0.33	-1.44	0.30	-1.44	0.29	-1.43	0.29	-1.41	0.29	-1.39

[a] Papadakis and Fowler (1971), used by permission of the American Institute of Physics.

| | 0.525 | | 0.625 | | 0.725 | | 0.825 | | 0.925 |
C	δ	C	δ	C	δ	C	δ	C	δ
1.14	−0.13	1.15	0.28	1.45	−0.10	1.20	−0.38	0.76	−0.16
1.13	−0.04	1.20	0.16	1.46	−0.16	1.18	−0.18	0.76	−0.12
1.13	0.02	1.23	0.06	1.47	−0.16	1.17	−0.23	0.77	−0.27
1.14	0.04	1.24	0.05	1.47	−0.18	1.16	−0.28	0.76	−0.28
1.15	0.11	1.28	0.04	1.47	−0.23	1.15	−0.37	0.76	−0.28
1.20	0.16	1.31	0.00	1.47	−0.22	1.15	−0.38	0.76	−0.29
1.22	0.13	1.33	−0.03	1.45	−0.19	1.14	−0.38	0.75	−0.30
1.25	0.12	1.35	−0.05	1.45	−0.17	1.12	−0.37	0.75	−0.30
1.32	0.21	1.39	−0.06	1.44	−0.17	1.10	−0.38	0.75	−0.31
1.35	0.22	1.40	−0.06	1.42	−0.19	1.10	−0.38	0.74	−0.30
1.38	0.16	1.42	−0.05	1.41	−0.21	1.08	−0.37	0.74	−0.31
1.43	0.09	1.45	−0.06	1.40	−0.23	1.06	−0.32	0.72	−0.30
1.51	0.08	1.47	−0.17	1.35	−0.30	1.00	−0.34	0.72	−0.23
1.54	0.01	1.47	−0.24	1.29	−0.24	1.00	−0.38	0.72	−0.24
1.59	−0.01	1.44	−0.13	1.29	−0.37	0.95	−0.28	0.72	−0.22
1.65	−0.01	1.33	−0.27	1.23	−0.24	0.90	−0.39	0.73	−0.17
1.67	−0.12	1.26	−0.18	1.20	−0.36	0.87	−0.30	0.75	−0.23
1.65	−0.16	1.23	−0.20	1.18	−0.38	0.84	−0.28	0.81	−0.28
1.58	−0.08	1.26	−0.34	1.19	−0.26	0.80	−0.42	0.76	−0.13
1.50	−0.08	1.31	−0.33	1.16	−0.34	0.82	−0.25	0.77	−0.30
1.42	−0.26	1.38	−0.33	1.03	−0.41	0.83	−0.34	0.77	−0.12
1.43	−0.32	1.38	−0.28	1.00	−0.30	0.83	−0.42	0.76	−0.14
1.43	−0.32	1.33	−0.35	0.90	−0.26	0.84	−0.40	0.71	−0.24
1.42	−0.30	1.29	−0.39	0.88	−0.26	0.83	−0.34	0.63	−0.30
1.40	−0.25	1.23	−0.43	0.90	−0.31	0.85	−0.25	0.64	−0.34
1.36	−0.28	1.05	−0.38	1.10	−0.46	0.88	−0.26	0.60	−0.10
1.46	−0.30	1.13	−0.26	0.85	−0.35	0.80	−0.44	0.72	−0.23
1.46	−0.36	1.20	−0.27	0.85	−0.28	0.75	−0.41	0.74	−0.30
1.45	−0.40	1.23	−0.30	0.90	−0.25	0.70	−0.36	0.68	−0.34
1.33	−0.47	1.22	−0.38	0.94	−0.26	0.69	−0.35	0.60	−0.34
1.12	−0.50	1.10	−0.51	1.07	−0.38	0.84	−0.28	0.55	−0.08
1.06	−0.48	1.05	−0.56	1.01	−0.46	0.86	−0.26	0.65	−0.06
0.90	−0.42	0.85	−0.56	1.05	−0.53	1.10	−0.35	1.10	−0.10
1.00	−0.28	0.79	−0.41	0.70	−0.52	0.69	−0.50	0.69	−0.34
1.09	−0.24	0.83	−0.28	0.67	−0.40	0.62	−0.49	0.62	−0.45
1.15	−0.26	0.88	−0.24	0.67	−0.28	0.56	−0.39	0.54	−0.44
1.17	−0.31	0.90	−0.28	0.69	−0.25	0.55	−0.30	0.51	−0.39
1.21	−0.32	0.96	−0.24	0.74	−0.19	0.57	−0.21	0.47	−0.28
1.23	−0.37	1.00	−0.27	0.79	−0.18	0.60	−0.13	0.47	−0.15
1.23	−0.41	1.03	−0.29	0.82	−0.19	0.64	−0.11	0.49	−0.08
1.23	−0.49	1.06	−0.37	0 89	−0.24	0.72	−0.11	0.56	−0.01
1.17	−0.65	0.98	−0.53	0.94	−0.39	0.82	−0.24	0.69	−0.08
1.04	−0.81	0.98	−0.71	0.91	−0.59	0.83	−0.45	0.76	−0.30
0.92	−0.93	0.89	−0.84	0.84	−0.73	0.79	−0.61	0.74	−0.48
0.85	−0.99	0.82	−0.91	0.79	−0.81	0.75	−0.70	0.71	−0.58
0.73	−1.08	0.71	−1.01	0.70	−0.93	0.67	−0.83	0.65	−0.73
0.60	−1.18	0.60	−1.12	0.59	−1.06	0.59	−0.98	0.58	−0.90
0.50	−1.24	0.50	−1.19	0.50	−1.14	0.50	−1.07	0.50	−1.00
0.29	−1.37	0.29	−1.34	0.29	−1.31	0.29	−1.27	0.29	−1.23

—4—

Elastic Surface Wave Devices[1]

J. DE KLERK

Westinghouse Research Laboratories,
Pittsburgh, Pennsylvania

I. Introduction 213
II. Phase Coded Signals 215
III. 13 Bit Barker Code Correlator 223
IV. Programmable Sequence Generator 231
V. Pulse Compression Filters 236
References 242

I. Introduction

Elastic surface wave filters are capable of performing many of the complex functions used in modern radar, ECM, and communications systems. In pulse compression radar, for example, a long coded pulse is transmitted and the received signal is processed to compress it into a relatively narrow pulse. In this way large average powers can be transmitted, which avoids the need for transmitting high peak powers. By this means, long range high target resolution can be achieved. The method of compression is determined by the type of code used in the long transmitted pulse. Codes employed can be phase modulated, frequency modulated, or time-frequency modulated. All of these codes enable the radar receivers to operate in "noisy" electromagnetic environments, by discriminating against all signals other than those containing the desired code. The discriminating element is a filter which matches the transmitted code.

A coded signal may be represented by either the time response $f(t)$ or the frequency response $F(\omega)$ of the encoding filter. One method of generating a coded signal is by applying an electrical impulse to the encoding filter as shown in Fig. 1. The return signal reflected from a target is down converted to the I.F. frequency and fed to the matched filter before amplification, detection, and display. The frequency response of the matched filter is the complex conjugate $F^*(\omega)$ of the encoding filter response. The matched filter output is the desired compressed pulse, which can be represented by the inverse

[1] This work was supported in part by the Office of Naval Research.

213

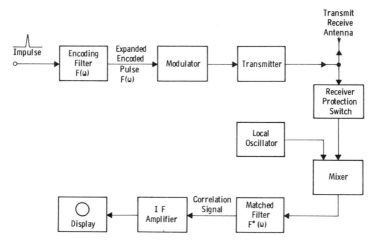

FIG. 1. Encoding and decoding by matched filters.

Fourier transform of the product of the matched filter frequency response and the signal input response:

$$g(t) = (1/2\pi)\int_{-\infty}^{+\infty} |F(\omega)|^2\, e^{i\omega t}\, d\omega. \tag{1}$$

Another way of representing the output of the matched filter is by using the time response $f(t)$ of the encoding filter. The same filter can be used for encoding and decoding, or expansion and compression, if the time inverse of the return signal is applied to the matched filter as shown in Fig. 2. The matched

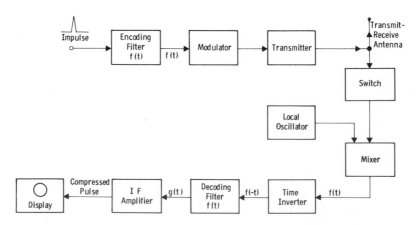

FIG. 2. Encoding and decoding by same filter using time inversion.

filter output can then be expressed by the convolution of the signal with the impulse response of the matched filter:

$$g(t) = \int_{-\infty}^{+\infty} f(T)f(-T)\, dT. \tag{2}$$

II. Phase Coded Signals

A long phase coded pulse consists of a number of subpulses all of equal length and amplitude. The phase of each subpulse is determined by the sequence of the chosen code. The most frequently used codes employ binary phase coding, and consist of sequences of rf bursts, at constant frequency, phased so that a 180° shift occurs each time the sign of the subpulse is changed. This is illustrated in Fig. 3. Any signal returning from a target is

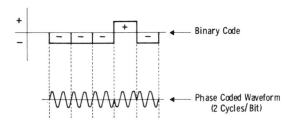

FIG. 3. Binary phase coded signal.

compressed or decoded by means of a matched filter. The compression ratio is identical to the number of subpulses or bits in the transmitted phase code. The width of the compressed pulse, and hence the range resolution, is equal to the length of one subpulse or bit.

A special group of binary phase codes, which allow the maximum peak compressed pulse amplitude to sidelobe ratio of $N{:}1$, where N is the number of bits in the code, was developed by Barker (1953). There are only seven codes in this group, the maximum value of N being 13, which provides a peak-to-sidelobe ratio of 22.3 dB and a compression ratio of 13, where the width of the peak is measured at half amplitude. Longer codes, which do not fall into the Barker group, would provide a greater compression ratio but a lower peak to sidelobe ratio.

Phase coded surface elastic waves can readily be generated by means of interdigital transducers (de Klerk, 1971a) as illustrated in Fig. 4. The code generated in this case would be the 5 bit code of Fig. 3. The 180° phase reversal is achieved by connecting two adjacent fingers of the I.D. structure to the same comb rail. As these connected fingers are at the same electric potential, no elastic strains will be generated between them. The field components generated in the substrate below the I.D. grid at the phase change location is illustrated in Fig. 5.

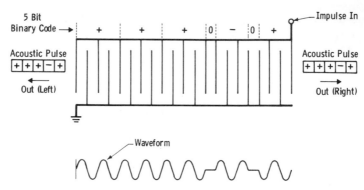

FIG. 4. Generation of phase coded elastic surface waves (5 bit Barker code).

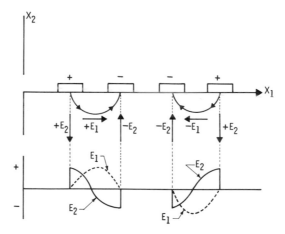

FIG. 5. E field components generated in the substrate below the I.D. grid at the phase change.

If an electrical impulse is applied to the I.D. comb filter of Fig. 4, surface acoustic pulses will propagate along the surface in both directions normal to the fingers. The phase codes of these two acoustic pulses are shown in block form to the left or to the right of the "encoding filter." The leading edge of the pulse propagating to the left would generate the electromagnetic signal, in a single finger pair detector, plotted directly below on the left in Fig. 4. Similar plots are shown for the pulse propagating to the right. It will be seen that the electric signal on the left is the time reversal of that on the right. Furthermore, timewise the last two bits on the left and the first two on the right have each lost one half cycle due to this method of acoustic phase reversal. In contrast, an electronically generated code would have its full complement of cycles per bit. Both encoded signals can be used with the phase coded filter shown in Fig. 4.

The decoding or correlating filter operates in the reverse order to that of the encoding filter. The phase coded electromagnetic signal is applied to a launcher I.D. grid, usually consisting of a few pairs or even a single pair of fingers. The resultant elastic waves will propagate in both directions away from the launcher grid. The backward propagating wave can be damped by means of some wax on the substrate surface behind the launcher grid. As the forward propagating surface elastic wave passes under the decoding or correlating I.D. grid structure, it will generate electric signals in this structure. Figure 6 illustrates the operation of the decoding or correlating filter. The electrical input to the launcher grid and the layout of the device are shown in Fig. 6a. Figure 6b shows the progress of the elastic waves in discreet steps under the correlating grid structure and the total electrical output from this grid at each step. Figure 6c shows the ideal envelope of the correlation signal. The signal-to-noise or peak-to-sidelobe ratio is 5:1 or 14 dB. This figure also shows that the peak width at half amplitude is just one bit width long yielding a compression ratio of 5:1, as stated above.

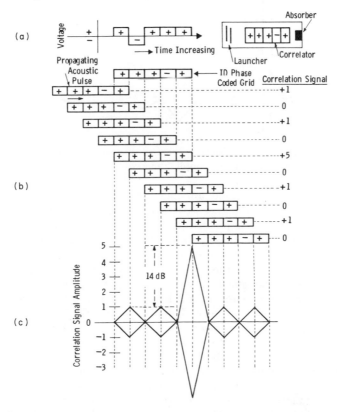

Fig. 6. Generation of a correlation signal by a phase coded I.D. grid (a), as the surface wave propagates below the grid (b), resulting in the idealized signal (c).

In this analysis it was assumed that no reflections of the elastic waves or regeneration (de Klerk, 1971a) took place at the fingers of the interdigital grid structure. This assumption is approximately valid for low coupling materials such as quartz, but is invalid for high and medium coupling materials such as lithium niobate and bismuth germanium oxide. By including regeneration, due to the back piezoelectric effect, for the 5 bit sequence used in Fig. 6, the output of the correlator grid structure on lithium niobate, assuming a regeneration factor of −0.2 for characteristic impedance termination, would be that shown in Fig. 7a. The interference between the launched phase coded

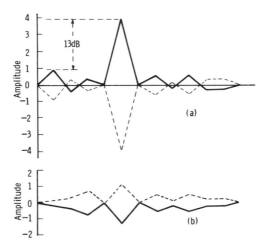

FIG. 7. Degradation of correlation signal due to second-order effects such as reflections and regeneration in high coupling materials.

elastic signal and regenerated antiphase elastic waves due to interaction between the signal and the grid structure reduces the peak-to-sidelobe ratio to 4.6 or 13 dB and the compressed peak amplitude by 1 dB. The regenerated signal, always 180° out of phase with the launched signal, is shown in Fig. 7b. Table I gives the analytical correlation grid signal outputs for Figs. 6 and 7. For longer code sequences, the interference would be greater.

These second-order effects have been computed (Jones *et al.*, 1972) for a 13 bit Barker code correlator on YZ lithium niobate by a different method, using equivalent circuit models. Their experimental 13 bit Barker code derived by applying an impulse to an interdigital encoder sampled every half wavelength and detected by an I.D. grid with half the number of finger pairs contained in one whole bit, for forward and reverse directions is shown in Fig. 8. The computed responses for both directions are given in Fig. 9. These experimental and computed impulse responses show remarkable qualitative similarities. However, comparison of theory with experiment for the correlated signals, shows the existence of fairly large quantitative discrepancies. Theoretical and experimental values of peak-to-sidelobe ratios

TABLE I

SIGNAL AMPLITUDES IN 5 BIT BARKER DECODER AS A FUNCTION OF BIT
COINCIDENCE NUMBER

Bit coincidence number	Correlation grid output		
	Ideal (no interference) Fig. 6c	Regenerated signal phase shifted (π) Fig. 7b	Degraded signal Fig. 7a
0	0	0	0
1	1	−0.16	+0.84
2	0	−0.32	−0.32
3	1	−0.64	+0.36
4	0	0	0
5	5	−1.12	+3.88
6	0	0	0
7	1	−0.45	+0.55
8	0	−0.13	−0.13
9	1	−0.42	+0.58
10	0	−0.26	−0.26
11	0	−0.26	−0.26
12	0	0	0

reported by Jones *et al.* (1972) are given in Table II. Columns 5 and 6 reveal substantial differences between theory and experiment. While the measured values of peak-to-sidelobe ratio for identical conditions except for direction of propagation, i.e. the order of the code sequence, differ by only 1 dB, computed values differ by 7 dB. These figures show that the equivalent circuit model used was inaccurate, and requires modification of the assumptions made for reflection and regeneration.

Some of the surface wave power contained in a wave incident upon an interdigital grid will be reflected back due to the periodic change in acoustic impedance of the substrate below and between fingers. As the velocity *below* the metal fingers is lower than that *between* the fingers, due to the effective shorting of the piezoelectric modulus by the conducting layer (Campbell and Jones, 1968), the acoustic impedance ρv differs in these two locations. This is illustrated in Fig. 10. Using Campbell and Jones' values of $v_1 = 3.487 \times 10^5$ cm/sec^{-1} and $v_2 = 3.401 \times 10^5$ cm/sec^{-1} for YZ LiNbO$_3$ and taking the density of LiNbO$_3$ substrate as 4.7, the values of Z_1 and Z_2 are

$$Z_1 = 16.39 \quad \text{and} \quad Z_2 = 15.98 \quad \text{gm cm}^{-2}\text{sec}^{-1} \times 10^5.$$

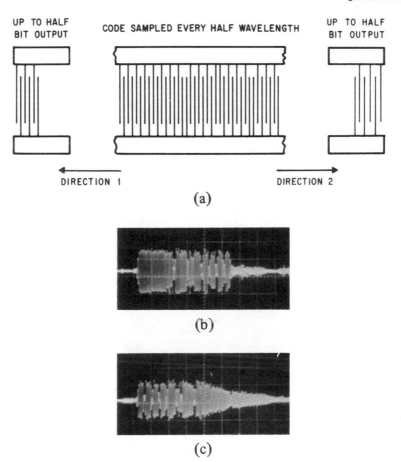

FIG. 8. Impulse responses of Barker code filter in forward and reverse directions (after Jones *et al.*, 1972).

Using these values the reflection coefficient at the edges of the fingers can be calculated:

$$R_{12} = \frac{Z_2 - Z_1}{Z_2 + Z_1} = \frac{-0.41}{32.37} = -1.26 \times 10^{-2},$$

$$R_{21} = \frac{Z_1 - Z_2}{Z_1 + Z_2} = +1.26 \times 10^{-2}.$$

Hence 1.26% of the energy is reflected when propagating from the unmetallized surface to the metallized surface, with a phase change of π rad. When the forward propagating wave emerges from below the metallized surface 1.26% is reflected back without any phase change. The resultant reflected

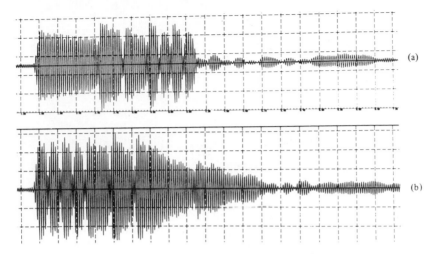

Fig. 9. Computer impulse response for both directions of propagation (after Jones *et al.*, 1972).

Table II

Peak-to-Sidelobe Ratios for 13 Bit Barker Code Filters on LiNbO$_3$[a]

Number of sampling points		Mode of code gen.	Propagation direction	Theory (dB)	Expt. (dB)	Jones *et al.* Fig. No.
Correlator grid	Detector grid					
N[b]	N/2	Electronic (perfect)	1	20	18	13
N	N/2	Electronic (perfect)	2	13	17	14
N	N/2	Surface wave impulse	2	10	13	15
1/N	N	Surface wave impulse	1	c	19.6	16, 17

[a] After Jones *et al.* (1972).
[b] N = number of cycles per bit.
[c] No theoretical value reported.

wave can be regarded as having two components, viz. those reflected from two interlaced diffraction gratings. One grating would comprise the π phase shift edges and the other the zero phase drift edges. The component reflected from the π grating is advanced in phase by π rad while the component reflected from the zero grating is retarded by π due to the extra forward and

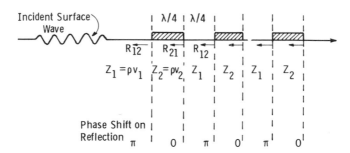

FIG. 10. Reflections at acoustic impedance mismatch due to velocity changes below interdigital grating.

backward propagation distance of $\lambda/2$ under the metal finger. The two components are thus in phase and will propagate as a backward reflected wave.

Regeneration, as distinct from reflection, occurs because both direct and inverse piezoelectric effects operate simultaneously as the surface wave propagates under the I.D. grid structure. As the first half cycle arrives under the first finger pair the direct piezoelectric effect causes electric polarization in the form of bound charges in the piezoelectric medium. This polarization induces free electric charges of opposite polarity on the metal fingers, which charges become instantaneously distributed over the whole metal grid structure. Thus, due to the inverse piezoelectric effect, another surface wave is launched under the complete grid and propagates in both directions. The regenerated wave is out of phase with the incident surface waves by π rad. While the incident surface wave propagates under the I.D. grid, new waves will be regenerated and will propagate in both directions. For the case of a phase coded I.D. grid, the backward wave propagates with the inverse phase code and the forward wave propagates with the code of the incident wave. These regenerated waves interfere with the incident wave and cause distortion of the electric output signal from the I.D. grid. The amount of distortion is strongly dependent upon the electromechanical coupling of the piezoelectric material and the electrical termination used.

The influence of both electrical termination and electromechanical coupling on reflection and regeneration[2] (de Klerk, 1971a) is shown in Fig. 11. When the capacitance of the reflecting grid is shunted by a variable inductance, the value of the latter can be chosen to resonate at the driving frequency. With this critical termination, much more energy is regenerated than when the " reflecting " grid is terminated by the characteristic impedance or left open-circuited. Reflection coefficient measurements were made for lithium niobate and for quartz with all three terminations. The lithium niobate curve shows that only 10 finger pairs are required to get the maximum reflection coefficient of 0.82 when critically terminated as described above.

[2] These measurements were made by M. R. Daniel at Westinghouse Research Laboratories.

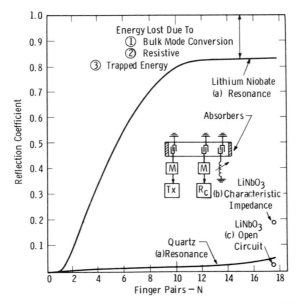

Fig. 11. Reflection coefficient of surface waves on LiNbO₃ and SiO₂ as a function of the number of finger pairs in the reflecting I.D. grid for various electrical terminations.

The remaining 18% can be accounted for by conversion to bulk modes, regeneration in the forward direction from the trailing end of the I.D. grid, resistive heating of the metal fingers, and energy trapped under the grid due to multiple reflections from the fingers.

Reference to Fig. 11 will confirm that less than 20% of the incident energy is reflected back when the "reflecting" grid consisting of 17 finger pairs is terminated by the characteristic impedance of this I.D. grid on lithium niobate. When this grid is left at open circuit, less than 2% of the incident energy is regenerated and reflected. Hence, "mass loading" due to the interdigital grid was less than 2%.

III. 13 Bit Barker Code Correlator

A 30 MHz, 13 bit, 14 cycles/bit Barker code correlator which is very stable and precise in frequency was developed at Westinghouse Research Laboratories and is being mass produced for three different radar systems at Westinghouse Aerospace and Electronic Systems Division (Thomas *et al.*, 1972). This correlator was developed to replace a bulk wave correlator consisting of 13 bulk delay lines for radar pulse compression. All three radar systems for which the correlator was developed require a correlation peak-to-maximum-sidelobe ratio of at least 21 dB, while the theoretical maximum peak-to-sidelobe ratio for a 13 bit Barker code is 22.3 dB. In a practical device the 21 dB ratio can only be achieved by very careful fabrication of transducers with very accurate amplitude, frequency, and phase responses. The required 21 dB ratio had to be maintained over the temperature range −40°C to +90°C. For this reason the correlator was fabricated on ST cut quartz and required no temperature compensation over the whole temperature range.

In the radar systems which use this correlator or decoder, the Barker code is generated by a conventional electronic digital phase encoder which operates independently of the surface wave decoder. Hence, it was necessary to fabricate the surface wave decoders to operate at a precisely controlled center frequency. In practice, it was found that in order to obtain a 21 dB ratio of correlation peak to maximum sidelobes, the center frequency of the decoder must be within approximately 8 kHz of the frequency of the encoder. This can be seen in Fig. 12, which shows the response at the center frequency

(a)

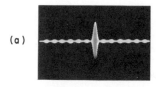

13 Bit Barker at f_c

(b)

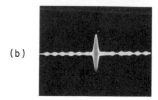

13 Bit Barker at f_c + 9 kHz

Fig. 12. Correlation peak-to-sidelobe ratio for 13 bit Barker code decoder at (a) center frequency f_c and at (b) f_c + 9 kHz (after Thomas *et al.*, 1972).

and at 9 kHz away from the center frequency. The correlation peak does not decrease significantly for this frequency deviation, but the valley between the correlation peak and the first sidelobe on either side increases until it is higher than the first sidelobe, thus decreasing the peak-to-sidelobe ratio to below 21 dB.

The manner in which the correlation signal deteriorates with even further frequency deviation is shown in Fig. 13. The response for frequency increments of approximately 12 kHz is shown, starting from center frequency at the bottom. The sidelobes increase rapidly and the main lobe decreases

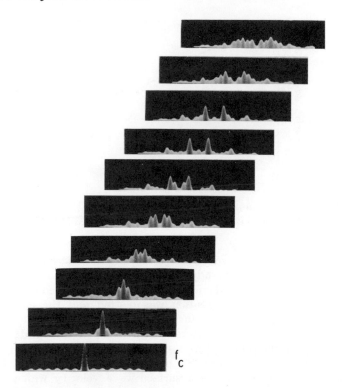

FIG. 13. Response of surface wave 13 bit Barker decoder at f_c ($=30$ MHz) and at frequency increments of 12 kHz (after Thomas *et al.*, 1972).

until the peak-to-sidelobe ratio reaches unity at approximately 35 kHz deviation from the center frequency.

Figure 14 shows the variation of peak-to-sidelobe ratio as a function of bit-to-bit phase errors. The 21 dB level occurs when the phase error is approximately 10 deg. Phase error can be controlled by proper care in making the original artwork, but frequency errors depend on the size of the transducers, which in turn depends on the amount of photoreduction in making the mask for etching the interdigital transducers.

In order to obtain a device with a center frequency within 8 kHz of the desired center frequency, the following procedure was used. The original mask was made on a computer-controlled plotter at 20 times the estimated final size. Since the velocity of surface waves on the quartz used was not known to the desired accuracy, the exact photoreduction could not be determined without making devices and measuring the center frequency. The camera was set up for a reduction of 20:1 and a series of photoreductions was made near this setting by moving the camera mount in 5 mil increments to produce a slightly larger and smaller reductions. Care was taken to control

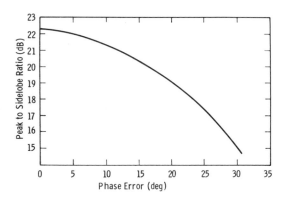

Fig. 14. Degradation of correlated Barker response with respect to bit-to-bit phase errors (after Thomas *et al.*, 1972).

the photoreduction process to yield a series of photomasks with an equal frequency difference between adjacent photoreductions in the series. By measuring the frequency of a device made from each photoreduction, a graph of frequency versus camera position was plotted and the exact camera position for the desired center frequency was taken from this plot.

In order to produce correlators which maintain the 21 dB peak-to-sidelobe ratio over the temperature range −40°C to +90°C without using a constant temperature enclosure, ST cut X propagating natural Brazilian quartz was used because it has a surface wave velocity which is nearly constant over this temperature range. This cut is known as a +42.75° rotated Y cut in "right-handed" quartz, as indicated in Fig. 15. In Fig. 16 is plotted the frequency of best correlation as a function of temperature for a typical device made with ST cut X propagating natural Brazilian quartz. The graph shows that the peak-to-sidelobe ratio remains above 21 dB over a temperature range of −60°C to +100°C which is a greater temperature range than is required. It was found that the elastic constants, and hence the surface wave velocity, of natural Brazilian and synthetic quartz differed so significantly that the photomask made for natural Brazilian quartz could not be used on synthetic quartz for fabricating correlators with the desired center frequency accuracy.

The correlators are made with 5000 Å thick aluminum transducers, etched using a solution of nitric, acetic, and phosphoric acids. Aluminum was chosen because its acoustic impedance is near that of quartz, which is necessary to prevent excessive reflections from the transducer fingers.

The photoresist is filtered as it is applied to the quartz substrate to prevent imperfections in the pattern due to impurities in the photoresist. Occasional shorts between fingers are eliminated by applying a small drop of etching solution to the area of the short and, with proper care, the short is removed before the remaining fingers are damaged by undercutting. Exposures of the photoresist are made through a high resolution photographic

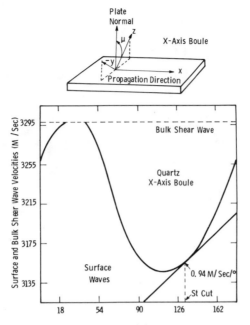

FIG. 15. Location of axes for 42.75° rotated Y cut quartz with respect to substrate surface (after Slobodnik and Conway, 1970).

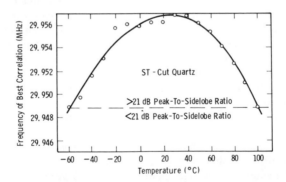

FIG. 16. Frequency of best correlation as a function of temperature (after Thomas *et al.*, 1972).

mask with collimated light. A 0.005 in separation is maintained between the mask and the substrate in order to prevent scratching the photoresist or the photographic emulsion on the mask.

A photograph of the decoder which uses 7 finger pairs per bit is shown in Fig. 17, together with certain important characteristics. The quartz substrate is glued to a printed circuit board to aid in assembly into a package.

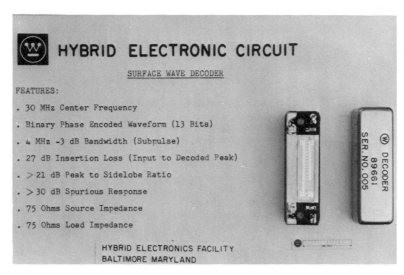

FIG. 17. Photograph of Barker decoder with some characteristics (after Thomas *et al.*, 1972).

Gold leads are bonded to the aluminum transducer pads using a thermocompression technique in which the bond area of the lead is repeatedly tacked to the pad using a hot capillary loaded gold wire. The other end of the lead is bonded to the printed circuit board which is, in turn, connected to hermetically sealed pins mounted in the package. Torroidal transformers and miniature tunable capacitors are used at the input and output to match the impedance of the device to 75 ohms. A metal shield is placed above the substrate to prevent electromagnetic feedthrough from input to output, which would produce a spurious signal. A metal lid is finally solder-sealed to the package to prevent long term degradation of the performance due to corrosion of the aluminum or condensation of water vapor on the device surface.

The errors inherent in this manufacturing process were estimated in order to determine whether it would be possible to produce correlators at the correct frequency with a high yield. There are several possible sources of error, which are listed in Table III. They are classified either as errors in surface wave velocity or errors in the size of the transducer pattern.

Velocity errors are due to misorientation of the transducer pattern with respect to the crystal axes. Since the pattern is registered with respect to a reference edge of the substrate, an error in the orientation of the pattern or the crystal axes with respect to this edge can cause velocity and hence frequency errors.

The change in velocity for small changes in angle between the propagation direction and the X axis is shown in Fig. 18. The curve is nearly para-

TABLE III

SOURCES OF FREQUENCY ERROR

Velocity errors
1. X axis misorientation
2. Mask misregistration
3. Substrate normal misorientation
Size errors
1. Thermal expansion of mask
2. Projection height changes

bolic up to several degrees change and the velocity is given by the expression

$$\Delta v = 0.22\ \theta^2, \tag{3}$$

where θ is in degrees and v is in meters per second. Hence, the frequency error due to misalignment of the transducer axis with the X axis is given by the expression

$$\Delta \nu = 0.22\ \theta^2/\lambda_s \tag{4}$$

where λ_s is the acoustic wavelength.

Suppliers of quartz crystals can align a reference edge parallel to the X axis within 10 min of arc and the quartz substrate and the photomask can

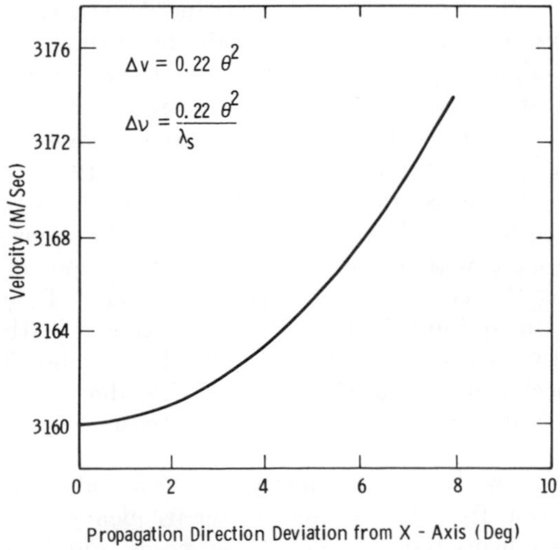

FIG. 18. Surface wave velocity as a function of propagation angle near the X-axis of ST cut quartz (after Thomas *et al.*, 1972).

be oriented with respect to the reference edge to within approximately 2 min of arc using a specially designed alignment fixture. Substituting these angular errors into the equation, it will be seen that the maximum frequency error due to X axis and mask misalignments is 84 Hz.

A more serious frequency error results from the deviation of the normal to the crystal face from the desired direction. The surface wave velocity as a function of the direction of the plate normal is shown in Fig. 15 (Slobodnik and Conway). ST cut quartz corresponds to an angle of 132.75° on this curve. The slope of velocity as a function of angle at this orientation is approximately 0.94 m/sec per angular degree. Crystal manufacturers can hold the face normal to $\pm$ 15 min of arc for substrates cut from different crystals and to less than 15 min for substrates cut from the same crystal. This angular error corresponds to a frequency error of 2.3 kHz. This error is acceptable assuming that other error sources do not increase the total error to the 8 kHz limit.

In addition to velocity errors, there are errors due to incorrect size of the transducer pattern. Differential thermal expansion between the photomask and the substrate is one possible source of size error. For high resolution glass photographic plates, the frequency error is only 24 Hz/°C temperature difference between the mask and the substrate. The temperature of the mask will not deviate greatly from that of the substrate due to their close proximity. Hence, this error is a minor one.

Another minor frequency error is due to slight variations in the size of the pattern because the photomask is not in direct contact with the quartz substrate during the exposing of the photoresist to the light source. However, using carefully collimated light, the maximum error due to changes in height of the mask above the substrate has been reduced to an estimated 33 Hz.

In order to verify that all errors are small except those due to orientation of the reference edge with respect to the crystal axes by the manufacturer of the quartz substrate, a long substrate was cut in half and a decoder was fabricated on each piece using the same reference edge for aligning both substrates with respect to the photomask. Thus, the substrates had the same surface normal and X axis orientation with respect to the transducers. The only differences between the decoders were possible differences in orientation of the reference edge with respect to the mask and possible temperature differences during the two exposures. The frequencies of the two decoders were measured with equipment which had a resolution of 1 kHz. No difference in frequency between these two decoders could be found. Thus, the main errors to be expected during production are those due to misorientation of the face normal, and they should be within approximately 2–3 kHz as indicated earlier.

At the time of writing approximately 1300 decoders have been fabricated. The deviation from the desired center frequency has been less than 2.5 kHz, which confirms the above error estimates. Furthermore, the yield so far has been close to 100%, which completely justifies the fabrication procedure adopted.

IV. Programmable Sequence Generator

Certain signal processing applications require the ability to change the
sequence of bits in a biphase code, instead of using the fixed code in the
Barker decoder just described. That this can readily be achieved useing
acoustic surface waves has been demonstrated (O'Clock *et al.*, 1971, 1972). In
this case a combination of acoustic surface waves and semiconductor device
technology was utilized to provide electronically switchable coding techniques
for acoustic surface wave signal processing devices.

Code flexibility was achieved by integrating the sequence generator
matched filter combination with a suitable array of semiconductor switching
elements. Phase control of the biphase code was achieved by using diode
transmission gates between each interdigital finger and the rails of ac sum
lines of the surface wave devices. The circuit is arranged as shown in Fig. 19.

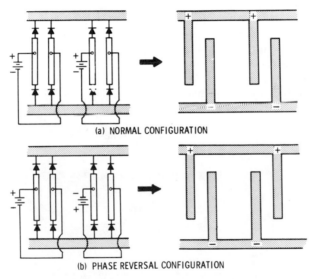

(a) NORMAL CONFIGURATION

(b) PHASE REVERSAL CONFIGURATION

FIG. 19. Programmable sequence generator—normal and phase-reversed diode
configurations (after O'Clock *et al.*, 1971).

The phase of any finger pair is controlled by the polarity of the dc bias
applied to four diodes as shown. Each finger is connected to a rail at each end
via a switching diode. By reversing the bias applied to a finger pair, the phase
of the finger pair is changed by 180°.

Several different semiconductor device technologies could be employed
to construct transmission gates. For instance, bipolar transistors will pass
rf, have very short switching time, and are readily available in a variety of
configurations, including integrated circuit arrays at reasonable cost. Unfortunately, however, their forward resistance is rather large and this type of
device uses relatively large amounts of power. Monolithic MOS transistors on

silicon are also economical but have large capacitances which will not effici-
ently pass signals in the vhf range. Complementary MOS transistors in
silicon-on-sapphire (SOS) format overcome the disadvantages of monolithic
MOS transistors. Although they will pass the rf and present a suitably large
impedance in the " off " state, they exhibit large " on " resistance as do other
transistors. One class of integrated circuit device *does* possess the necessary
frequency response, low capacitance, and low forward resistance, viz. diodes
in the MOS/SOS format. These diodes have a forward resistance of approximately 10 ohms at 10 mA bias current. Other types, particularly Schottky
barrier and $p–i–n$ diodes with beam leads, have attractive features which
make them suitable for use as transmission gates. Particularly well suited
are $p–i–n$ diodes with " on " resistance of approximately 5 ohms and extremely low capacitance. Further attractions are their reliability, ready availability, and low cost.

In their first reported sequence generator (O'Clock *et al.*, 1971) the
authors used beam lead diode switches as shown in Fig. 20. This hybrid
structure consisted of an acoustic surface wave tapped delay line on $LiNbO_3$,
where some of the interdigital transducer taps were connected through beam
lead diode switches to the rf sum lines. Each miniature diode was approximately 0.25 mm wide and 0.50 mm long, including leads. The complete
hybrid integrated-circuit package was approximately 3.2 cm by 2.5 cm. This

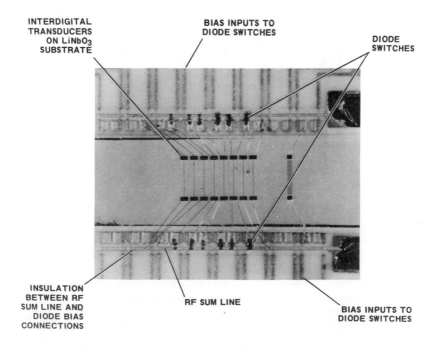

FIG. 20. Switchable sequence generator (after O'Clock *et al.*, 1971).

package incorporated the rf sum lines, bias pad connections for the diodes, and provision for mounting the acoustic delay line. Close examination of the hybrid circuit of Fig. 20. will reveal that some taps were connected to the rf sum lines through diode switches, while others were permanently wired directly to the rf sum lines. This arrangement allowed comparisons to be made between bits generated by taps that were permanently wired to the rf sum lines and those connected by diodes. Two types of diodes were used in this hybrid sequence generator. The smaller beam lead diodes in Fig. 20 were Hewlett Packard 2740 beam lead hot carrier diodes, while the larger beam lead diodes were Texas Instrument MD90 $p–i–n$ diodes.

The 8 bit sequence generator shown in Fig. 20 operated at 60 MHz with a 5 MHz bit rate or 12 cycles/bit. An electrical impulse, consisting of either a 0.2 μsec 60 MHz rf pulse or a 0.2 μsec video pulse was used as the impulse signal to launch the programmed biphase code. The launched surface acoustic signal was detected by an interdigital transducer half a bit in length. The theoretical insertion loss of this 8 bit switchable device was estimated to be 25 dB unmatched, whereas the measured value was 29 dB.

Approximately nine months after reporting their first device, the same authors (O'Clock *et al.*, 1972) reported two similar but much more sophisticated devices, viz. a 16 bit switchable and a 32 bit fixed sequence generator/correlator.

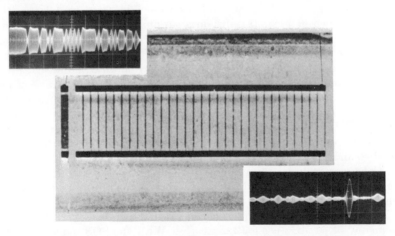

FIG. 21. 32 bit fixed sequence generator/correlator (after O'Clock *et al.*, 1972).

The fixed code sequence generator/correlator is shown in Fig. 21. This 32 bit device was fabricated on a quartz substrate and operated at 60 MHz. The upper left inset waveform shows the generated binary code while the lower right oscillogram shows the correlation signal output obtained from the correlator when the generated code was launched by the 1 bit long launcher grid on the left. The correlation peak-to-maximum sidelobe ratio was 12 dB.

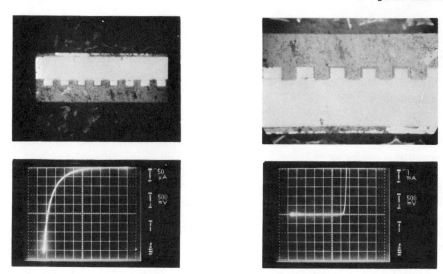

FIG. 22. SOS diodes and characteristics (after Allison *et al.*, 1969).

The switching diodes used in the 16 bit sequence generator/correlator are shown in Fig. 22. These SOS diodes were fabricated in a 1.7 μ thick silicon film grown heteroepitaxially on a sapphire substrate in the following manner (Allison *et al.*, 1969). Phosphorus was diffused into the entire wafer, resulting in n^+ silicon doped to about 10^{19} cm^{-3}. The wafer was next thermally oxidized, holes were etched in the oxide, and boron was diffused through the holes into the silicon. The silicon was next etched into arrays of eight p^+–n^+ diodes. Each diode was 0.008 in wide. The devices were then metallized with 1 μ of aluminum. The resulting diodes had a reverse voltage limited by Zener breakdown to -4 V, a forward bias resistance r_0 of 7 ohms, and a interelectrode capacity (C_p) of approximately 0.5 pF. Each diode was bonded to an interdigital finger on the acoustic delay line by a 0.025 mm Al wire. Care was required to minimize the length of the bonding leads as a 2 cm length of this wire could introduce an additional attenuation of 8 dB at vhf.

The 16 bit switchable sequence generator shown in Fig. 23 operated at a frequency of 60 MHz using 12 cycles/bit. With the SOS diode array used only 1.2 to 1.4 mW/bit was consumed. The entire 16 bit array used less than 20 mW. The processing loss between the 16 bit code input (shown in upper left inset) and the output correlation peak in the lower right inset was approximately 19 dB from the device as a correlator. The measured correlation peak-to-sidelobe ratio for this device was 11 dB compared to the theoretical maximum of 12 dB. Figure 24 shows four different codes generated by the 16 bit sequence generator of Fig. 23. The range of correlation peak-to-sidelobe ratios for the codes shown in Fig. 24 was found to be from 8 dB to 10 dB.

For the diode and switching arrangement used in the 16 bit sequence

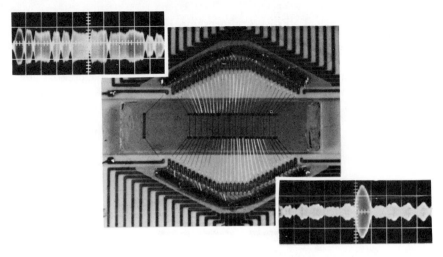

Fig. 23. 16 bit switchable sequence generator/correlator (after O'Clock *et al.*, 1972).

generator/correlator, it was found that isolation between bits was not perfect. Switching one diode pair on or off was found to cause a $\pm$ 5% variation in the amplitude of another bit. Long bias lines were found to have a similar effect. It was found that if bias lead lengths were restricted to less than $\lambda/20$ electromagnetic, problems associated with electrical mismatch and the effects of adjacent bits influencing each other could be minimized. In addition, it was found that the resistance of the diode switching elements themselves should be less than 50 ohms to minimize insertion loss and the power required per bit.

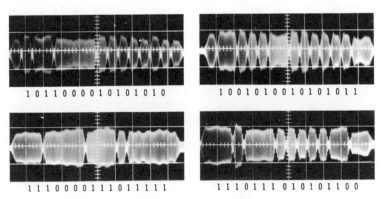

Fig. 24. Biphase coded waveforms from 16 bit switchable sequence generator (after O'Clock *et al.*, 1972).

V. Pulse Compression Filters

Elastic surface waves éxhibit no velocity dispersion, i.e. all frequencies propagate at the same velocity, on any sample which has a thickness of many acoustic wavelengths. Velocity dispersion, however, can be created for a surface wave device by allowing the distance of propagation to vary with frequency (de Klerk, 1972). Figure 25 illustrates this method of creating

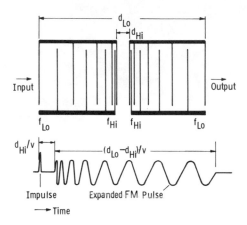

FIG. 25. Dispersive filter for generating an expanded frequency modulated pulse.

velocity dispersions by employing two interdigital transducers which are linearly frequency modulated and mirror images of each other. When an impulse, short compared with half a period at the highest frequency, is applied to the left grid, an expanded linear frequency modulated pulse is detected at the right grid. The length of the expanded pulse can be expressed, in terms of propagation distances and velocity, by

$$\Delta t = (d_{Lo} - d_{Hi})/v \tag{5}$$

where d_{Lo} is the propagation distance for the lowest frequency, d_{Hi} the propagation distance for the highest frequency, and v is the surface wave velocity. The bandwidth of the expanded FM pulse will be

$$\Delta f = (f_{Hi} - f_{Lo}). \tag{6}$$

By applying the expanded FM pulse, derived as shown from the right-hand grid of Fig. 25, to the left-hand grid shown in Fig. 26, the signal detected at the right-hand grid will be compressed by a factor which is the product of the bandwidth and pulse length, or

$$\text{pulse compression ratio} = \Delta f\, \Delta t$$
$$= (f_{Hi} - f_{Lo})(d_{Lo} - d_{Hi})/v. \tag{7}$$

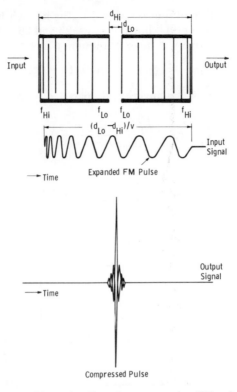

FIG. 26. Dispersive filter for compressing FM pulses.

Correlation gain obtained during compression of the pulse can be expressed as

$$\text{relative correlation gain} = \tfrac{1}{2}(f_{\text{Hi}} + f_{\text{Lo}})\,\Delta t$$
$$= f_c\,\Delta t = \text{RCG} \tag{8}$$

where f_c is the center frequency or arithmetic mean frequency.

$$\text{Absolute gain} = \text{RCG} - \text{conversion loss.}$$

In designing frequency modulated interdigital grids for pulse expansion or compression, the number and placement of finger pairs required can be calculated as follows.

From a knowledge of the dispersion time width, Δt, and the center frequency, f_c, the number of complete cycles in the FM or chirp signal is obtained from the following expression:

$$n = f_c\,\Delta t. \tag{9}$$

Hence, the total number of quarter wavelengths in each half of the pair of expansion or compression grids shown in Figs. 25 and 26 will be $2n + 1$. Each successive finger and space will have a frequency increment δf where

$$\delta f = \Delta f / (2n + 1). \tag{10}$$

In this case the widths of the fingers and spaces, as indicated in Fig. 27, are given by

$$w_m = \frac{v}{4} \left(\frac{1}{f_m + (m - 1)f} \right) \tag{11}$$

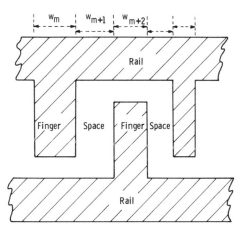

Fig. 27. Detail of frequency modulated grid showing change in width of fingers and spaces to simulate velocity dispersion.

where v is the surface wave velocity; $m = 1, 3, 5, \ldots (2n + 1)$ for fingers, and $m = 2, 4, 6, \ldots 2n$ for spaces. Equation (11) is valid for low piezoelectric coupling materials. For strong coupling materials, however, it is necessary to use the appropriate surface wave velocities for the fingers and for the spaces. Thus, $v = v_o$ for the fingers, i.e., the unstiffened velocity (Campbell and Jones, 1968) and $v = v_\infty$ for the spaces, i.e., the velocity stiffened by the piezoelectric effect.

It may also be necessary to amplitude modulate the frequency modulated pulse expansion or compression filter to compensate for frequency dependent effects such as propagation loss, beam spread, electrode resistive losses, and electrical matching. Amplitude modulation is achieved by varying the finger lengths of the interdigital grid. Each effect mentioned above will require its own unique amplitude modulation function for effective compensation.

Experimental evidence (Tancrell and Williamson, 1971) has shown that removing the excess finger lengths from the surface can lead to serious

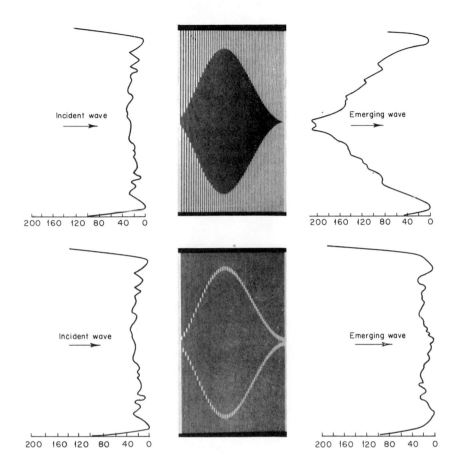

FIG. 28. Wavefront distortion of a surface wave due to amplitude modulation (after Tancrell and Williamson, 1971).

wavefront distortion as indicated in the upper half of each of Figs. 28 and 29. By allowing the excess to remain on the surface, which can be readily done by creating a gap approximately one wavelength long in the affected fingers, the wavefront remains relatively undistorted. The distortion arises from the $(\Delta v/v)$ effect (Campbell and Jones, 1968) caused by the nonuniform metallization of the surface of the piezoelectric medium when the "excess" finger lengths are removed. This type of distortion is greatest for high coupling materials and directly proportional to the number of fingers involved.

Surface wave pulse expansion and compression filters of the type just described are subject to several second-order effects which can seriously

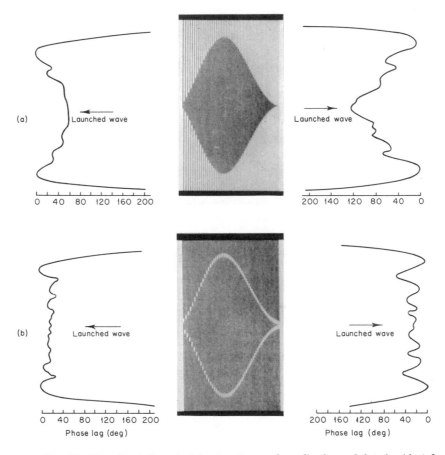

FIG. 29. Wavefronts launched by two types of amplitude modulated grids (after Tancrell and Williamson, 1971).

degrade the filter performance when using high coupling piezoelectric substrates, such as $LiNbO_3$. The second-order effects involve wavefront distortion (Figs. 28 and 29), multiple reflections between fingers and grids, bulk mode generation, and reradiation (de Klerk, 1971b). These undesirable effects can be avoided by reflecting the surface wave through two 90° angles (Williamson and Smith, 1973). This technique is illustrated in Fig. 30. The device consists of an input and an output interdigital transducer, each consisting of a few evenly spaced finger pairs, the spacing being determined by the wavelength at center frequency, and two sets of obliquely placed etched gratings. The spacings between slots in these gratings is varied uniformly, the closest spacings being adjacent to the input and output grids. This type of structure will selectively reflect frequencies according to the spacings between slots. Thus, the portions of the gratings nearest the transducers will most strongly reflect the high frequencies, while those furthest away will favor the low frequencies, thus simulating velocity dispersion due to the different path lengths for the various frequencies. Ideally, the depths of the elements

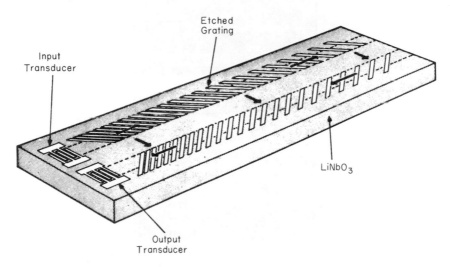

Fig. 30. Schematic diagram of a reflective array surface wave compression filter. The propagation paths at different frequencies are indicated (after Williamson and Smith, 1973).

of the grating should also vary in a manner directly proportional to their spacing to achieve uniform amplitude of reflection over the complete passband. Fabrication of such a device is naturally rather more complicated than conventional surface wave pulse compression filters. The process involves three separate photolithographic steps. In the first, the input and output grids are fabricated by conventional positive photoresist techniques (Smith *et al.*, 1971). In the second, the substrate is recoated with positive photoresist and the grating pattern is exposed and developed, leaving openings where the grating elements will be machined. These areas are ion beam etched to the desired depths by selectively exposing them to the ion beam for varying lengths of time. In the third step, a closed border of thin film aluminum is placed around each transducer. When grounded, these isolation shields reduce the amount of direct electromagnetic feedthrough from input to output transducer.

Several devices of this type, fabricated by Williamson and Smith (Williamson and Smith, 1973) on $LiNbO_3$, gave very encouraging characteristics. Figure 31 shows oscillograms of signals derived from two such

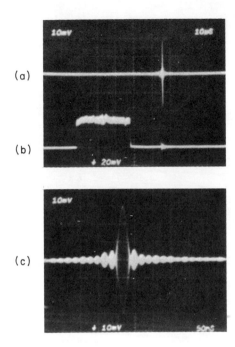

FIG. 31. Oscillograms obtained from two reflective array filters. One device generated an expanded pulse of 30 μsec duration with a bandwidth of 50 MHz, while the other
compressed the expanded pulse to approximately 20 nsec. (a) Compressed pulse. (b)
Expanded pulse. (c) Expanded display of compressed pulse (after Williamson and Smith,
1973).

reflective array filters; one generated an expanded linear FM pulse while the
other compressed the expanded pulse. The expanded pulse had a gated
length of 30 μsec shown in Fig. 31b. The 4 dB width of the compressed pulse
(see Fig. 31c) was approximately 20 nsec, resulting in a compression ratio
of 1500.

REFERENCES

Allison, J. F., Dumin, D. J., Heiman, R. P., Mueller, C. W., and Robinson, P. H. (1969).
  Proc. IEEE **57**, 1490.
Barker, R. H. (1953). "Group Synchronizing of Binary Digital Systems, in Communication Theory" (W. Jackson, ed.), Chapter 19, p. 273. Butterworth, London.
Campbell, J. J., and Jones, W. R. (1968). *IEEE Trans. Sonics Ultrason.* **15**, 209.
de Klerk, J. (1971a). *Ultrasonics* **9**, 35.
de Klerk, J. (1971b). "*Invited*" *Proc. 1970 Ultrason. Symp. IEEE* Cat. No. 70 C69 SU,
  p.94.
de Klerk, J. (1972). *Phys. Today* **25**, 32.

Jones, W. S., Hartmann, C. S., and Sturdivant, T. D. (1972). *IEEE Trans. Sonics Ultrason.* **19**, 368.

O'Clock, G. D., Jr., Grasse, C. L., and Gandolfo, D. A. (1971). *Proc. IEEE* **59**, 1536.

O'Clock, G. D., Jr., Gandolfo, D. A., and Sunshine, R. A. (1972). *Proc. IEEE* **60**, 732.

Slobodnik, A. J., Jr., and Conway, E. D. (1970). "Microwave Acoustics Handbook," Vol. 1, p. 59.

Smith, H. I., Bachner, F. J., and Efremow, N. (1971). *J. Electrochem. Soc.* **118**, 821.

Tancrell, R., and Williamson, R. C. (1971). *Appl. Phys. Lett.* **19**, 456.

Thomas, R. L., Vale, C. R., and Foster, T. M. (1972). *IEEE Trans. Sonics Ultrason.* **19**, 415.

Williamson, R. C., and Smith, H. I. (1973). *IEEE Trans. Microwave Theory Tech.* **21**, 195.

—5—

Nonlinear Effects in
Piezoelectric Quartz Crystals

J. J. GAGNEPAIN and R. BESSON

Ecole Nationale Supérieure de Chronométrie et de Micromécanique,
Besancon, France

I.	Introduction	245
II.	Fundamental Equations of Quartz	247
	A. Definition of Strains	247
	B. Equation of Equilibrium	248
	C. Stress–Strain Relation	250
	D. Fundamental Equations	251
III.	Characteristic Coefficients	252
	A. Elastic Coefficients	252
	B. Piezoelectric Coefficients	255
	C. Dielectric Coefficients	261
	D. Damping and Conductivity Coefficients	265
IV.	Nonlinear Effects in Shear Vibrating Quartz Crystal Resonators	266
	A. Fundamental Equations	266
	B. Wave Propagation Equation	268
	C. Boundary Conditions	270
	D. Solutions	271
	E. Amplitude Frequency Effect	272
V.	Equivalent Electrical Circuit of a Quartz Resonator	278
	A. The Linear Equivalent Electrical Circuit	278
	B. Electrical Conductivity Influence	279
	C. Nonlinear Equivalent Circuit	281
VI.	Influence of an Applied dc Electric Field	283
	A. Diffusion Phenomena	283
	B. Frequency Applied Field Relation	284
VII.	Conclusion	287
	References	287

I. Introduction

Since the appearance of Cady's first quartz vibrator, the evolution of quartz crystal units has continued over a long period of time. The crystal units appear to be very important in frequency and time measurements, in radio communication, and recently in electronic watchmaking since the introduction of quartz watches.

Quartz crystal units cannot be considered for use in primary frequency standards because of their long term drift. Nevertheless they are very useful for short term studies because of their excellent spectral purity and short term stability; in each passive atomic standard a quartz crystal unit is included so as to achieve at the same time the low long term drift of the atomic device and the short term stability of the quartz unit.

A quartz crystal unit is also often used when a fair degree of frequency stability is needed under set conditions (weight, dimensions, low cost, low environmental sensitivity).

In fact, the best stabilities for a quartz crystal unit are in the neighborhood of 10^{-12} or 5×10^{-13} over a few seconds and 10^{-10} to 10^{-11} per day. These results have been obtained by new technical methods (electrode plating, thermocompression bonding, cold welding, etc.). Higher Q factors, better spectral purity, and lower frequency drift are also obtained. In oscillators, components are carefully chosen and electronic circuitry is specially studied.

Long term frequency stability is closely tied to the aging phenomena of resonators and components. Short term stability is associated with three different types of noise (Uebersfeld *et al.*, 1973; Rutman, 1972): internal noise of the oscillating loop, additional external noise principally caused by output amplifiers, and flicker noise which is a low frequency noise with a power spectral density represented by a f^{-1} curve.

The signal to noise ratio may be improved by increasing the oscillation level. As a result the short term stability is improved, but the long term stability is partly destroyed.

Increasing the oscillation level (i.e., increasing the power dissipated in the crystal) is limited by the nonlinear resonator effects apparent in the frequency amplitudes, by distorsions in the response curves, and by the harmonic frequencies generated. These phenomena may be important enough to prevent the normal use of resonators, and the origin of nonlinear phenomena should certainly be determined in order to lower nonlinear effects as far as possible.

Nevertheless, nonlinearities are also a matter of positive interest in surface wave devices; nonlinear elasticity may be used to perform convolution and correlation in real time (Quate and Thompson, 1970).

Nonlinear phenomena in quartz crystal are numerous and associated with different kinds of interaction processes. We shall consider nonlinear elastic, piezoelectric, electric, and damping effects (damping and elastic effects are related). These phenomena are not in agreement with classical laws (Hooke's law for instance), and correspond to very low time constants. We wish to distinguish between these effects and indirect effects, such as temperature effects. Temperature effects may also generate amplitude frequency effects during crystal heating when the oscillation level is increased, which corresponds to a long time constant. Coupling phenomena between two vibration modes close to one another may also be confused with non-linear effects. Coupling phenomena will not be included here, because such an important subject requires a complete study of its own.

Strictly speaking, crystal plating and fixation influence should also be accounted for; this has yet to be investigated.

Nonlinear phenomena have already been demonstrated experimentally. The AT and BT cut response curves have been obtained by Seed (1962b) revealing jump phenomena for a high level of excitation. Similar work has been done by Hammond *et al.* (1963). Experiments by Warner (1960) have shown that resonant frequency is a quadratic function with respect to current for an AT cut crystal. Smolarski (1965) and Gagnepain (1968) determined the amplitude frequency effect in a free oscillating quartz crystal from phase variation measurements.

Theoretical results are few and far between. An approximate expression for the AT cut voltage frequency effect has been given by Seed (1962a). McMahon (1968) proposed basic equations but did not solve them. A similar approach to the AT cut nonlinear problem has been made by Franx (1967), and Gerber (1951) and Smolarski (1966) have determined phenomenologically nonlinear effects.

II. Fundamental Equations of Quartz

A. DEFINITION OF STRAINS

We suppose that all strains are reversible. Therefore we consider perfectly elastic phenomena, although we are dealing with high amplitudes. The coordinates x, y, z of one particle before deformation become x_1, y_1, z_1 after deformation.

The components of the displacement vector are

$$u = x_1 - x, \quad v = y_1 - y, \quad w = z_1 - z. \tag{1}$$

We consider the Jacobian matrix K given by

$$K = \frac{u, v, w}{x, y, z} \tag{2}$$

and we denote the components of the strain by η_{ij}

$$\eta_{ij} = \tfrac{1}{2}(K + K^*) + \tfrac{1}{2}KK^*, \tag{3}$$

where K^* is the transpose of K.

KK^* is a nonlinear term introducing squared magnitudes of the strains. These nonlinearities are not related to the intrinsic crystal properties, but result from the definition of strain. In the infinitesimal theory of strain (Hooke's approximation) these products are negligible.

Many different symbols are used for strains. The η_{ij} notation is used by Murnaghan (1951). One can also quote the most commonly used notations:

the tensor notation with two subscripts S_{ij} and, the abbreviated notation with half the number of subscripts S_i.

Factors of two are introduced between the single and double subscript shear strains. We have the following relations:

$$S_{11} = S_1, \; S_{22} = S_2, \; S_{33} = S_3, \; 2S_{23} = S_4, \; 2S_{13} = S_5, \text{ and } 2S_{12} = S_6, \quad (4)$$

the Voigt notation which is analogous to the single subscript notation, $x_x, y_y, z_z, y_z, z_x, x_y$ (Voigt, 1928).

Using the single subscript notation S_i, we obtain the following expressions for the strain:

$$
\begin{aligned}
S_1 &= \frac{\partial u}{\partial x} + \frac{1}{2}\left[\left(\frac{\partial u}{\partial x}\right)^2 + \left(\frac{\partial v}{\partial x}\right)^2 + \left(\frac{\partial w}{\partial x}\right)^2\right], \\[2mm]
S_2 &= \frac{\partial v}{\partial y} + \frac{1}{2}\left[\left(\frac{\partial u}{\partial y}\right)^2 + \left(\frac{\partial v}{\partial y}\right)^2 + \left(\frac{\partial w}{\partial y}\right)^2\right], \\[2mm]
S_3 &= \frac{\partial w}{\partial z} + \frac{1}{2}\left[\left(\frac{\partial u}{\partial z}\right)^2 + \left(\frac{\partial v}{\partial z}\right)^2 + \left(\frac{\partial w}{\partial z}\right)^2\right], \\[2mm]
S_4 &= \frac{\partial w}{\partial y} + \frac{\partial v}{\partial z} + \left[\frac{\partial u}{\partial y}\frac{\partial u}{\partial z} + \frac{\partial v}{\partial y}\frac{\partial v}{\partial z} + \frac{\partial w}{\partial y}\frac{\partial w}{\partial z}\right], \\[2mm]
S_5 &= \frac{\partial u}{\partial z} + \frac{\partial w}{\partial x} + \left[\frac{\partial u}{\partial x}\frac{\partial u}{\partial z} + \frac{\partial v}{\partial x}\frac{\partial v}{\partial z} + \frac{\partial w}{\partial x}\frac{\partial w}{\partial z}\right], \\[2mm]
S_6 &= \frac{\partial u}{\partial y} + \frac{\partial v}{\partial x} + \left[\frac{\partial u}{\partial x}\frac{\partial u}{\partial y} + \frac{\partial v}{\partial x}\frac{\partial v}{\partial y} + \frac{\partial w}{\partial x}\frac{\partial w}{\partial y}\right].
\end{aligned}
\tag{5}
$$

B. Equation of Equilibrium

In the deformed state a portion of the medium is supposed to be in equilibrium under the action of various forces which consist of the surface forces given by the stress matrix T_i

$$
T_i = \begin{bmatrix} T_1 & T_6 & T_5 \\ T_6 & T_2 & T_4 \\ T_5 & T_4 & T_3 \end{bmatrix}, \tag{6}
$$

where T_i is related to one unit of surface, and the body forces F_j

$$
F_j = \begin{bmatrix} F_1 \\ F_2 \\ F_3 \end{bmatrix}. \tag{6a}
$$

Weight is a typical example of a body force.

Considering a body in equilibrium, the resultant of all forces must be zero and

$$\iint_S T_i \, dA_h + \iiint_V F_j \, dV = 0. \tag{7}$$

dA_h is the matrix of the surfaces and is given by

$$dA_h = \begin{bmatrix} dA_1 \\ dA_2 \\ dA_3 \end{bmatrix}. \tag{8}$$

Using Ostrogradski's relation we obtain

$$\iiint_V [(\mathrm{div}\ T_i)^* + F_j] \, dV = 0. \tag{9}$$

This relation can be extended over every arbitrary portion V of the medium. Hence

$$(\mathrm{div}\ T_i)^* + F_j = 0. \tag{10}$$

In the case of a dynamical system the body would be in equilibrium if the inertial forces were added. Hence the equations of motion are given by

$$(\mathrm{div}\ T_i)^* + F_j = \rho \frac{\partial^2 u_j}{\partial t^2}, \tag{11}$$

where u_j is the matrix of the displacements $\begin{bmatrix} u \\ v \\ w \end{bmatrix}$. In this case we can neglect

the body forces and obtain the equation of equilibrium in its final form:

$$\rho \, \partial^2 u_j / \partial t^2 = (\mathrm{div}\ T_i)^*. \tag{12}$$

Developing this relation we obtain the well known system:

$$\begin{aligned}
\rho \frac{\partial^2 u}{\partial t^2} &= \frac{\partial T_1}{\partial x_1} + \frac{\partial T_6}{\partial y_1} + \frac{\partial T_5}{\partial z_1}, \\[2mm]
\rho \frac{\partial^2 v}{\partial t^2} &= \frac{\partial T_6}{\partial x_1} + \frac{\partial T_2}{\partial y_1} + \frac{\partial T_4}{\partial z_1}, \\[2mm]
\rho \frac{\partial^2 w}{\partial t^2} &= \frac{\partial T_5}{\partial x_1} + \frac{\partial T_4}{\partial y_1} + \frac{\partial T_3}{\partial z_1}.
\end{aligned} \tag{13}$$

ρ is the specific mass of the crystal after deformation.

C. STRESS–STRAIN RELATION

The relation proposed by Murnaghan (1951) is the following:

$$T_i = \frac{1}{\det J} J \frac{\partial \Phi}{\partial S_i} J^*, \tag{14}$$

where Φ is the energy per unit initial volume, J the Jacobian matrix

$$\frac{x_1, y_1, z_1}{x, y, z}$$

and J^* is the transpose of J. Using Eq. (1), we have $J = K + E_3$ where E_3 is the unit matrix of dimension 3.

In the case of infinitesimal deformations, the stress–strain relation becomes

$$\bar{T}_i = \frac{\partial \Phi}{\partial S_i}. \tag{15}$$

We distinguish between the stress $\bar{T}_i$ of the linear theory and the stress T_i of the nonlinear theory. These two stresses are related as follows:

$$T_i = \frac{1}{\det J} J \bar{T}_i J^*. \tag{16}$$

On expanding this relation we usually obtain rather long expressions. However, it is only necessary to proceed as far as the fourth order nonlinear terms.

The order of nonlinearities is defined by using the energy expression. Consequently the usual linear terms are second order terms, and the first nonlinearities correspond to third order terms, etc.

In the particular case of a one-dimensional vibration, corresponding to a plane wave propagation, the relations are much simpler. For an X-cut rod, lengthwise vibrating, the principal stress is

$$T_2 = \left(1 + \frac{\partial v}{\partial y}\right) \bar{T}_2. \tag{17}$$

For an X-cut plate and a thickness compression vibration the stress will be

$$T_1 = \left(1 + \frac{\partial u}{\partial x}\right) \bar{T}_1. \tag{18}$$

For a vibrating plate in thickness shear the stress occurring is T_6. It can be readily shown, considering the order of the approximation, that

$$T_6 = \bar{T}_6. \tag{19}$$

D. Fundamental Equations

Quartz crystal properties are described by a few equations that we shall call the "fundamental equations of quartz." These equations take into account linear and nonlinear elastic, electric, and piezoelectric phenomena, the latter being due to the interaction between elastic and electric phenomena.

The first two equations, called "piezoelectric equations," are obtained by derivation of the energy with respect to the strains S_i [Eq. (20)] and with respect to the electric fields E_m [Eq. (21)].

$$\bar{T}_1 = C_{ij}S_j + \tfrac{1}{2}C_{ijk}S_jS_k + \tfrac{1}{6}C_{ijkl}S_jS_kS_l + \cdots + r_{ij}\frac{\partial S_j}{\partial t}$$

$$+ \frac{1}{2}r_{ijk}\frac{\partial}{\partial t}(S_jS_k) + \frac{1}{6}r_{ijkl}\frac{\partial}{\partial t}(S_jS_kS_l) + \cdots - e_{mi}E_m$$

$$- \tfrac{1}{2}e_{mn,i}E_mE_n - e_{m,ij}E_mS_j - \tfrac{1}{2}e_{m,ijk}E_mS_jS_k$$

$$- \tfrac{1}{2}e_{mn,ij}E_mE_nS_j - \tfrac{1}{6}e_{mnp,i}E_mE_nE_p\cdots. \tag{20}$$

C_{ij}, C_{ijk}, and C_{ijkl} are the elastic constants of the second, third, and fourth order, respectively. r_{ij}, r_{ijk}, and r_{ijkl} are the corresponding damping terms. e_{mi} is the piezoelectric constant of the second order, $e_{mn,i}$ and $e_{m,ij}$ those of the third order, and $e_{m,ijk}$, $e_{mn,ij}$, and $e_{mnp,i}$ those of the fourth order.

The second relation gives the electric displacement D_h in terms of the strains and of the electric field.

$$D_h = e_{hj}S_j + e_{hm,j}E_mS_j + \tfrac{1}{2}e_{h,jk}S_jS_k + \tfrac{1}{6}e_{h,jkl}S_jS_kS_l$$

$$+ \tfrac{1}{2}e_{hm,jk}E_mS_jS_k + \tfrac{1}{2}e_{hmn,j}E_mE_nS_j + \cdots + \varepsilon_{hm}E_m$$

$$+ \tfrac{1}{2}\varepsilon_{hmn}E_mE_n + \tfrac{1}{6}\varepsilon_{hmnp}E_mE_nE_p + \cdots. \tag{21}$$

ε_{hm}, ε_{hmn}, and ε_{hmnp} are the dielectric coefficients of the second, third, and fourth order. (The different coefficients will be studied in the next section.)

Using the fundamental dynamic equation we obtain

$$\rho\frac{\partial^2 u_j}{\partial t^2} = (\text{div }T_i)^*, \tag{22}$$

where div T_i is the column matrix $(\partial/\partial x_1, \partial/\partial y_1, \partial/\partial z_1)$ and T_i the matrix of the stresses given by Eq. (6):

$$T_i = \begin{bmatrix} T_1 & T_6 & T_5 \\ T_6 & T_2 & T_4 \\ T_5 & T_4 & T_3 \end{bmatrix}. \tag{23}$$

Boundary conditions are to be associated with Eq. (22); they express that the fixation points are nodal points and that the stresses are null at the ends of the crystal plate, if the surfaces are free.

The continuity condition has to be added. On the major surfaces of the crystal the normal component of the electric displacement is equal to the superficial density of charge. Therefore, the current expression is readily obtained:

$$I = \frac{d}{dt} \iint_A D_h \, dA_h, \qquad (24)$$

where A_h is the electrode surface.

When conductibility is negligible, there is no free charge in the crystal (the flux of the electric displacement being conservative). Thus we can write

$$\text{div } D_h = 0. \qquad (25)$$

If the resonator is used in forced oscillation conditions, the voltage V across the crystal is fixed. On the other hand, in a freely oscillating plate the voltage appears as a supplementary unknown value. Voltage and electric field are related by

$$E_m = -\text{grad } V. \qquad (26)$$

If the conductibility is not negligible, we must introduce the density of charge q:

$$\text{div } D_h = q. \qquad (27)$$

Introducing the current density J_m and the coefficient of conductibility σ_{mn} we obtain

$$J_m = \sigma_{mn} E_n. \qquad (28)$$

Finally, we present an equation governing the law of conservation of electricity:

$$\text{div } J_m = -\frac{\partial q}{dt}. \qquad (29)$$

III. Characteristic Coefficients

A complete description of quartz crystal properties implies the use of characteristic elastic, piezoelectric, and electric coefficients (Cady, 1946). Most of the time, the coefficients are defined by the free energy expression

$$\Phi = \Phi_{\text{elastic}} + \Phi_{\text{piezoelectric}} + \Phi_{\text{electric}}. \qquad (30)$$

Elastic energy will be written in terms of the power terms of the strains S_i. Similarly electric energy will be written using the power terms of the fields E_m, and piezoelectric energy using products of electric fields and strains.

A. Elastic Coefficients

Elastic energy can be written in the following way:

$$\Phi = (1/2)C_{ij}S_iS_j + (1/6)C_{ijk}S_iS_jS_k + (1/24)C_{ijkl}S_iS_jS_kS_l + \cdots$$
$$(1/n!)C_{ij\ldots n}S_iS_j \cdots S_n. \qquad (31)$$

C_{ij}, C_{ijk}, ... $C_{ijk \cdots n}$ are respectively the elastic coefficients of the second, third, up to the nth order. The subscripts i, j, k, ... may have values from 1 to 6 according to Einstein's convention. This convention introduces before each energy term a numerical factor of $1/n!$ type (Brugger, 1964, 1965).

The first order terms in energy have not been considered because they are not very significant. They simply indicate that certain initial stresses are applied.

Elastic energy may also be written in terms of strains. Compliance coefficients are then defined as s_{ij}, s_{ijk}, ... $s_{ijk \cdots n}$. When the possible symmetries in the crystal are not taken into account there are 21 second order, 56 third order, and 126 fourth order elastic coefficients.

Due to the cyclic threefold z-axis and the twofold x-axis of quartz crystal some coefficients are equal to zero while others are related. Therefore the matrix of the second order coefficients has only 6 independent coefficients.

Third order coefficients have been studied by Brugger (1964, 1965). There are 14 independent coefficients and 17 related ones, making 31 non-zero coefficients. All of the 14 independent coefficients have been measured by Thurston (1966).

The number of coefficients increases rapidly with their order. There are 126 fourth order coefficients when the symmetries are not taken into account. These coefficients were first studied by Seed (1962a).

TABLE I

FOURTH ORDER INDEPENDENT COEFFICIENTS OF QUARTZ

C_{1111}	C_{3333}	C_{4444}	C_{6666}	C_{1112}	C_{1113}	C_{1123}	C_{2214}	C_{3331}
C_{4456}	C_{5524}	C_{4443}	C_{1133}	C_{3344}	C_{1456}	C_{1155}	C_{1134}	C_{2356}
		C_{4423}	C_{4413}	C_{3314}	C_{6614}	C_{6624}		

In the case of quartz we have found 23 independent coefficients which are given in Table I. We have found that there were 46 relations between the coefficients which are given in Table II. Therefore there is a total of 69 non-zero coefficients.

The fourth order elastic coefficients have not been thoroughly studied, and not many numerical values exist. Two values have been given by Fowles (1967) concerning C_{1111} and C_{3333}:

$$C_{1111} = 1.59 \times 10^{13} \text{ N/m}^2 \pm 20\%,$$
$$C_{3333} = 1.84 \times 10^{13} \text{ N/m}^2 \pm 20\%.$$

For the AT-cut, we have measured the coefficient C_{6666}^{D} (superscript D indicating that C_{6666}^{D} is defined at constant or zero electric displacement). This measurement has been performed by means of the amplitude frequency effect in resonators which is completely different from the method of Thurston or Fowles, who measured wave velocities in crystal.

TABLE II

RELATIONS BETWEEN THE FOURTH ORDER ELASTIC COEFFICIENTS

$C_{2222} = C_{1111}$	$C_{2266} = \frac{1}{6}(C_{1111} - C_{1112})$	$C_{2223} = C_{1113}$
$C_{2221} = C_{1112}$	$C_{6612} = \frac{1}{6}(C_{1111} - 4C_{6666} - C_{1112})$	$C_{2213} = C_{1123}$
$C_{1166} = C_{2266}$	$C_{1122} = \frac{1}{3}(-C_{1111} + 4C_{1112} + 8C_{6666})$	$C_{6613} = \frac{1}{4}(C_{1113} - C_{1123})$
$C_{5555} = C_{4444}$	$C_{4455} = \frac{1}{3}C_{4444}$	$C_{6623} = C_{6613}$

$C_{1124} = -C_{2214} + C_{6614} + C_{6624}$	$C_{3312} = -C_{1133}$
$C_{1114} = 3(-C_{2214} + 2C_{6614} - 2C_{6624})$	$C_{2233} = C_{1133}$
$C_{2256} = \frac{1}{2}(-2C_{2214} + 3C_{6614} - 5C_{6624})$	$C_{6633} = C_{1133}$
$C_{2224} = 3(C_{2214} - 3C_{6614} + C_{6624})$	$C_{3355} = C_{3344}$
$C_{1156} = \frac{1}{2}(-2C_{2214} + 7C_{6614} - C_{6624})$	$C_{3332} = C_{3331}$
$C_{1256} = \frac{1}{2}(-2C_{2214} + 3C_{6614} - C_{6624})$	$C_{5534} = -C_{4443}$
$C_{6665} = \frac{3}{2}(C_{6614} - C_{6624})$	

$C_{4442} = -4C_{4456} - C_{5524}$	$C_{1234} = C_{1134} - 2C_{2356}$	$C_{2255} = C_{4412}$
$C_{5514} = 2C_{4456} + C_{5524}$	$C_{1356} = 2C_{1134} - 3C_{2356}$	$C_{5566} = C_{1456}$
$C_{5556} = 3C_{4456}$	$C_{2234} = 4C_{2356} - 3C_{1134}$	$C_{3324} = -C_{3314}$
$C_{4441} = 2C_{4456} - C_{5524}$	$C_{6634} = C_{1234}$	$C_{3356} = C_{3314}$

$C_{5512} = C_{4412}$	$C_{1144} = C_{4412}$	$C_{5523} = C_{4413}$
$C_{2456} = C_{1456}$	$C_{2244} = C_{1155}$	$C_{5513} = C_{4423}$
$C_{4466} = C_{1456}$	$C_{4412} = C_{1155} - 4C_{1456}$	$C_{3456} = \frac{1}{2}(C_{4423} - C_{4413})$

Certain coefficients have also been measured in our laboratory by using a special pressing device that applies stresses up to $6 \times 10^8 \ N/m^2$ on a quartz cylinder. The stresses parallel with the cylinder axis are measured. Subsequent strains are also measured and "apparent" coefficients are obtained from the stress–strain relation with the aid of a computer. Only coefficients of the C_{ii}, C_{iii}, C_{iiii} type can be measured this way. Nevertheless, certain shear coefficients may be obtained by using several different cuts of the crystal. The coefficients measured are those defined at a zero electric field. It should also be pointed out that the coefficients measured this way are "apparent" ones that differ slightly from the coefficients previously defined. In fact these last coefficients are calculated from

$$\bar{T}_i = C_{ij} S_j + \tfrac{1}{2} C'_{ijk} S_j S_k + \tfrac{1}{6} C_{ijkl} S_j S_k S_l. \tag{32}$$

This equation involves the main strain which is parallel to the applied stress, but it also involves other strain components which are nonzero. Therefore it is necessary either to measure the latter (and that is difficult for shear components), or to define "apparent" coefficients given by the following relation:

$$\bar{T}_i = C_{ii}^a S_i + \tfrac{1}{2} C_{iii}^a S_i{}^2 + \tfrac{1}{6} C_{iiii}^a S_i{}^3 \cdots. \tag{33}$$

In fact it is possible to relate these apparent coefficients to the coefficients defined at constant electric field if all the matrix values are known; but this would obviously be difficult.

This problem does not occur when converse coefficients such as s_{ii}, s_{iii}, s_{iiii} are measured, since for these the stress components different from the main one are null at the end of the crystal. Applying a stress along the y axis we obtain

$$C_{222}^{a} = -2.1 \times 10^{11} \text{ N/m}^2,$$
$$C_{2222}^{a} = 4.9 \times 10^{13} \text{ N/m}^2.$$

B. Piezoelectric Coefficients

1. Definition

From the energy expression, piezoelectric linear second order coefficients, direct or converse, e_{ij} or d_{ij}, are defined (the process is similar to the one used previously for elastic coefficients). Nonlinear third order and fourth order coefficients are also defined. They will be written using the following conventions: third order direct coefficients: $e_{ij,m}$, $e_{i,mn}$; third order converse coefficients: $d_{ij,m}$, $d_{i,mn}$; fourth order direct coefficients: $e_{ijk,m}$, $e_{ij,mn}$, $e_{i,mnp}$; fourth order converse coefficients: $d_{ijk,m}$, $d_{ij,mn}$, $d_{i,mnp}$.

In order to avoid possible confusion between the subscripts that correspond to strains or stresses and those which correspond to fields, we separated them by a comma; the subscripts to the left always relate to the fields. Therefore

$$e_{ij,m} = e_{ji,m} \quad \text{but} \quad e_{ij,m} \neq e_{j,im} \quad \text{and} \quad e_{i,mnp} = e_{i,nmp} = e_{i,pmn} \cdots.$$

Piezoelectric energy is written:

$$-\Phi = e_{im} E_i S_m + \tfrac{1}{2} e_{ij,m} E_i E_j S_m + \tfrac{1}{2} e_{i,mn} E_i S_m S_n + \cdots.$$

Independent third order coefficients and corresponding relations have been determined. There are 41 nonzero coefficients among a total of 99; 16 coefficients are independent and there are 25 relations. The independent coefficients and the relations are given in Tables III and IV.

The $d_{i,mn}$ type coefficients are usually called electrostriction coefficients and they are found in any crystal.

2. Measurement

The second order coefficients have often been measured over a long period of time. The variations between different authors' experimental values, pointed out especially by Bottom (1969), have been due mainly to three causes: (i) different samples exhibit different characteristic coefficients (Langevin, 1939; Besson, 1970); (ii) nonlinear third order or higher order coefficients have very often been ignored in second order coefficient measurements; and (iii) "hysteresis" effects appear in the strain versus electric field

TABLE III

INDEPENDENT THIRD ORDER PIEZOELECTRIC
COEFFICIENTS

$e_{11,1}$	$e_{11,2}$	$e_{11,3}$	$e_{11,4}$	$e_{33,1}$	$e_{13,5}$	$e_{23,1}$	$e_{33,3}$
$e_{1,11}$	$e_{1,22}$	$e_{1,44}$	$e_{1,13}$	$e_{1,14}$	$e_{1,34}$	$e_{3,15}$	$e_{1,24}$

TABLE IV

RELATIONS BETWEEN THE THIRD ORDER PIEZO-
ELECTRIC COEFFICIENTS

$$e_{22,1} = e_{11,2} \qquad\qquad e_{1,55} = -e_{1,44}$$
$$e_{22,2} = e_{11,1} \qquad\qquad e_{1,56} = \tfrac{1}{2}(e_{1,24} - e_{1,14})$$
$$e_{22,3} = e_{11,3} \qquad\qquad e_{2,16} = \tfrac{1}{4}(e_{1,11} + 3e_{1,22})$$
$$e_{22,4} = -e_{11,4} \qquad\qquad e_{1,66} = -\tfrac{1}{2}(e_{1,11} + e_{1,22})$$
$$e_{33,2} = e_{33,1} \qquad\qquad e_{2,35} = -e_{1,34}$$
$$e_{12,5} = e_{11,4} \qquad\qquad e_{2,36} = -e_{1,13}$$
$$e_{12,6} = \tfrac{1}{2}(e_{11,1} - e_{11,2}) \qquad e_{3,25} = -e_{3,15}$$
$$e_{13,6} = e_{23,1} \qquad\qquad e_{2,45} = e_{1,44}$$
$$e_{23,2} = -e_{23,1} \qquad\qquad e_{3,46} = -e_{3,15}$$
$$e_{23,4} = e_{13,5} \qquad\qquad e_{2,46} = \tfrac{1}{2}(e_{1,14} - e_{1,24})$$
$$e_{1,12} = -\tfrac{1}{2}(e_{1,11} + e_{1,22})$$
$$e_{1,23} = -e_{1,13}$$
$$e_{2,15} = -e_{1,24}$$
$$e_{2,25} = -e_{1,14}$$
$$e_{2,26} = -\tfrac{1}{4}(e_{1,22} + 3e_{1,11})$$

curves so that the measured strains depend on the fields previously applied (Besson, 1971).

Third order coefficient values have been proposed by Hruska and Kazda (1968) using the frequency variation of a resonator with dc applied fields.

Measurements have been performed in our laboratory using an original device. Since the best accuracies are obtained with frequency measurements, length variations have been changed into capacitance variations, and then into frequency variations. A 0.05 Å resolution apparatus has been used to perform static length variation measurements up to ± 500 Å with a 5×10^{-5} linearity.

The 0.05 Å resolution is small compared to atomic distances and must therefore be explained. Frequency variations are obtained through capacitance variations; then the variations in position of an equipotential surface Σ_M are actually measured. Even if Σ_M is very different from a plane, it is perfectly possible to consider a small translation without deformation, as is suggested by Fig. 1.

Σ_M actually stands for an average position with respect to space and time variables as well. In fact, thermal agitation causes vibrations of the sample

FIG. 1. Translation of the conductor equipotential surface in a displacement measurement.

surface. The vibration frequency corresponds mainly to the sample resonance frequency, and the vibration amplitude is approximately 0.2 Å (Mayer, 1959).

Static processes appear highly interesting because they eliminate part of the noise; then the previously given resolution is possible if the phenomenon studied simply causes a translation of Σ_M into $\Sigma'_{M'}$.

The measuring device is represented by the scheme of Fig. 2. A grounded

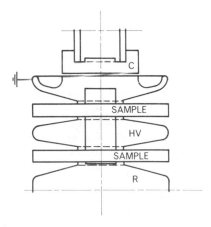

FIG. 2. Principle of the piezoelectric coefficients measurement. C: fixed conductor. Sample: quartz crystal or fused silica disks. HV: high voltage electrode. R: grounded electrode.

electrode R, a first sample, a high voltage electrode, a second sample, and a measurement electrode are piled up. The symmetry axis is vertical; the electrode R is a mechanical and electrical reference. The displacement of the measurement electrode is determined by measuring the variation of its capacitance with respect to a fixed conductor C.

The conductor C is insulated by a fused silica cylinder. It remains fixed during the measurement but, of course, its position can be adjusted beforehand. The measurement capacitance is the variable element of an LC sinusoidal oscillator using a unijunction transistor. The frequency of this oscillator is compared to the frequency of a synthesizer.

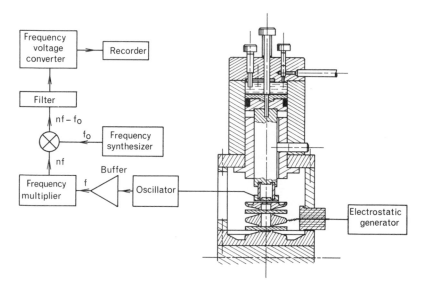

FIG. 3. Complete device used for the measurement of piezoelectric coefficients.

The complete measuring device is represented by Fig. 3. The complete device placed on a massive concrete table is insulated from vibrations as far as possible: its temperature stability is approximately 0.02°C per day. The stability of the LC oscillator is about 10^{-7} over 20 min. The most troublesome frequency variations are long term drifts, but actually they do not cause much interference. The electric field applied across the samples is supplied by a high stability dc electrostatic generator.

Two different samples may be used (Fig. 2) if they are sufficiently identical. In order to separate even effects from odd effects the samples are then successively placed according to the four existing possibilities. In fact, one of the samples is usually fused silica, the electrostriction coefficient of which has been previously measured. The successive coefficients are then obtained from the following strain field relation with the aid of a computer.

$$S_1 = d_{11}E_1 + \tfrac{1}{2}d_{11,1}E_1^2 + \tfrac{1}{6}d_{111,1}E_1^3. \tag{34}$$

For example, the results for a Brazil quartz X cut crystal are

$$d_{11} = (-2.232 \pm 0.006) \times 10^{-12} \text{ m/V},$$
$$d_{11,1} = (5.0 \pm 0.1) \times 10^{-20} \text{ m}^2/\text{V}^2,$$
$$d_{111,1} = 2.7 \times 10^{-26} \text{ m}^3/\text{V}^3.$$

About twenty independent measurements performed with the same sample exhibited a very low standard deviation.

Using a Y cut we obtained

$$d_{12} = (2.329 \pm 0.006) \times 10^{-12} \text{ m/V},$$
$$d_{11,2} = (1.42 \pm 0.02) \times 10^{-19} \text{ m}^2/\text{V}^2,$$
$$d_{111,2} = -5.7 \times 10^{-25} \text{ m}^3/\text{V}^3.$$

Numerous experiments have been performed with X cuts of different origin. Table V lists typical piezoelectric coefficients for Madagascar natural quartz some exhibiting large defects, and Table VI for synthetic high Q Sawyer quartz.

TABLE V

MADAGASCAR NATURAL QUARTZ

d_{11} (10^{-12} m/V)	$d_{11,1}$ (10^{-20} m^2/V^2)	
-0.86	8	With large defects
-2.36	7	Without apparent defects

TABLE VI

HIGH Q SAWYER QUARTZ

d_{11} (10^{-12} m/V)	$d_{11,1}$ (10^{-20} m^2/V^2)	$d_{111,1}$ (10^{-27} m^3/V^3)	
-2.220	5.8		Sample including the seed
-2.252	3.6	-1.5	Sample that does not include the seed

3 " Hysteresis " Effect

Independently of the previous influence of the crystal origin, significant differences may be observed in experiments concerning a single sample. These variations are related to a dependence between the properties of the crystal and the previously applied fields; this dependence will be called the "hysteresis" effect.

An X-cut is used and electric field is supposed to have never been previously applied. The thickness of the crystal will be denoted e and its relative variation $\Delta e/e$. For positive electric fields the curves of Fig. 4 will be obtained.

Experimental values, obtained when the electric field increases for the first time from 0 up to E_M, correspond to OVB (see Fig. 4). After that, experimental values correspond to the OCB part of the curve. It appears that applying E_M causes a permanent modification of the crystal. This modification

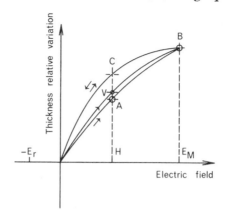

FIG. 4. A dc electric positive field is applied on a quartz sample.

TABLE VII

RESULTS OF TWO SERIES OF EXPERIMENTS

	1	2
E_M (V/m)	3.2×10^6	1.6×10^7
E_r (V/m)	9.7×10^5	9.9×10^5
e (m)	2.5×10^{-3}	3×10^{-4}

remains constant with time and to lower it it is necessary to apply a negative electric field, the amplitude of which is greater than E_r. Two series of experiments have been performed, giving the results of table VII. If a negative E_M field is applied then, by using positive increasing fields, the OAB part will be obtained.

A similar phenomenon is observed for measurements with negative fields; the complete results are illustrated in Fig. 5. The field variations in Fig. 5 correspond to Table VIII. If the sample width is 2.5 mm the ratio AC/HC will be about 0.01. The coefficients in the development of the strain–

TABLE VIII

FIELD VARIATIONS IN FIG. 5

	OVB	BCO	ODE	EFO	OAB	BCO
E	E_M ↗ 0	E_M ↘ 0	0 ↘ $-E_M$	0 ↗ $-E_M$	E_M ↗ 0	E_M ↘ 0

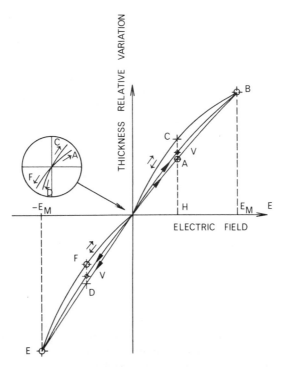

FIG. 5. "Hysteresis" effect when dc positive and negative electric fields are applied

field relation are of course different according to the part of the curve being considered. The d_{11} variation is very small, but the $d_{11,1}$ coefficient typically varies from 1.0×10^{-20} to 6×10^{-20} m^2/V^2.

Finally, it is necessary to point out that the sample modification disappears by heating the crystal at 480° for some hours.

C. DIELECTRIC COEFFICIENTS

Electrical relations may be obtained either by expressing the electric polarization P in terms of the electric field E, or by expressing the electric displacement D in terms of E. Converse relations may also be used.

It has been chosen to express D in terms of E. Linear coefficients will be denoted by ε_{ij} and will correspond to permittivity coefficients. The electric energy may be written:

$$-\Phi_{\text{elect.}} = \tfrac{1}{2}\varepsilon_{ij}E_i E_j + \tfrac{1}{6}\varepsilon_{ijk}E_i E_j E_k + \tfrac{1}{24}\varepsilon_{ijkl}E_i E_j E_k E_l + \cdots, \quad (35)$$

where ε_{ij}, ε_{ijk}, ε_{ijkl} are respectively the second, third, and fourth order coefficients.

Second order coefficients are well known; three of them are nonzero coefficients: ε_{11}, $\varepsilon_{22} = \varepsilon_{11}$, and ε_{33}. All the other permittivity matrix coefficients are null.

A study of the third and fourth order coefficients gives the following results: Third order coefficients consist of the independent coefficient ε_{111} and the relation: $\varepsilon_{221} = -\varepsilon_{111}$. Fourth order coefficients consist of the independent coefficients ε_{1111}, ε_{3333}, ε_{1133}, ε_{2223} and the relations $\varepsilon_{2222} = \varepsilon_{1111}$, $\varepsilon_{1122} = \frac{1}{3}\varepsilon_{1111}$, $\varepsilon_{2233} = \varepsilon_{1133}$, and $\varepsilon_{1123} = -\varepsilon_{2223}$.

Complete relations up to the fourth order may therefore be written:

$$
\begin{aligned}
D_1 = \varepsilon_{11}E_1 + \tfrac{1}{2}\varepsilon_{111}E_1{}^2 - \tfrac{1}{2}\varepsilon_{111}E_2{}^2 + \tfrac{1}{6}\varepsilon_{1111}E_1{}^3 + \tfrac{1}{6}\varepsilon_{1111}E_1E_2{}^2 \\
+ \tfrac{1}{2}\varepsilon_{1133}E_1E_3{}^2 - \varepsilon_{2223}E_1E_2E_3 ,
\end{aligned}
\tag{36}
$$

$$
\begin{aligned}
D_2 = \varepsilon_{11}E_2 - \varepsilon_{111}E_1E_2 + \tfrac{1}{6}\varepsilon_{1111}E_2{}^3 + \tfrac{1}{2}\varepsilon_{2223}E_2{}^2E_3 \\
+ \tfrac{1}{6}\varepsilon_{1111}E_1{}^2E_2 + \tfrac{1}{2}\varepsilon_{1133}E_2E_3{}^2 - \tfrac{1}{2}\varepsilon_{2223}E_1{}^2E_3 ,
\end{aligned}
\tag{37}
$$

$$
\begin{aligned}
D_3 = \varepsilon_{33}E_3 + \tfrac{1}{6}\varepsilon_{3333}E_3{}^3 + \tfrac{1}{6}\varepsilon_{2223}E_2{}^3 + \tfrac{1}{2}\varepsilon_{1133}E_1{}^2E_3 \\
+ \tfrac{1}{2}\varepsilon_{1133}E_2{}^2E_3 - \tfrac{1}{2}\varepsilon_{2223}E_1{}^2E_2 .
\end{aligned}
\tag{38}
$$

The different experimental methods usually employed to perform the measurements of ε_{11} and ε_{33} are not accurate enough to exhibit nonlinear dielectric coefficients. An original method is used in our laboratory to obtain directly these nonlinear coefficients (it may be used for any dielectric sample). An *LC* oscillator, similar to the one that has been previously described for piezoelectric coefficient measurements, has been designed (Besson and Gagnepain, 1972). The variable capacitance of this oscillator is obtained by using two quartz disks, as identical as possible. A dc voltage is applied across the samples (see Fig. 6).

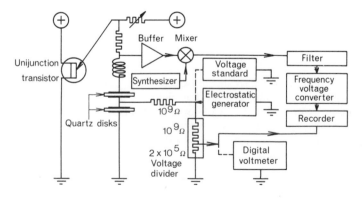

FIG. 6. Experimental device used for the measurement of the nonlinear dielectric coefficients.

A recorder is used to obtain at the same time the applied field and the capacitance variation. It can be readily shown by calculating the oscillator frequency that the coefficient measured this way corresponds to $\partial D/\partial E$. Linear dielectric coefficients are obtained by substituting a standard capacitance for the samples. Piezoelectric or electrostrictive strains must be taken into account as well as thermal effects.

1. Results for an X-Cut

The electric displacement D_1 is written:

$$D_1 = \varepsilon_{11}E_1 + \tfrac{1}{2}\varepsilon_{111}E_1{}^2 + \tfrac{1}{6}\varepsilon_{1111}E_1{}^3. \tag{39}$$

Two experimental cases are possible, which allow the separation of effects due to ε_{111} and to ε_{1111}. If the applied fields have the same sense for the two disks, the relative frequency variation is

$$\frac{f-f_0}{f_0} = -\frac{1}{2}\frac{\varepsilon_{111}}{\varepsilon_{11}}E_1 - \frac{1}{4}\frac{\varepsilon_{1111}}{\varepsilon_{11}}E_1{}^2. \tag{40}$$

On the other hand, if the applied fields have different senses for the two disks, then

$$\frac{f-f_0}{f_0} = -\frac{1}{4}\frac{\varepsilon_{1111}}{\varepsilon_{11}}E_1{}^2. \tag{41}$$

In fact, the observed capacitance variation is not only due to permittivity variations, but also to the variations in dimension of the crystal (influence of d_{11}, d_{12}, and d_{14}). The crystal thickness will be denoted e and the plated surface S. For a maximum field of 3×10^5 V/m we have

$$\frac{de}{e} - \frac{dS}{S} = 1.4 \times 10^{-6} \quad \text{(due to } d_{11} \text{ and } d_{12}\text{)}.$$

Moreover, d_{14} causes a surface shear in the yz plane. Thus the plated surfaces which were circular become elliptic, but the surface relative variation is smaller than 10^{-14} and may be ignored. Finally, the results obtained are

$$\varepsilon_{11} = (4.00 \pm 0.05) \times 10^{-11} \text{ F/m},$$

$$\varepsilon_{111} = (2.4 \pm 0.4) \times 10^{-22} \text{ F/V},$$

$$\varepsilon_{1111} < 8 \times 10^{-31} \text{ F} \cdot \text{m/V}^2.$$

2. Results for a Y-Cut

The field is applied along the y-axis.

$$D_2 = \varepsilon_{11}E_2 + \tfrac{1}{6}\varepsilon_{1111}E_2{}^3. \tag{42}$$

Therefore, the relative frequency variation is

$$\frac{f-f_0}{f_0} = -\frac{1}{4}\frac{\varepsilon_{1111}}{\varepsilon_{11}}E_2{}^2. \tag{43}$$

The influence of piezoelectric strains due to d_{25} and d_{26} may be ignored since they introduce a relative variation of about 10^{-14} for a maximum field of 3×10^5 V/m. It is again found that

$$\varepsilon_{1111} < 8 \times 10^{-31} \text{ F} \cdot \text{m/V}^2.$$

3. *Results for a Z-Cut*

The only applied field is E_3. Then D_3 may be written:

$$D_3 = \varepsilon_{33} E_3 + \tfrac{1}{6}\varepsilon_{3333} E_3{}^3. \tag{44}$$

Hence, the relative frequency variation is

$$\frac{f - f_0}{f_0} = -\frac{1}{4}\frac{\varepsilon_{3333}}{\varepsilon_{33}} E_3{}^2. \tag{45}$$

Since the second order coefficients are zero, the strains due to them are null. Therefore the only strains are a result of the third order coefficients $d_{33,1}$, $d_{33,2}$, and $d_{33,3}$. The relative frequency variation due to these strains is about 7.5×10^{-11} for a maximum field of 3×10^5 V/m; it is negligible compared to the observed relative frequency variation. Then the results obtained are

$$\varepsilon_{33} = (4.10 \pm 0.02) \times 10^{-11} \text{ F/m},$$
$$\varepsilon_{3333} = (1.5 \pm 0.3) \times 10^{-24} \text{ F} \cdot \text{m/V}^2.$$

ε_{333} is confirmed to be null.

4. *Results for AT- and BT-Cuts*

AT- and BT-cuts are obtained by a rotation θ of the Y-cut about the x-axis. The AT cut corresponds to $\theta = 35° \, 15'$ and the BT cut to $\theta = -49°$. By applying a field E_2' along the axis of the considered cut a displacement D_2' is obtained, given by

$$D_2' = \varepsilon_{22}' E_2' + \tfrac{1}{6}\varepsilon_{2222}' E_2'^3, \tag{46}$$

where

$$\varepsilon_{22}' = \varepsilon_{22} \cos^2\theta + \varepsilon_{33} \sin^2\theta \tag{47}$$

and

$$\varepsilon_{2222}' = \varepsilon_{3333} \sin^4 \theta + \cos^2 \theta \sin \theta (4\varepsilon_{2223} \cos \theta + 6\varepsilon_{2233} \sin \theta) + \varepsilon_{2222} \cos^4 \theta. \tag{48}$$

Measuring ε_{2222}' for the AT-and BT-cuts permits a determination of the two independent coefficients ε_{2223} and ε_{1133}.

It may be readily shown that piezoelectric strains cause a negligible effect. For an AT-cut the observed variation is very low and approaches the resolution of the measuring device, and

$$\varepsilon_{2222}'\text{AT} \simeq 1.5 \times 10^{-26} \text{ F} \cdot \text{m/V}^2.$$

For a BT cut

$$\varepsilon'_{2222}\text{BT} = (5.2 \pm 0.8) \times 10^{-26} \text{ F} \cdot \text{m/V}^2.$$

Hence

$$\varepsilon_{2223} \simeq 1.2 \times 10^{-25} \text{ F} \cdot \text{m/V}^2.$$
$$\varepsilon_{1133} \simeq -2.3 \times 10^{-25} \text{ F} \cdot \text{m/V}^2.$$

D. Damping and Conductivity Coefficients

1. *Damping Coefficients*

Damping is partly attributed to imperfections and impurities in the crystal, when it is called internal friction. But damping may also be caused by plating and crystal fixation especially when the crystal mass is low. Measuring the internal friction is not interesting for our purpose, and total damping coefficients will be considered; they may be obtained from resonator properties.

Linear and nonlinear damping coefficients are defined in a general form. Second, third, and fourth order coefficients are respectively denoted by r_{ij}, r_{ijk}, and r_{ijkl}. The new stress–strain relation, including only elastic quantities, is written:

$$\bar{T}_i = C_{ij}S_j + r_{ij}(\partial S_j/\partial t) + \tfrac{1}{2}C_{ijk}S_jS_k + \tfrac{1}{2}r_{ijk}(\partial/\partial t)(S_jS_k)$$
$$+ \tfrac{1}{6}C_{ijkl}S_jS_kS_l + \tfrac{1}{6}r_{ijkl}(\partial/\partial t)(S_jS_kS_l) \cdots. \tag{49}$$

r_{ij} may be calculated from the measured values of the Q factor, while r_{ijk} and r_{ijkl} are obtained from the Q factor variations with the excitation level. These Q factor variations will be obtained theoretically and experimentally in the next sections.

2. *Electric Conductivity*

Quartz, being a very good insulator, has low conductibility. Therefore it has always been ignored in calculations dealing with quartz resonators. It will be shown in the next sections that this assumption is a correct one. However, it would be interesting to know whether conductibility influences the damping of a crystal or not (i.e. its Q factor), because it is possible to obtain considerable conductibility variations by heating the crystal.

Conductibility has been measured for several cuts of different samples by applying a dc voltage of 1000 V and measuring the subsequent current. The following values have been obtained for a temperature of $23\,°\text{C}$

$$\text{for an X-cut: } \sigma_{11} = 4.10^{-14}\Omega^{-1}\text{m}^{-1} \pm 10\%,$$
$$\text{for a Y-cut: } \sigma_{22} = 4.10^{-15}\Omega^{-1}\text{m}^{-1} \pm 10\%,$$
$$\text{for a Z-cut: } \sigma_{33} = 2{,}8.10^{-12}\Omega^{-1}\text{m}^{-1} \pm 10\%.$$

Conclusion: Various coefficients have been theoretically studied. Their total has been given, and relations between them have been obtained.

From the experimental point of view measurements have mainly concerned the coefficients occuring in theoretical expressions. The measured values and the values from the literature will allow, in the next Section, a comparison between theoretical and experimental results.

IV. Nonlinear Effects in Shear Vibrating Quartz Crystal Resonators

A. FUNDAMENTAL EQUATIONS

Vibrating plates in thickness shear are now mostly used in the medium and high frequency ranges, because of their good stability and their high Q factors (2.5×10^6 for the 5 MHz fifth overtone resonators).

Since a theoretical study is being made, simple conditions are assumed. We shall consider a pure vibrating mode which has a variable excitation level. The frequency amplitude effect will be investigated without calculating the resonant frequency of the resonator from its dimensions, and without stating precisely the accurate ratios of the different overtone frequencies on the fundamental frequency. Only the main deformation will be considered. Our model is unidimensional; therefore a plane wave propagation will be considered (Gagnepain, 1972).

We assume infinite plates of thickness e. The cut is $Y + \theta°$, obtained from the Y-cut by rotation of an angle θ about the x-axis as shown in Fig. 7.

We consider the displacement u which is a function of y, the corresponding strain S_6, and the main stress T_6. Here the electric field is perpendicular to the plane wave, and it will depend on both time and the space

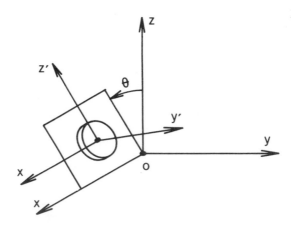

FIG. 7. $Y + \theta$ cut obtained by a θ rotation of a Y cut about the X-axis.

variable y. There lies the difference between the thickness shear mode and the lengthwise vibrating mode, because in the latter the electric field, being parallel to the wave plane, depends only on the time variable.

The fundamental equations are

$$\bar{T}_6 = C_{66} S_6 + \frac{1}{6} C_{6666} S_6{}^3 + r_{66} \frac{\partial S_6}{\partial t} + \frac{1}{2} r_{6666} S_6{}^2 \frac{\partial S_6}{\partial t} - e_{26} E_2$$

$$- \tfrac{1}{6} e_{222,6} E_2{}^3 - \tfrac{1}{2} e_{22,66} E_2{}^2 S_6 - \tfrac{1}{2} e_{2,666} E_2 S_6{}^2, \tag{50}$$

$$\bar{T}_6 = T_6, \tag{51}$$

$$D_2 = e_{26} S_6 + \tfrac{1}{2} e_{222,6} E_2{}^2 S_6 + \tfrac{1}{2} e_{22,66} E_2 S_6{}^2 + \tfrac{1}{6} e_{2,666} S_6{}^3$$

$$+ \varepsilon_{22} E_2 + \tfrac{1}{6} \varepsilon_{2222} E_2{}^3, \tag{52}$$

$$\rho \frac{\partial^2 u}{\partial t^2} = \frac{\partial T_6}{\partial y}, \tag{53}$$

$$I = S \frac{\partial D_2}{\partial t}, \tag{54}$$

$$\frac{\partial D_2}{\partial y} = 0. \tag{55}$$

Equations (50) and (52) illustrate that nonlinear occurring coefficients are fourth order coefficients. Third order coefficients are zero in the case considered:

$$C_{666} = 0, \qquad e_{2,66} = 0, \qquad e_{22,6} = 0, \qquad \varepsilon_{222} = 0. \tag{56}$$

Equation (55) means that space charges are ignored (i.e. that crystal conductivity is considered null). It will be seen in Section V,B that this assumption is quite correct. Therefore the electric displacement D_2 is independent of y but depends only on the time variable.

Using Eqs. (50) and (52) we express the field E_2 and the stress T_6 in terms of the strain S_6 and the electric displacement D_2.

$$E_2 = \beta_{22} D_2 - h_{26} S_6 + \tfrac{1}{2} h_{222,6} D_2{}^2 S_6 + \tfrac{1}{2} h_{22,66} D_2 S_6{}^2$$

$$+ \tfrac{1}{6} h_{2,666} S_6{}^3 + \tfrac{1}{6} \beta_{2222} D_2{}^3, \tag{57}$$

$$T_6 = C_{66}^{\mathrm{D}} S_6 + \tfrac{1}{6} C_{6666}^{\mathrm{D}} S_6{}^3 + \tfrac{1}{2} h_{22,66} D_2{}^2 S_6 + \tfrac{1}{2} h_{2,666} D_2 S_6{}^2$$

$$+ \frac{1}{6} h_{222,6} D_2{}^3 - h_{26} D_2 + r_{66} \frac{\partial S_6}{\partial t} + \frac{1}{2} r_{6666} S_6{}^2 \frac{\partial S_6}{\partial t}, \tag{58}$$

where

$$\beta_{22} = \frac{1}{\varepsilon_{22}}, \tag{59}$$

$$h_{26} = \frac{e_{26}}{\varepsilon_{22}}, \tag{60}$$

$$\beta_{2222} = -\frac{\varepsilon_{2222}}{\varepsilon_{22}^4}, \tag{61}$$

$$h_{22,66} = 2e_{222,6}\frac{e_{26}}{\varepsilon_{22}^3} - e_{22,66}\frac{1}{\varepsilon_{22}^2} - \varepsilon_{2222}\frac{e_{26}^2}{\varepsilon_{22}^4}, \tag{62}$$

$$h_{2,666} = -3e_{222,6}\frac{e_{26}^2}{\varepsilon_{22}^3} + 3e_{22,66}\frac{e_{26}}{\varepsilon_{22}^2} - e_{2,666}\frac{1}{\varepsilon_{22}}$$

$$+ \varepsilon_{2222}\frac{e_{26}}{\varepsilon_{22}^4}, \tag{63}$$

$$h_{222,6} = \varepsilon_{2222}\frac{e_{26}}{\varepsilon_{22}^4} - \frac{e_{222,6}}{\varepsilon_{22}^3}. \tag{64}$$

C_{66}^{D} and C_{6666}^{D} are the elastic coefficients of the second and fourth order defined at constant electric displacement. They can be related to the C_{66}^{E} and C_{6666}^{E} coefficients defined at constant electric field. (Most of the time, for simplicity, these latter coefficients will be denoted as C_{66} and C_{6666}.)

$$C_{66}^{\mathrm{D}} = C_{66}^{\mathrm{E}} + \frac{e_{26}^2}{\varepsilon_{22}}, \tag{65}$$

$$C_{6666}^{\mathrm{D}} = C_{6666}^{\mathrm{E}} - \varepsilon_{2222}h_{26}^4 + 4e_{222,6}h_{26}^3 + 4e_{2,666}h_{26}$$

$$- 6e_{22,66}h_{26}^2. \tag{66}$$

B. Wave Propagation Equation

By introducing from Eq. (58) the mechanical stress T_6 into Eq. (53) and using the strain expressions, the following equation of motion is obtained:

$$\frac{\partial^2 u}{dt^2} = C^2\left[1 + \frac{1}{2}\frac{C_{6666}^{\mathrm{D}}}{C_{66}^{\mathrm{D}}}\left(\frac{\partial u}{\partial y}\right)^2 + \frac{1}{2}\frac{h_{22,66}}{C_{66}^{\mathrm{D}}}D_2{}^2 + \frac{h_{2,666}}{C_{66}^{\mathrm{D}}}D_2\frac{\partial u}{\partial y}\right]\frac{\partial^2 u}{\partial y^2}$$

$$+ \frac{r_{66}}{\rho}\frac{\partial^3 u}{\partial y^2\,\partial t} + \frac{1}{2}\frac{r_{6666}}{\rho}\left[2\frac{\partial u}{\partial y}\frac{\partial^2 u}{\partial y^2}\frac{\partial^2 u}{\partial y\,\partial t} + \left(\frac{\partial u}{\partial y}\right)^2\frac{\partial^3 u}{\partial y^2\,\partial t}\right]. \tag{67}$$

This last equation is obviously nonlinear. Nonlinearities are introduced in the velocity of propagation and in the damping terms. An analogous equation was proposed by McMahon (1968) for longitudinal wave propagations along the Z-axis and for shear waves polarized along the Y-axis and propagating along the Z-axis.

Solutions of the following form are chosen:

$$u = \sum_{n=0}^{\infty} [A_n(y) \cos n\omega t + B_n(y) \sin n\omega t], \tag{68}$$

$$D_2 = \sum_{n=0}^{\infty} (M_n \cos n\omega t + N_n \sin n\omega t), \tag{69}$$

where A_n and B_n are functions of the space variable y whereas M_n and N_n are constant, because the electric displacement D_2 is independent of y and depends only on the time variable t. The form of the solutions has been chosen in relation to the form of the boundary conditions.

By introducing Eqs. (68) and (69) into Eq. (67) and by identification, a new set of differential equations giving A_n, B_n, M_n, and N_n is obtained.

Writing these equations is a lengthy procedure so we only give the two first equations corresponding to the fundamental frequency.

$$
\begin{aligned}
\frac{d^2 A_1}{dy^2} + \frac{\omega^2}{C^2} A_1 = \quad & k^2 \left[\frac{3}{4} \left(\frac{dA_1}{dy}\right)^2 \frac{d^2 A_1}{dy^2} + \frac{1}{4} \left(\frac{dB_1}{dy}\right)^2 \frac{d^2 A_1}{dy^2} + \frac{1}{2} \frac{dA_1}{dy} \frac{dB_1}{dy} \frac{d^2 B_1}{dy^2} \right. \\
& + \alpha_1 \left(\frac{3}{4} M_1^2 \frac{d^2 A_1}{dy^2} + \frac{1}{4} N_1^2 \frac{d^2 A_1}{dy^2} + \frac{1}{2} M_1 N_1 \frac{d^2 B_1}{dy^2} \right) \\
& + \alpha_2 \left(\frac{3}{4} M_1 \frac{dA_1}{dy} \frac{d^2 A_1}{dy^2} + \frac{1}{4} M_1 \frac{dB_1}{dy} \frac{d^2 B_1}{dy^2} + \frac{1}{4} N_1 \frac{dA_1}{dy} \frac{d^2 B_1}{dy^2} \right. \\
& \left. + \frac{1}{4} N_1 \frac{dB_1}{dy} \frac{d^2 A_1}{dy^2} \right) + \frac{\alpha\omega}{C^2} \frac{d^2 B_1}{dy^2} \right] - \frac{1}{4} k^2 \mu \frac{\omega}{C^2} \left[-\left(\frac{dA_1}{dy}\right)^2 \frac{d^2 B_1}{dy^2} \right. \\
& \left. + \left(\frac{dB_1}{dy}\right)^2 \frac{d^2 B_1}{dy^2} + 2 \frac{dA_1}{dy} \frac{dB_1}{dy} \frac{d^2 A_1}{dy^2} \right], \tag{70}
\end{aligned}
$$

$$
\begin{aligned}
\frac{d^2 B_1}{dy^2} + \frac{\omega^2}{C^2} B_1 = -k^2 & \left[\frac{1}{4} \left(\frac{dA_1}{dy}\right)^2 \frac{d^2 B_1}{dy^2} + \frac{3}{4} \left(\frac{dB_1}{dy}\right)^2 \frac{d^2 B_1}{dy^2} + \frac{1}{2} \frac{dA_1}{dy} \frac{dB_1}{dy} \frac{d^2 A_1}{dy^2} \right. \\
& + \alpha_1 \left(\frac{1}{4} M_1^2 \frac{d^2 B_1}{dy^2} + \frac{3}{4} N_1^2 \frac{d^2 B_1}{dy^2} + \frac{1}{2} M_1 N_1 \frac{d^2 A_1}{dy^2} \right) \\
& + \alpha_2 \left(\frac{1}{4} M_1 \frac{dA_1}{dy} \frac{d^2 B_1}{dy^2} + \frac{1}{4} M_1 \frac{dB_1}{dy} \frac{d^2 A_1}{dy^2} + \frac{1}{4} N_1 \frac{dA_1}{dy} \frac{d^2 A_1}{dy^2} \right. \\
& \left. + \frac{3}{4} N_1 \frac{dB_1}{dy} \frac{d^2 B_1}{dy^2} \right) - \alpha \frac{\omega}{C^2} \frac{d^2 A_1}{dy^2} \right] - \frac{1}{4} k^2 \mu \frac{\omega}{C^2} \left[-\left(\frac{dA_1}{dy}\right)^2 \frac{d^2 A_1}{dy^2} \right. \\
& \left. + \left(\frac{dB_1}{dy}\right)^2 \frac{d^2 A_1}{dy^2} - 2 \frac{dA_1}{dy} \frac{dB_1}{dy} \frac{d^2 B_1}{dy^2} \right], \tag{71}
\end{aligned}
$$

with

$$k^2 = \frac{1}{2}\frac{C^D_{6666}}{C^D_{66}}, \qquad k^2\alpha_1 = \frac{1}{2}\frac{h_{22,66}}{C^D_{66}}, \qquad k^2\alpha_2 = \frac{h_{2,666}}{C^D_{66}},$$

$$k^2\alpha = \frac{r_{66}}{\rho}, \qquad k^2\mu = \frac{1}{2}\frac{r_{6666}}{\rho}, \qquad C^2 = \frac{C^D_{66}}{\rho} \tag{72}$$

k^2 is a small factor corresponding to the nonlinearities of the fourth order.

The solutions of these two differential equations are of the form

$$A_1 = a_1 \sin(\omega y/C) + f_1 \cos(\omega y/C) + k^2 F, \tag{73}$$

$$B_1 = b_1 \sin(\omega y/C) + g_1 \cos(\omega y/C) + k^2 G. \tag{74}$$

F and G are functions of a_1, b_1, f_1, g_1 (especially of cubic terms such as a_1^3, b_1^3, $a_1^2 b_1$, ...).

C. Boundary Conditions

Since a one-dimensional model is used in this study, two boundary conditions are found. First, the plate is supposed to be fixed at the thickness center which is a nodal plane (see Fig. 8). Therefore it can be written in the form

$$u = 0 \qquad \text{for} \qquad y = 0. \tag{75}$$

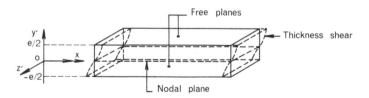

Fig. 8. Thickness shear vibrating plate.

The two surfaces at the ends of the plate are free, and the corresponding stresses are null. Thus

$$T_6 = 0 \qquad \text{for} \qquad y = \pm e/2, \tag{76}$$

where e is the thickness of the plate.

Using the first boundary condition one obtains:

$$f_1 = g_1 = 0.$$

The second boundary condition is obtained by using the T_6 expression given by Eq. (58) into which Eqs. (68) and (69) are introduced; identification is performed with respect to $\cos n\omega t$ and $\sin n\omega t$ terms. Thus two relations expressing dA_1/dy and dB_1/dy as functions of M_1 and N_1 are obtained for the fundamental frequency.

$$dA_1/dy = \alpha_4 M_1 + k^2 H, \tag{77}$$

$$dB_1/dy = \alpha_4 N_1 + k^2 J, \tag{78}$$

where

$$\alpha_4 = \frac{e_{26}}{\varepsilon_{22} C_{66}^D}.$$

H and J are functions of third power terms with respect to M_1 and N_1 (such as $M_1{}^3$, $N_1{}^3$, $M_1 N_1{}^2$, ...).

D. Solutions

The a_1, b_1, M_1, and N_1 constants occurring in the expression for the fundamental frequency vibration are obtained from Eqs. (73) and (74) using Eqs. (77) and (78). Two supplementary equations are needed; they are obtained by integrating the electric field with respect to the crystal thickness.

$$\int_{-e|2}^{e|2} E_2\, dy = V_0 \cos\omega t. \tag{79}$$

V_0 is the applied voltage (the electrodes being plated on the crystal surface).

By solving these sets of equations, the second overtone terms (which are a_2, b_2, M_2, and N_2) are shown to be zero since the third order nonlinear coefficients are zero.

The a_1 and b_1 fundamental frequency constants are given by the following simplified system:

$$
\begin{aligned}
(a_1{}^2 + b_1{}^2)(p_1 a_1 + p_1' b_1) - a_1 \delta_0 + \frac{1}{2Q} b_1 &= \frac{2e_{26}}{n^2 \pi^2 C_{66}^D} V_0, \\
(a_1{}^2 + b_1{}^2)(p_1 b_1 - p_1' a_1) - b_1 \delta_0 - \frac{1}{2Q} a_1 &= 0.
\end{aligned}
\tag{80}
$$

δ_0 is the relative difference between the excitation frequency and the resonant frequency. Q is the quality factor of the resonator and n is the overtone rank. The p_1 and p_1' coefficients depend respectively on the nonlinear elastic coefficient C_{6666}^D and the nonlinear damping coefficient r_{6666}.

$$p_1 = \frac{3n^2}{64} \frac{C_{6666}^D}{C_{66}^D} \frac{\pi^2}{e^2}, \tag{81}$$

$$p_1' = \frac{5n^2}{64} \frac{r_{6666}}{\rho C} \frac{\pi^3}{e^3}. \tag{82}$$

System (80) has been simplified by taking into account the different coefficient magnitudes. In particular, the influence of the piezoelectric and electric coefficients ($h_{2,666}$, $h_{22,66}$, $h_{222,6}$, ε_{2222}) is negligible compared to the C^{D}_{6666} influence (for these coefficient magnitudes see Section VI,B).

E. Amplitude Frequency Effect

The resonator current is obtained from the electric displacement, the normal component of which corresponds to surface charge density.

$$i = S \, \partial D_2 / \partial t. \tag{83}$$

S is the plated surface and D_2 in this case is perpendicular to this surface. I_0 the current amplitude, can be expressed in terms of a_1 and b_1, the mechanical vibration components.

$$I_0 = (2e_{26} S \omega_0 / e)(a_1{}^2 + b_1{}^2)^{1/2}. \tag{84}$$

In the same way, the phase angle ϕ between this current and the excitation voltage can be written

$$\phi = \mathrm{arctg}(a_1/b_1). \tag{85}$$

From system (80) and Eq. (84) the amplitude frequency effect expression can be deduced.

$$f = f_0 \left[1 + \frac{3n^2}{256} \frac{C^{\mathrm{D}}_{6666}}{C^{\mathrm{D}}_{66}} \frac{\pi^2 I_0{}^2}{e_{26}^2 S^2 \omega_0{}^2} \right]. \tag{86}$$

This expression exhibits a quadratic variation law for frequency difference as a function of current amplitude.

The theoretical resonance curves for amplitude and phase are presented for the case of a fundamental 5 MHz resonator (see Figs. 9a and 10a and) for the case of a fifth overtone 5 MHz resonator (see Figs. 11a and 12a). Theoretical curves are obtained using Eq. (80) and drawn by an analog computer.

Corresponding experimental curves have been obtained and are shown in Figs. 9b, 10b, 11b, and Fig. 12b. The amplitude frequency effect curve corresponds approximately to the maximum of the amplitude–resonance curves.

When the excitation level is important the amplitude frequency effect arises, and jump phenomena appear on resonance curves. [These phenomena had already been experimentally shown by Seed (1962b,) and Franx (1967), but a complete theoretical interpretation had not been given.] In this particular case, system (80) may exhibit three pairs ($a_1 b_1$) of solutions. One pair corresponds to instability, and the corresponding values (dotted line) cannot be experimentally obtained (Fig. 13).

It should be pointed out that the curves obtained are similar to the solutions of Düffing's differential equations or similar ones. They will be used in the following section when equivalent nonlinear circuits are proposed.

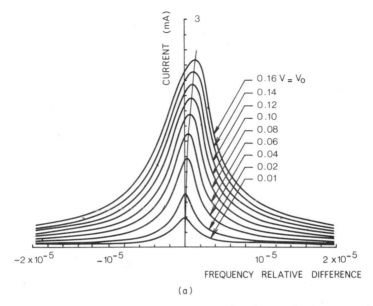

FIG. 9a. Theoretical response curves representing the amplitude versus the frequency for a 5 MHz fundamental resonator.

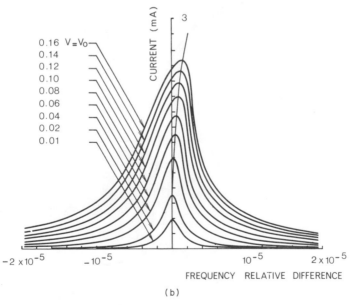

Fig. 9b. Experimental response curves representing the amplitude versus the frequency for a 5 MHz fundamental resonator.

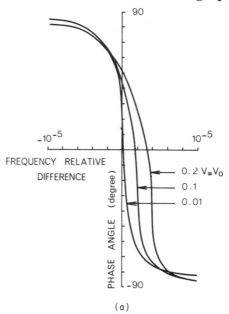

Fig. 10a. Theoretical response curves representing the phase versus the frequency for a 5 MHz fundamental resonator.

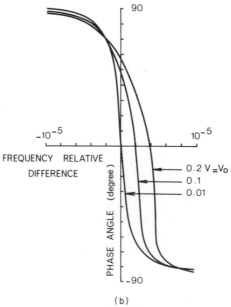

Fig. 10b. Experimental response curves representing the phase versus the frequency for a 5MHz fundamental resonator.

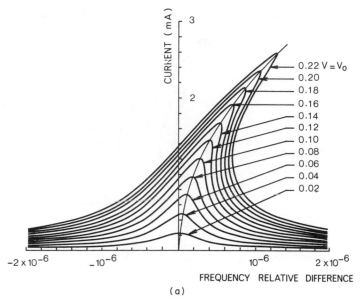

FIG. 11a. Theoretical response curves representing the amplitude versus the frequency for a 5 MHz fifth overtone resonator.

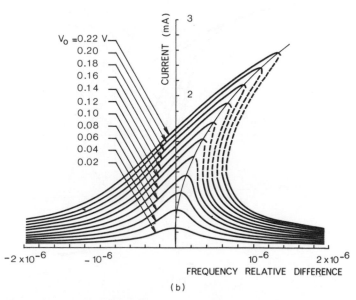

Fig. 11b. Experimental response curves representing the amplitude versus the frequency for a 5 MHz fifth overtone resonator.

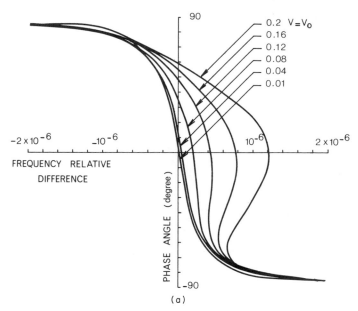

FIG. 12a. Theoretical response curves representing the phase versus the frequency for a 5 MHz fifth overtone resonator.

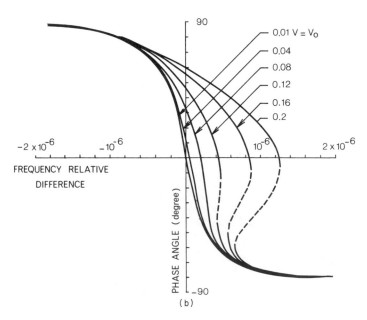

Fig. 12b. Experimental response curves representing the phase versus the frequency for a 5 MHz fifth overtone resonator.

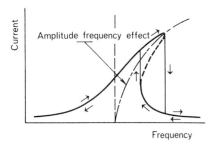

FIG. 13. Jump phenomenon obtained for a high excitation level.

The C^D_{6666} coefficient is calculated by means of experimental curves and the theoretical relation of Eq. (86). Then the following results are obtained: Using a 5 MHz fundamental resonator (first resonator, "Z" plating)

$$C^D_{6666} = (7 \pm 0.1)10^{13} \text{ N/m}^2.$$

Using a 5 MHz fifth overtone resonator (second resonator, rectangular plating)

$$C^D_{6666} = (1,5 \pm 0,1)10^{14} \text{ N/m}^2.$$

Using a 5 MHz fifth overtone resonator (third resonator, circular central plating)

$$C^D_{6666} = (1.3 \pm 0.1)10^{14} \text{ N/m}^2.$$

These three resonators are illustrated in Fig. 14.

The damping linear and nonlinear coefficients r_{66} and r_{6666} are measured respectively from the Q factor and from its variation with the excitation level. The following values have been obtained:
first resonator

$$r_{66} = 2.7 \times 10^{-3} \text{ N} \cdot \text{sec/m}^2 \pm 1\%,$$
$$r_{6666} = 3 \times 10^6 \text{ N} \cdot \text{sec/m}^2 \pm 1\%,$$

second resonator

$$r_{66} = 8.5 \times 10^{-4} \text{ N} \cdot \text{sec/m}^2 \pm 1\%,$$
$$r_{6666} = 3.3 \times 10^5 \text{ N} \cdot \text{sec/m}^2 \pm 4\%,$$

third resonator

$$r_{66} = 6.2 \times 10^{-4} \text{ N} \cdot \text{sec/m}^2 \pm 1\%,$$
$$r_{6666} = 1.2 \times 10^5 \text{ N} \cdot \text{sec/m}^2 \pm 4\%.$$

No significant difference is found between the two fifth overtone resonators. The difference for the first resonator is due to the small thickness of this crystal (0.33 mm), then plating and fixation influence become important

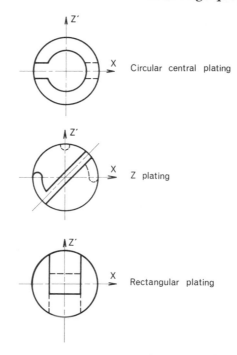

Circular central plating

Z plating

Rectangular plating

FIG. 14. Plated electrodes used for the experiments.

since damping coefficients arise. This influence is less important in the case of fifth overtone resonators because their thickness is five times greater (1.65 mm).

V. Equivalent Electrical Circuits of a Quartz Resonator

B. The Linear Equivalent Electrical Circuit

In the linear problem the fundamental equations are much simpler. They are written (again in the thickness shear case)

$$T_6 = C_{66}S_6 - e_{26}E_2 + r_{66}\,\partial S_6/dt, \tag{87}$$

$$D_2 = e_{26}S_6 + \varepsilon_{22}E_2. \tag{88}$$

The equilibrium equation is

$$\rho\,\frac{\partial^2 u}{\partial t^2} = C_{66}^{D}\,\frac{\partial^2 u}{\partial y^2} + r_{66}\,\frac{\partial^3 u}{\partial y^2\,dt}. \tag{89}$$

We choose a solution of the following form:

$$u = A(y)\cos\omega t + B(y)\sin\omega t. \tag{90}$$

Using the boundary conditions ($u = 0$ for $y = 0$) we obtain for A and B

$$A = a \sin \frac{\omega' y}{C} \operatorname{ch} \frac{\omega^2 \alpha y}{2C^3} - b \cos \frac{\omega' y}{C} \operatorname{sh} \frac{\omega^2 \alpha y}{2C^3}, \tag{91}$$

$$B = b \sin \frac{\omega' y}{C} \operatorname{ch} \frac{\omega^2 \alpha y}{2C^3} + a \cos \frac{\omega' y}{C} \operatorname{sh} \frac{\omega^2 \alpha y}{2C^3}, \tag{92}$$

where

$$\omega' = \omega(1 - \tfrac{3}{8} \omega^2 \alpha^2 / C^4), \qquad C^2 = C_{66}^{\mathrm{D}}/\rho, \quad \text{and} \quad \alpha = r_{66}/\rho. \tag{93}$$

Therefore a and b may be readily calculated using the second boundary condition ($T_6 = 0$ for $y = \pm e/2$) and integrating the electric field with respect to plate thickness.

$$\int_{-e/2}^{e/2} E_2 \, dy = V_0 \cos\omega t.$$

By using the previous relations, the electric displacement D_2 may be calculated and the current deduced. It appears that the resonator may be represented by the equivalent circuit of Fig. 15. It is well known that the

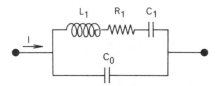

FIG. 15. Equivalent electrical circuit of a quartz resonator.

expressions for the L_1, C_1, R_1, C_0 elements of the equivalent network are (Van Dyke, 1925, 1928, 1932)

$$L_1 = \frac{\rho e^3}{8e_{26}^2 S}, \qquad C_1 = \frac{8e_{26}^2 S}{\pi^2 C_{66}^{\mathrm{D}} \, en^2},$$

$$R_1 = \frac{\pi^2 er_{66} \, n^2}{8e_{26}^2 S}, \qquad C_0 = \frac{\varepsilon_{22} S}{e}. \tag{94}$$

The resonant frequency is given by

$$\omega_0 = \frac{\pi C}{e} - \frac{4e_{26}^2 \, C}{C_{66}^{\mathrm{D}} \, \varepsilon_{22} \, \pi e}. \tag{95}$$

B. Electric Conductivity Influence

It will be shown that electric conductivity may be ignored because its influence is insignificant. Of course, this is irrelevant in piezoelectric semiconductors which are used to amplify or attenuate acoustic waves by applying a dc electric field. This field generates electron–phonon interactions as was shown by Hutson (1960), Hutson *et al.* (1961), and Hutson and White (1962).

In the problem that is considered here, the electric conductivity generates space charges of volume density q given by

$$\operatorname{div} D_2 = \partial D_2/\partial y = q. \tag{96}$$

J_2, the current density, is written in the form

$$J_2 = \sigma_{22} E_2, \tag{97}$$

where σ_{22} is the conductivity coefficient part of the matrix σ_{ij} in which σ_{11}, σ_{22}, and σ_{33} are the only occurring coefficients. It can be shown that

$$\sigma_{22} = \sigma_{11}.$$

The expression for the conservation of electricity is written:

$$\operatorname{div} J_2 = -\partial q/\partial t. \tag{98}$$

Then, the following system is obtained:

$$\rho\, \frac{\partial^2 u}{\partial t^2} = C_{66}\, \frac{\partial^2 u}{\partial y^2} + r_{66}\, \frac{\partial^3 u}{\partial y^2\, \partial t} + \frac{e_{26}}{\sigma_{22}}\, \frac{\partial^2 D_2}{\partial y\, \partial t}, \tag{99}$$

$$D_2 + \frac{\varepsilon_{22}}{\sigma_{22}}\, \frac{\partial D_2}{\partial t} = e_{26} S_6. \tag{100}$$

Solutions of the following form are chosen:

$$u = A(y)\, \cos\omega t + B(y)\, \sin\omega t, \tag{101}$$

$$D_2 = M(y)\, \cos\omega t + N(y)\, \sin\omega t. \tag{102}$$

The difference between this and the usual case where $\sigma_{22} = 0$ lies in the relation of Eq. (100); M and N are no longer constants but functions of the space variable y. By introducing these solutions into Eqs. (99) and (100), and expressing M and N as functions of A and B, a new set of differential equation is obtained:

$$C'^2 (d^2 A/dy^2) + \omega^2 A + \omega \alpha' (d^2 B/dy^2) = 0, \tag{103}$$

$$C'^2 (d^2 B/dy^2) + \omega^2 B - \omega \alpha' (d^2 A/dy^2) = 0. \tag{104}$$

It appears that the propagation velocity C and the damping coefficient [given by Eq. (93)] become

$$C'^2 = C^2 - \frac{\sigma_{22}^2 e_{26}^2}{\varepsilon_{22}^3 \omega_0{}^2 \rho}, \tag{105}$$

$$\alpha' = \alpha + \frac{e_{26}^2 \sigma_{22}}{\varepsilon_{22}^2 \omega_0{}^2 \rho}. \tag{106}$$

Since σ_{22} is approximately $4 \times 10^{-15}\ \Omega^{-1}\ m^{-1}$, the supplementary terms introduced by nonnull conductivity are very low and may perfectly well be ignored even if the temperature variation of σ_{22} is taken into account.

C. Nonlinear Equivalent Circuit

Considering the elements of the previous linear equivalent circuit as constants, this circuit does not give any account of nonlinear phenomena: it can only be used for low excitation levels. For higher excitation levels the elements of the equivalent circuit will have values depending on the applied resonator voltage or the resonator current.

We shall express L_1, C_1, R_1, and C_0 in terms of powers of the current amplitude I.

$$L_1 = L_0(1 + \alpha_1 I + \alpha_2 I^2 + \cdots), \tag{107}$$

$$R_1 = R_0(1 + \beta_1 I + \beta_2 I^2 + \cdots), \tag{108}$$

$$1/C_1 = (1/C_{10})(1 + \gamma_1 I + \gamma_2 I^2 + \cdots), \tag{109}$$

$$1/C_0 = (1/C_{00})(1 + \lambda_1 I + \lambda_2 I^2 + \cdots). \tag{110}$$

It is not necessary to introduce terms higher than the second order because in the previous study of nonlinear resonator properties only terms up to the fourth order coefficients were used. Moreover, it has been seen that when a thickness shear motion is considered, all the third order coefficients are null. Therefore it can be readily shown by identification that $\alpha_1 = \beta_1 = \gamma_1 = \lambda_1 = 0$.

From relations (110) the resonant series frequency and the Q factor expressions can be obtained:

$$f = f_0[1 - \tfrac{1}{2}(\alpha_2 - \gamma_2)I^2], \tag{111}$$

$$Q = Q_0[1 - (\beta_2 - \tfrac{1}{2}\alpha_2 - \tfrac{1}{2}\gamma_2)I^2], \tag{112}$$

where

$$f_0{}^2 = \frac{1}{4\pi^2 L_0 C_{10}} \quad \text{and} \quad Q_0 = \frac{L_0\omega_0}{R_0}. \tag{113}$$

Relating the α_2, β_2, γ_2 coefficients to nonlinear elastic and damping coefficients will now be achieved. The frequency expression as a function of current is given by Eq. (86). The Q factor may be calculated from the system (80) that gives the a_1 and b_1 mechanical vibrations components, and from the expression for the current [Eq. (84)]. Thus the resonance curve bandwidth at -3dB will be calculated. This calculation has been performed when Eq. (80) has one pair of solutions (a_1, b_1); (i.e. when the excitation level is not too high).

By successive approximations the following expression is obtained:

$$Q = Q_0[1 - (Q_0 e^2/4e_{26}^2 S^2 \omega_0{}^2)(p_1 + 3p_1')I^2]. \tag{114}$$

For the case of a fifth overtone resonator p_1 and p_1' are given by

$$p_1 = \frac{75}{64} \frac{C_{6666}^D}{C_{66}^D} \frac{\pi^2}{e^2}, \qquad p_1' = \frac{125}{64} \frac{r_{6666}}{\rho C} \frac{\pi^3}{e^3}. \tag{115}$$

It has been shown (Gagnepain, 1972) that nonlinear damping coefficients do not generate amplitude frequency effects. This means that α_2 and γ_2 are independent of p_1', and conversely that β_2 does not depend on p_1. Therefore it is possible to calculate α_2, γ_2, and β_2 thus:

$$\alpha_2 = -\frac{75}{256} \frac{C_{6666}^D}{C_{66}^D} \frac{\pi^2}{e_{26}^2 S^2 \omega_0^2} (1 + Q_0), \tag{116}$$

$$\gamma_2 = \frac{75}{256} \frac{C_{6666}^D}{C_{66}^D} \frac{\pi^2}{e_{26}^2 S^2 \omega_0^2} (1 - Q_0), \tag{117}$$

$$\beta_2 = \frac{375}{256} \frac{r_{6666}}{r_{66}} \frac{\pi^2}{e_{26}^2 S^2 \omega_0^2}. \tag{118}$$

In Eqs. (116) and (117) Q_0 is much greater than 1 ($Q \simeq 2 \times 10^6$), but nevertheless the $(1 + Q_0)$ and $(1 - Q_0)$ factors are introduced because in the frequency expression the quantity $(\alpha_2 - \gamma_2)$ occurs.

In fact, experimental results from resonance curves correspond to a quadratic law for the amplitude–frequency effect in good agreement with relation (111). The laws concerning the resistance are more difficult to obtain because of the fixation influence. Actually it is usually rather difficult to distinguish between the internal friction damping and the fixation damping.

By using the values previously determined for C_{6666}^D, r_{6666}, and r_{66} the following theoretical values are obtained:

$$\alpha_2 \simeq \gamma_2 \simeq -2 \times 10^5 \, A^{-2} \qquad \text{and} \qquad \beta_2 \simeq 2.3 \times 10^4 \, A^{-2}.$$

From several resonators the following average experimental values have been obtained:

$$\alpha_2 \simeq \gamma_2 \simeq -9 \times 10^4 \, A^{-2} \qquad \text{and} \qquad \beta_2 \simeq 5 \times 10^4 \, A^{-2}.$$

The C_0 parallel capacitance variations with current may be attributed to the fourth order nonlinear permittivity coefficient ε_{2222}, which occurs in the electric displacement expression. Using only this nonlinear coefficient we shall write

$$D_2 = e_{26} S_6 + \varepsilon_{22} E_2 + \tfrac{1}{6} \varepsilon_{2222} E_2^3. \tag{119}$$

The other equations are unchanged. By using a calculation similar to the previous ones, the C_0 expression for a fifth overtone resonator is obtained:

$$C_0 = C_{00} \left[1 + \frac{625}{512} \frac{\varepsilon_{2222}}{\varepsilon_{22}} \frac{\pi^4 C_{66}^{D\,2}}{e_{26}^2 Q_0^2 S^2 \omega_0^2} I^2 \right]. \tag{120}$$

λ_2 is obtained by identification:

$$\lambda_2 = -\frac{625}{512} \frac{\varepsilon_{2222}}{\varepsilon_{22}} \frac{\pi^4 C_{66}^{D\,2}}{e_{26}^2 Q_0{}^2 S^2 \omega_0{}^2}. \tag{121}$$

Using the previously given values and $\varepsilon_{2222} = 8 \times 10^{-31}$ F·m/V^2 we then obtain

$$\lambda_2 = -2.6 \times 10^{-14} A^{-2}.$$

Since the value is very low, this coefficient may be ignored. Therefore in the equivalent circuit, near resonant frequency, C_0 may be considered independent of current.

It should be pointed out that the nonlinearities due to ε_{2222} also have an influence on the motional element values. The corresponding expressions are long, but the correcting terms are always very low and may be ignored.

VI. Influence of an Applied dc Electric Field

An applied electric field may cause a frequency variation which is due on the one hand to crystal dimension variations—changing the resonant frequency through width and specific mass variations—(This first phenomenon is a linear one and can be considered as due to usual piezoelectric effect) and, on the other hand to a quadratic variation (Kusters, 1970; Gagnepain, 1973) which is due to nonlinear crystal coefficients. In the thickness shear mode fourth order coefficients are to be considered.

These phenomena may be used to study and measure nonlinear coefficients in a crystal, which has been one of the aims of this study. The practical interest of this frequency variation should also be pointed out. It becomes possible to achieve accurate control of a crystal resonant frequency or even to capture a frequency for instance in a quartz crystal unit. This is usually done using a varicap diode, but that may be a source of noise and therefore a cause of instability. The direct polarization realised by means of the very crystal electrodes or auxilliary electrodes had already been proposed by Hruska (1962). It seems that this polarization method is less inconvenient than using diodes because it becomes possible to act completely outside of the oscillating loop.

A. Diffusion Phenomena

When a dc voltage is applied to a resonator a very fast frequency variation followed by a slow frequency variation is observed. This last troublesome variation must, if possible, be eliminated.

An exponential law may be established (Fig. 16) for this last variation (Kusters, 1970) which corresponds to a frequency lowering or raising according to the shape and position of the electrodes. As it depends strongly on temperature a diffusion process of ions or a defect inside the crystal is of concern.

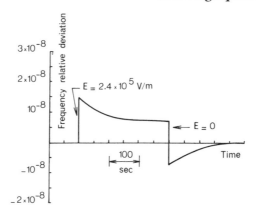

FIG. 16. Diffusion phenomenon in a quartz resonator.

This diffusion generates a field gradient and a mass displacement which leads to a frequency variation.

The time constant τ of the phenomenon varies as a function of temperature according to

$$\tau = \tau_0 \exp(\mathscr{E}/kT) \tag{122}$$

where $\mathscr{E}$ is the activation energy, k the Boltzmann constant, and T the absolute temperature. The τ variations (logarithmic scale) versus $1/T$ are drawn on Fig. 17, for several crystals. The following average values have been obtained for $\mathscr{E}$ and τ_0: $\mathscr{E} = 0.75$ eV and $\tau_0 = 1.2 \times 10^{-11}$ sec.

In order to be free of the diffusion phenomena, resonators were studied in liquid azot; then diffusion phenomena can be neglected since the time constant τ becomes very large.

B. FREQUENCY APPLIED FIELD RELATION

The frequency variation versus the applied dc field E is shown in Fig. 18 and can be expressed by

$$df/f = 6 \times 10^{-14} E + 3 \times 10^{-20} E^2. \tag{123}$$

Note that in the particular case of a lengthwise vibrating crystal studied by Hruska (1962), the frequency variation is purely linear. This variation is due to piezoelectric second order coefficients and to nonlinear third order elastic and piezoelectric coefficients.

The three components of the field E in Eq. (123) along the three axes will be denoted E_1, E_2, E_3. A rotation of the reference trihedral by an angle $\theta = 35° \ 15'$ about the x-axis has been performed in the AT-cut case. For a thickness shear motion the resonant frequency of the resonator is written

$$f = (1/2e)(C_{66}^D/\rho)^{1/2}. \tag{124}$$

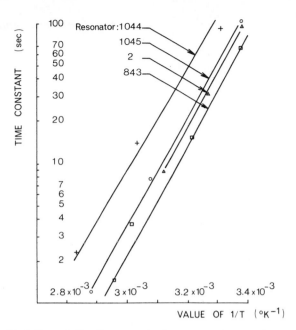

FIG. 17. Variation of the time constant τ versus inverse temperature using a logarithmic scale.

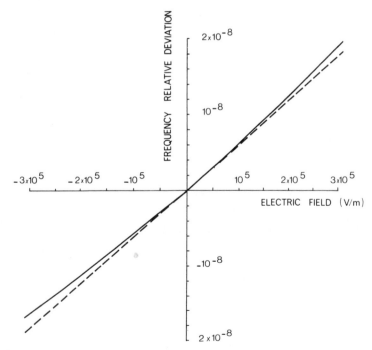

FIG. 18. Dc electric field influence on the resonant frequency of quartz resonators.

The linear part (i.e. the first term) in Eq. (123) corresponds to longitudinal strains caused by E_1 according to

$$df/f = -(de/e) - \tfrac{1}{2}(d\rho/\rho)$$
$$= (d_{11} \cos^2 \theta - d_{14} \sin\theta \, \cos\theta)E_1. \tag{125}$$

A quadratic variation is superposed to this linear variation; it corresponds to the second term in Eq. (123) and is attributed to:

1. Shear strains due to the E_2 component and the d_{25} and d_{26} coefficients. It can be readily shown that the quantity $(1/E_2{}^2) \, (df/f)$ is negligible since

$$(1/E_2{}^2)(df/f) = 6 \times 10^{-24} \; m^2/V^2.$$

2. Piezoelectric third order coefficients of the $d_{ij,k}$ type when they are not zero. Then $(1/E_2{}^2)(df/f)$, being approximately 5×10^{-22} m²/V², is also negligible.

3. Nonlinear fourth order piezoelectric coefficients $h_{2,666}$, $h_{22,66}$, and $h_{222,6}$, elastic coefficient C^D_{6666}, and nonlinear permittivity coefficient ε_{2222}. Actually piezoelectric coefficients only cannot be considered. Elastic coefficients also have an influence since applying a dc field also causes a strain which brings about a frequency variation from nonlinear elasticity.

The propagation equation and the boundary conditions are given by Eqs. (67), (75), and (76). In order to simplify the calculation, damping will be ignored.

Solutions of the following form are chosen:

$$u = A_0(y) + A_1(y) \cos\omega t, \tag{126}$$

$$D_2 = M_0 + M_1 \cos\omega t. \tag{127}$$

u and D_2 are introduced into the propagation equation, and identification is performed with respect to time.

E_2 is obtained from Eq. (57) and integrated over the crystal thickness:

$$\int_{-e/2}^{+e/2} E_2 \, dy = V_0 \cos\omega t + eE, \tag{128}$$

where $V_0 \cos \omega t$ is the time dependent part of the applied voltage and eE the dc part.

After lengthy calculations similar to the previous ones, an approximate expression is obtained for the case of an AT-cut:

$$\frac{1}{E^2} \frac{df}{f} \simeq \frac{1}{4} \left[C^D_{6666} \frac{e_{26}^2}{C_{66}^{D\,3}} + h_{22,66}^* \frac{\varepsilon_{22}^2}{C_{66}^D} \right], \tag{129}$$

where

$$h_{22,66}^* = h_{22,66} + 2h_{2,666} \frac{e_{26}}{\varepsilon_{22}C_{66}^D} + 16h_{222,6} \frac{e_{26}}{\pi^2}$$

$$- 8\varepsilon_{2222} \frac{e_{26}^2}{\pi^2 \varepsilon_{22}^4}. \tag{130}$$

$h_{22,66}$, $h_{2,666}$, and $h_{222,6}$ are given by Eqs. (62)–(64) and C^D_{6666} is given by Eq. (66).

From Eqs. (123) and (129) $h^*_{22,66}$ may be obtained:

$$h^*_{22,66} \simeq 7.5 \times 10^{11} \text{ m/F}.$$

$h_{22,66}$, $h_{2,666}$, and $h_{222,6}$ are independent coefficients; their approximative magnitudes cannot be higher than that of $h^*_{22,66}$. Therefore

$$|h_{22,66}| < 7.5 \times 10^{11} \text{ m/F}, \quad |h_{2,666}| < 3.10^{12} \text{ V/m}, \quad |h_{222,6}| < 3.10^{12} \text{ F}^2\text{V}.$$

These last results confirm that the approximations used in Section IV to obtain Eq. (80) are quite correct.

VII. Conclusion

At the end of this study, theoretical results should be compared with experimental results.

All the nonlinearity causes that seemed necessary to be considered have been introduced into the theoretical model. The values of certain coefficients have been measured directly from a quartz crystal by static methods.

Other coefficient values have been determined from experimental results through the amplitude frequency effect or the influence of a dc applied field on the resonator frequency. Several resonators gave coherent results. Experimental results are in good agreement with theoretical results as can be readily seen by comparing the experimental and theoretical curves.

Differences concerning the damping phenomena have been found. Since it is difficult to separate the internal friction of a crystal from damping due to plating and fixation, the influence of this is quite difficult to study theoretically.

In conclusion, it appears that it would be interesting to extend the result to a three-dimensional model.

REFERENCES

Besson, R. (1970). Thèse No. 48, Ecole Nat. Super. Chronomet. Micromec., Besançon.
Besson, R. (1971). *C. R. Acad. Sci.* **273**, 1078.
Besson, R., and Gagnepain, J. J. (1972). *C. R. Acad. Sci.* **274**, 835.
Bottom, V. (1969). *Proc. Annu. Frequency Contr. Symp. 23th, Atlantic City*, p. 21.
Brugger, K. (1964). *Phys. Rev. A* **133**, 1611.
Brugger, K. (1965). *J. Appl. Phys.* **36**, 759.
Cady, W. G. (1946). "Piezoelectricity." McGraw-Hill, New York.
Fowles, R. (1967). *Geophys. Res.* **72**, 5729.
Franx, C. (1967). *Proc. Annu. Frequency Contr. Symp., 21th, Atlantic City*, p. 436.
Gagnepain, J. J. (1968). *C. R. Acad. Sci.* **266**, 711.
Gagnepain, J. J. (1972). Thèse No. 59, Ecole Nat. Super. Chronomet. Micromec., Besançon.
Gagnepain, J. J. (1973). *C. R. Acad. Sci.* **276**, 491.
Gerber, E. A. (1951). *Electronics* **24**, 142.
Hammond, D., Adams, C., and Cutler, L. (1963). *Proc. Annu. Frequency Contr. Symp., 17th, Atlantic City*, p. 215.

Hruska, K. (1962). *Czech. J. Phys.* **12**, 338
Hruska, K., and Kazda, V. (1968). *Czech. J. Phys.* **18**, 500.
Hutson, A. R. (1960). *Phys. Lett.* **4**, 505.
Hutson, A. R., and White, D. L. (1962). *J. Appl. Phys.* **33**, 40.
Hutson, A. R., McFee, J. H., and White, D. L. (1961). *Phys. Rev. Lett.* **7**, 237.
Kusters, J. (1970). *Proc. Annu. Frequency Contr. Symp., 24th, Atlantic City*, p. 46.
Langevin, A. (1939). *C. R. Acad. Sci.* **270** 994.
McMahon, D. N. (1968). *J. Acoust. Soc. Amer.* **44**, 1007.
Mayer, G. (1959). Thèse No. 294, Paris.
Murnaghan, F. D. (1951). "Finite Deformation of an Elastic Solid." Wiley, New York.
Quate, C. F., and Thompson, R. B. (1970). *Appl. Phys. Lett.* **16**, 494.
Rutman, J. (1972). *Off. Nat. Etud. Rech. Aerosp., Publ.* No. 142.
Seed, A. (1962a). Ph. D. Thesis, London.
Seed, A. (1962b). *Proc. Int. Congr. Acoust., 4th, Copenhagen.*
Smolarski, A. (1965). *Bull. Acad. Pol. Sci., Ser. Sci. Tech.* **13**, 61.
Smolarski, A. (1966). *Bull. Acad. Pol. Sci., Ser. Sci. Tech.* **14**, 485.
Thurston, R. N. (1966). *J. Appl. Phys.* **37**, (1).
Uebersfeld, J., Olivier, M., and Groslambert, J. (1973). *Bull. BNM* **4**,(11), 9.
Van Dyke, K. S. (1925). *Phys. Rev.* **25**, 895.
Van Dyke, K. S. (1928). *Proc. IRE* **16**, 742.
Van Dyke, K. S. (1932). *Phys. Rev.* **40**, 1026.
Voigt, W. (1928). "Lehrbuch der Kristall Physik.", Stuttgart.
Warner, A. W. (1960). *Bell Syst. Tech. J.* **39**, 1193.

Acoustic Emission

ARTHUR E. LORD, JR.

Department of Physics and Atmospheric Science
Drexel University, Philadelphia, Pennsylvania

I.	Introduction	290
II.	Historical Work	291
III.	Early Work and General Background	294
	A. Initial Modern Work	294
	B. Early Work of Schofield	294
	C. Early Work of Tatro	295
	D. Early Structural Integrity Work	296
	E. Modern Instrumentation	297
	F. Types and Models of Acoustic Emissions	299
IV.	Materials Investigated with Acoustic Emission	301
	A. Unflawed Metal Specimens	301
	B. Flawed Metal Specimens	303
	C. Rocks	306
	D. Composite Materials	310
	E. Concrete	311
	F. Ceramics	311
	G. Ice	312
	H. Soils	312
	I. Wood	317
V.	Processes Studied with Acoustic Emission	320
	A. Welding	320
	B. Martensitic Transformations	321
	C. Slope Stability	322
	D. Magnetic Effects	324
VI.	Structural Integrity	330
VII.	Potpourri of Topics (Brief Descriptions)	333
	A. Kaiser Effect	334
	B. Emissions During Unloading	334
	C. Effects of Rate of Loading	334
	D. Application of Acoustic Emission to Civil Engineering Structures	334
	E. Effects of Nuclear Reactor Irradiation	334
	F. Measurement of Surface Coating Thickness from Acoustic Emission	335
	G. Detection of Boiling	335

H. Investigation of Honeycomb Material with Acoustic Emission 335
I. Determination of Bond Quality from Acoustic Emission Measurements 335
J. Creep Effects and Acoustic Emission 335
K. Fatigue Effects and Acoustic Emission 335
L. Effect of Temperature on Acoustic Emission 336
VIII. Conclusions and Suggestions for Further Work 336
A. Fundamental Area 336
B. Applied Area.. 337
IX. Appendix 338
A. Sources of Commercially Available Acoustic Emission and Related 338
Equipment 338
B. Acoustic Emission Working Group 338
References 339
Bibliography 345

I. Introduction

Acoustic emission involves stress waves internally generated during dynamic processes in various materials. The dynamic processes may be the result of and externally applied stress or the result of some other unstable situation, e.g. a phase transition or, on a larger scale, a shifting mine slope. The actual source of the stress waves depends on the material. It may be dislocation or crack motion in a metal, interparticle movement in a soil, or fiber breaking in a composite or wood. The term acoustic emission will be generally used here. The phenomenon has been called by many other names, e.g. stress wave emission, sonic pulse, elastic shock; in rocks—microseisms, microseismic activity, subaudible noise, rock noise, and seismoacoustic activity. Acoustic emission may be a somewhat misleading term, because in some materials the primary frequencies of the waves generated may be well above the audio range.

Figure 1 shows a detailed experimental situation for detecting and analyzing acoustic emission (Spanner, 1970). The stress wave pulses are detected with an electromechanical transducer of some sort, an amplifier (possibly a

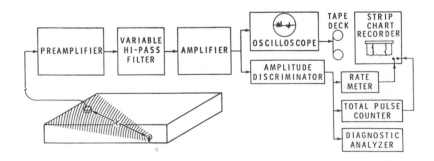

Fig. 1. A typical acoustic emission measurement and analysis system.

filter), and a display device. A simple display device can be a conventional counter or oscilloscope. Much more complicated display and recording devices are generally used.

A wide variety of materials have been studied using acoustic emission techniques. Metals and metallic structures have received the most attention. Of all other materials, rocks have been studied the most. Some of the other materials include composites, ice, wood, soils, ceramics, and concrete. Processes studied include phase transitions, welding, magnetic processes, and slope stability. A great deal of work has gone into determining the integrity of structures fabricated from metals.

There are some excellent review articles on acoustic emission already in print (Liptai *et al.*, 1971; Dunegan and Tatro, 1971; Green, 1969a; Liptai and Harris, 1971; Hardy, 1972; Knill *et al.*, 1968). The first four deal primarily with metals and metal structures, whereas Hardy and Knill *et al.*, deal with rocks. A recent bibliography has been collected by Drouillard (1974).

The philosophy of this review article will be to attempt to review the work in metals and rocks somewhat more concisely than the above authors, to spend some time dealing with acoustic emission in soils, wood, and magnetic processes (areas where the author has some research interests), to stress the early original work in mines, and to present as complete a bibliography as possible.

II. Historical Work

Hodgson (1943, 1958) in Canada and Obert (1941) and Obert and Duvall (1942, 1945ab, 1957, 1961) in the United States were all interested in predicting rock bursts in mines using the subaudible "microseisms" generated in the rocks. Their work started in the late 1930's. Hodgson, after some early development work, essentially used Obert's apparatus. The transducer, called a geophone, was a bimorphic piezoelectric crystal $2\frac{1}{2}$ in. long by $\frac{3}{4}$ in. wide by $\frac{1}{4}$ in. thick mounted as a cantelever in a steel tube $1\frac{1}{4}$ in. in diameter and about 8 in. long. It was designed to be the size of a stick of powder so it could be inserted in a rock-drill hole. A block diagram of Obert's apparatus is presented in Hardy (1972)]. When the geophone is subjected to a subaudible mechanical impulse, the crystal suffers a slight flexure which results in the generation of a transient voltage between the two terminals of the crystal. The output of the geophone was connected to an impedence-matching transformer and then traveled (for safety reasons) through as much as 1000 ft of cable to an amplifier. The amplifier was a conventional three-stage resistance coupled unit having a gain of a few hundred thousand with a flat frequency response from 150 to 10,000 Hz. Filters were available for high-, low-, or bandpass operation. A logarithmic amplifier was used at times. The recorder used a coil mounted on a pivoted stylus in the field of a permanent magnet. The mechanical energy of the impulse is thus coverted to a throw of the stylus. Recording was done by means of a small current carried by the stylus to and through the special record strip to a stationary metal platen.

The special paper made the recorder trace immediately visible. Each piece of equipment was essentially hand made by the workers themselves, whereas today one can purchase acoustic emission equipment of any desired degree of sophistication "off-the-shelf." (Appendix I lists sources of commercially available equipment.)

It was found during this work that microseisms (subaudible rock noises) do not occur in shallow mines, evidently because the pressure is not high enough there. Usually observations had to be made at a depth of at least 2000 ft to observe microseisms. The ability to use microseisms as a means of predicting rock bursts depended very strongly on the type of mine investigated. Due to dispersion in the attenuation of elastic waves in rock, the high frequency components of a microseism were damped more rapidly than the low frequency parts. A listener therefore could estimate crudely, from the frequency of the microseism, the distance to the source.

The predictability of rock bursts via microseism count has been found to be reasonably good in certain mines. Figure 2 shows the results for one

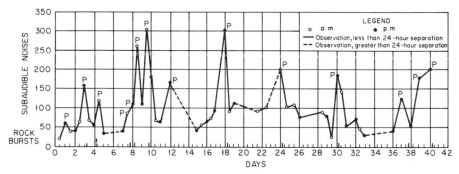

Fig. 2. Chart of subaudible noises (per two hour interval) in Ahmeek Copper Mine. Rock burst predictions are indicated by P's and actual Rock bursts by Marks along abscissa.

location in the Ahmeek Copper Mine in Michigan (Obert and Duvall, 1942). On this chart 14 rock burst predictions were made; 9 were followed by rock bursts within 12 hr or less. There were 5 predictions that were not followed by bursts, and 2 unpredicted bursts. The criteria for rock burst prediction was the following: "When the number of recorded noises increased in any interval (not exceeding 24 hr) by a factor of 2 or more a dangerous condition is indicated. Furthermore if after such an increase, the number of noises continue to increase, the state of danger is presumed to persist."

Obert and Duvall (1945b) contributed an extremely important early piece of information to the acoustic emission literature. They went into the laboratory, with controlled external stresses, to ascertain whether the microseisms originate from intermovement along fissures, seams, or fractures of geological origin or from the homogeneous rock itself. (Of course, there are microcracks in the "homogeneous" material.) It was definitely shown that microseisms can originate from initially homogeneous material. Obert and Duvall also observed a general behavior in emission rate versus stress, which subsequent workers have come to regard as 'normal" behavior in a wide variety of materials. Namely, if emission rate is plotted against stress, a high rate is found to start at some value of stress significantly below the crushing strength. The rate then decreases with increasing stress and, finally, near or at failure, increases once again. The microseisms observed in the laboratory were similar to those observed in the mines.

Triangulation studies (with multiple geophones) were attempted by Obert in order to locate the source of the microseisms. Obert also looked at the problem of underground opening stability and roof control (Obert and Duvall, 1957, 1961).

It is interesting to note in passing that Hodgson, a seismologist, in addition to an interest in predicting rock bursts, also was interested in seismic velocities in the vicinity of Ottowa, Canada. Some rock bursts at the Lake Shore Gold Mines in Kirkland Lake were severe enough to show up on the vertical Benioff seismograph in Ottawa. In fact, papers were published on seismic velocities between Kirkland Lake and Ottawa (Hodgson, 1942, 1947). One burst from Kirkland Lake was so severe that it registered in Weston, Massachusetts—a distance of 581 miles.

It is proper to point out here that this early work in "acoustic emission" contained most of the precursors of the future work (albeit crude though it may have been by today's standards). Instrumentation development was important—transducers, amplifiers, etc.; the use of filtering was tried; laboratory studies confirmed the sources of emissions in the materials themselves; triangulation for source location was tried; the work had strong practical interest; and there was strong collaboration among the people involved.

The Russians have been very active in using rock noise (seismoacoustics) to predict rock bursts (Antsyferov, 1966). Their work started in 1952, considerably later than the American and Canadian workers. The Russian workers favored electrodynamic geophones, with some work in piezoelectric types. A few of the interesting projects undertaken with seismo-acoustic techniques were: location of lost boreholes, correlation of measured rock pressure with seismo-acoustic activity, origin of emissions ahead of working face, source location via triangulation techniques, and determination of efficiency of pressure relief due to the drilling of gas relief boreholes.

A permanent seismo-acoustic warning station gave the following statistics: 80 danger zones were diagnosed—these produced 21 rock bursts and 9 face falls, and 15 faults were discovered. The statement is made, "The problem of predicting whether a mine working area is near a danger zone for a rock burst can now be considered as solved." In view of the prediction statistics given here and also by the American and Canadian workers, it appears that the above statement is somewhat extreme. Final decisions concerning the predictive use of seismo-acoustic techniques will certainly have to be made based on many factors including economics.

III. Early Work and General Background

A. Initial Modern Work

Mason *et al.* (1948; Mason, 1950) observed what appeared to be acoustic emissions in the ultrasonic frequency region during the mechanical twinning of a very small tin specimen. A plane quartz cerystal was used as the detector. This work gave some of the first indirect evidence for the existence of dislocations in a mechanical process (twinning). At that time, preliminary measurements showed no effect in aluminum.

Kaiser (1950, 1953) is credited with the first serious work in the field of acoustic emission. He worked with polycrystalline zinc, steel, aluminum, copper, and lead. Emissions were found in all the materials tested and the source of the emissions was presumed to be grain boundary motion induced by the applied stress. This hypothesis was subsequently discounted, because later workers found single crystals to be very active emitters. Kaiser also observed that the acoustic emission activity was irreversible. Emissions are not generated during the reloading of a material until the stress exceeds its previous high value. This fact has become quite useful in acoustic emission studies and is known as the "Kaiser Effect." The Kaiser Effect applies to most metals, but not generally to other materials. Details of the Kaiser Effect will be discussed later.

B. Early Work of Schofield

Schofield (1961) of Lessells and Associates (now Teledyne Materials Research) of Waltham, Massachusetts performed some very important early acoustic emission experiments. In this report may be found a good historical review of sounds associated with mechanical deformation. Work was done with aluminum and zinc single crystals, commercial copper, 24ST-4 aluminum, lead, and 70-30 brass. The primary purpose of this early work was to determine the source of the emissions, and the single crystal work showed conclusively that grain boundary effects were not the only source of emissions.

Hydraulic loading was used on the tensile specimens and measurements were performed in a soundproof room in order to eliminate machine and ambient noise.

Figure 3 shows Schofield's basic instrumentation. The data were analyzed in terms of total emission, emission rate, and amplitude and frequency of the emission. Two transducer operations were used to locate the sources. Schofield realized, as did Tatro, that the emissions were modified by both electrical and mechanical filtering before being recorded. To assess this modification, "known" pulses were introduced into the specimen and recorded as usual for emissions.

In this work, Schofield was the first to make a real distinction between burst type (discrete) and continuous type emissions. The "Kaiser Effect"

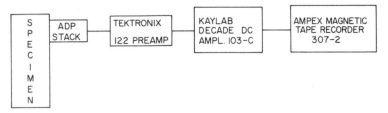

F‍IG. 3. The equipment used in the early work of Schofield.

was also verified. He was able to show that twin production and grain boundary reorientation (in a bicrystal) produced extremely large noises.

Schofield showed conclusively that the emissions were generated within the test section of the tensile specimen and that the emissions come from the deformation process itself.

Using simple dislocation-deformation arguments, it was estimated that each acoustic emission pulse resulted from the motion of some 5–50 dislocations. Schofield was able to show in some specimens a reasonably good one-to-one correspondence between acoustic emissions and slip lines.

Later work by Schofield (1963a,b, 1964) was related to surface and volume effects in regard to acoustic emission. He used oriented single crystals of aluminum and gold in this work. The aluminum could be tested with and without the oxide layer, and gold was used because it does not form a brittle oxide as does the aluminum. The oxide coating on the aluminum single crystals was not a source of acoustic emissions. The surface and its condition seemed to play a secondary role in influencing emission response. The surface induced modification was primarily in the strain, where acoustic emissions started, and the quantity of large burst activity. Schofield's later work produced the major contribution that acoustic emission is mainly a volume and not a surface effect.

C. E‍ARLY W‍ORK OF T‍ATRO

Tatro (1959) became interested in acoustic emission in 1956, hoping to use the technique as a yield detector in materials. Tatro and Liptai (1962) worked with polycrystalline aluminum (2024-T4 and 2011-T3) and steel (C 1018). The acoustic emission count versus strain was strongly dependent on the material and the surface condition of the specimen. On the basis of their results it was concluded that acoustic emission was a surface phenomenon. (At this time Schofield was also of the same opinion.) It was hypothesized that slip line formation, crack formation, or both were the cause of the acoustic emissions. Figure 4 shows a block diagram of Tatro and Liptai's experimental apparatus, which is typical of the early modern work.

Tatro and Liptai (1963) and Liptai (1963) also worked with single crystals. These were anodized in such a way as to enhance the large burst component of the acoustic emission. The oxide films were stripped from the crystals and examined with the electron microscope.

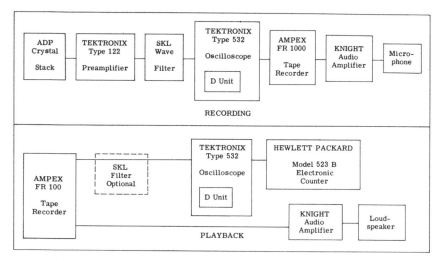

FIG. 4. Instrumentation used in the early work of Tatro and Liptai.

Tatro and Kroll (1964) and Egle and Tatro (1967) were concerned with the effect of instrumentation and propagation in the specimen on the resulting measured acoustic emissions. In the first mentioned study acoustic emission pulses were simulated by implanting a small cylindrical piezoelectric transducer in an oversized tensile test sample and attempting to excite it with a short duration pulse. The total system of specimen plus detection instrumentation was analyzed theoretically. Then a comparison was made between the resulting "emissions" received and theory to observe how the system altered the natural emissions. In the second study, the overall (mechanical and electrical) system response to both longitudinal and bending waves was evaluated. Then, the Fourier transforms of the acoustic emission response pulses were compared with the system's response so that something could be said about the nature of the actual emissions. They used the normal geometry for acoustic emission measurements (tensile test type specimen) and found that the bending response was actually larger than the longitudinal part.

D. Early Structural Integrity Work

The workers at Aerojet-General Corporation were the first to employ acoustic emission to verify structural integrity (Green *et al.*, 1964). They were working with the Polaris filament-wound, solid rocket motor cases. Due to the nature of the material, conventional techniques of experimental stress analysis were not satisfactory. In early test, the rocket motor cases had emitted potentially useful "popping noises." Conventional accelerometers were mounted at ten locations. The usual frequency response of the accelerometers was to 8000 Hz. The amplified output of the accelerometers was tape recorded for later analysis. Various methods of data presentation were tried in order to be able to predict burst pressure. One novel approach was the use of a "missile print," analogous to the "voice print." Here the output of an accelerometer is characterized by plotting frequency, pressure, and acceleration in a two-dimensional topographic map.

E. Modern Instrumentation

Work in Metals: Figure 5 shows a typical instrumentation system used for work in metals or metal systems (Dunegan and Harris, 1969). Other groups have used reasonably equivalent instrumentational setups (Nakamura *et al.*, 1972; Hutton and Parry, 1971; Green, 1969a). The sensing transducer

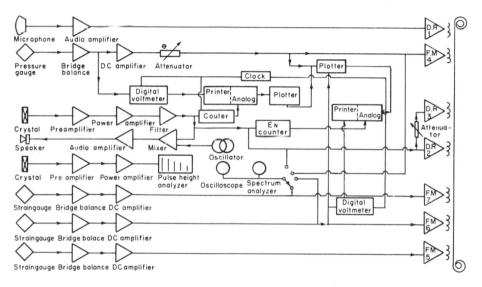

Fig. 5. A modern acoustic emission instrumental setup, as used by Dunegan and Harris.

is usually PZT with a fundamental thickness mode resonance in the hundreds of kilohertz. [Other types of transducers have been used—Knill *et al.* (1968) list electromagnetic, capacitive, and resistive. Hardy (1972) describes work with semiconducting strain gages. Magnetostrictive transducers have been used by Vetrano and Jolly (1972) and Lynnworth and Bradshaw (1971)]. One important area in instrumentation advance is the development of transducers with flatter frequency response, broader bandwidth, and high temperature capability (Engle and Dunegan, 1969). The low noise preamplifier has a gain of some 80 dB. The bandpass filter chooses a segment of the high kilohertz or low megahertz frequency range. The amplifier has a gain of some 10 dB.

Early work in acoustic emission was concerned with very much lower frequencies than those mentioned above. Hence ambient noise of all types had been a very significant problem in the emission experiments. Dunegan *et al.* (1964), were the first to work in this higher frequency regime and subsequently ambient noise was no longer the significant problem it once had been. The resulting amplified signal from the acoustic emission is fed into a counter (usually with a preset trigger level) or a tape recorder for subsequent detailed analysis. Counts per unit time and the accumulated counts seems to be of most utility, so it is a must to have a counter with a time gate and reset. If the data from the counter is to be displayed on an *xy* type recorder, then some digital to analog device must be used. A common method of presenting data is to plot acoustic emission versus some engineering parameter, say strain or stress, so strain gages, load cells, and associated circuitry are also incorporated in the experiment. A good treatment of instrumentation is found in Liptai *et al.* (1969), and Tatro (1971, 1972) describes the total experimental situation.

Efforts to derive information from pulse height and spectral analysis have not often proved fruitful. One reason for this is indicated in the work of Malone (1965). He showed that the apparatus and experimental technique used can alter the received emissions significantly. Certain frequencies, related to the piezoelectric crystal and specimen resonances, are enhanced "artificially." Some of these problems are discussed in Liptai and Harris (1971). Tatro and Kroll (1964) and Egle and Tatro (1967), whose work was discussed earlier, were concerned with this very problem. There is probably little hope of ever recording the *true* nature of an emission. The use of total emission counts and count rate usually seems the most rewarding approach.

Work in Rocks: There are two schools of study of acoustic emissions from

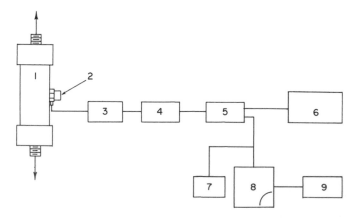

FIG. 6. Experimental setup used by Hardy. 1—Rock specimen, 2—accelerometer, 3—preamplifier, 4—variable gain amplifier, 5—bandpass filter, 6—magnetic tape recorder, 7—speaker system or earphones, 8—cathode ray oscilloscope, 9—electronic counter.

stressed rock. Mogi (1968) and Scholz (1968a,b,c,d) worked in the frequency range from 10–100 kHz to 1 MHz, while Hardy *et al.* (1972) and Knill *et al.* (1968) used the lower frequency region from 2 to 10,000 Hz. Figure 6 shows Hardy's setup. Knill *et al.* used an accelerometer and charge amplifier, the output of the charge amplifier going into a high speed uv recorder or a tape recorder. Mogi and Scholz used ceramic transducers (Ba TiO$_3$ or PZT), preamplifiers, amplifiers, and a tape recorder (Scholz) or photography of oscilloscope tracings (Mogi).

Work in Other Materials: The instrumentation used for acoustic emission work in other materials is usually based on the type described above. If there is any difference, it will be noted in the section on the individual material or process.

F. Types and Models of Acoustic Emissions

Continuous Emissions: These emissions appear in metals as a phenomenon similar to noise but are actually traceable to the deformation process. They occur rather steadily, are of high frequency, and very small magnitude. Their amplitude increases with increasing tensile stress and then decreases once yielding occurs. Continuous emissions supposedly arise from dislocations moving through the crystal and possibly slip movements. There seems to be no work at all concerning this type of emission in nonmetals.

Burst Emissions: As the name implies, this component of acoustic emission does not occur continuously, but in bursts of higher amplitude than the continuous emissions. These emissions, which are supposedly associated with failures such as twinning, microcracks, and blocks of dislocations breaking away from obstacles, occur at larger plastic strains. Burst emissions can also be associated with growth of an existing crack. The source of emissions in certain nonmetals depends on the particular material and will be discussed in the appropriate section.

Models of Emissive Sources: It is safe to say that at present there is no rigorous theory for the actual internal mechanism of acoustic emissions in any material. There are some heuristic models and a very few specialized rigorous attempts have been made. Liptai *et al.* (1969) and Engle and Dunegan (1969) are of the opinion that breakaway of dislocations at obstacles is a major source of acoustic emissions in metals.

Pollack (1968) used a simple argument of a stretched spring coupled to a spring. If, during deformation, the spring moves fast enough, oscillations of the spring can develop and it is assumed that these oscillations can radiate elastic waves. If the system moves slowly, then no significant oscillations will develop. The string could be a dislocation or even possibly a surface of a crack.

Models of the sources of acoustic emissions are given in Gillis (1971, 1972) and Tetelman and Chow (1972). Frederick and Felbeck (1972) developed a model based on the activation of dislocation sources and the subsequent shutting off of the sources by the back stress of piled-up dislocations. Their model was able to explain the effect of microstructure of the metal on the acoustic emissions.

Engle (1966) used single crystals of lithium fluoride oriented for easy glide under direct shear. Acoustic emissions and specimen displacement pulses were monitored. The largest displacement pulses (10^{-5} in.) were observed in coincidence with large acoustic pulses. Smaller displacement pulses (2×10^{-7} in.) and emissions were observed together but not in coincidence. Some large acoustic emission pulses had no displacement pulses associated with them. On the basis of his result, he was able to determine group velocities of the dislocation which agreed well with previous measurements in lithium fluoride. The model proposed for the emissions involves an interaction between groups of piled-up dislocations and the obstacles that cause the pile-up. "The pinning causes an increase in local strain energy stored in the region of the obstacle. When the driving stress on the leading dislocation, composed of the applied stress and the additional stress due to the pileup itself, is large enough to cause breakaway and acceleration of part of the group, the local strain energy is availabe to excite lattice vibrations that appear as acoustic emissions." He considers other sources, including ones at high stresses when the stress concentrations at the leading edge of dislocation groups cause crack nucleation.

Sedgwick (1968) worked with LiF and KCl loaded in compression. He concluded that the main acoustic emission mechanism in ionic single crystals within the elastic region was the operation of fast dislocation sources (e.g. a Frank-Read mechanism). According to Sedgwick, the LiF had much more acoustic emission than the KCl because LiF is a "hard" material (fast dislocation movement) whereas KCl is a "soft" material (slow dislocation movement).

Liptai *et al.* (1971) model one source on a certain grain boundary source, where during straining the upper half of the grain slips over the lower half by a certain distance. Using simple elasticity theory, the energy change accompanying the slip is calculated. They consider the initial event as an impulse which sets the grain into resonant vibration, with the upper half shearing over the lower half. With a grain diameter of 5×10^{-3} in. and using the physical properties of aluminum, this frequency of vibration is found to be about 2 MHz. Other estimates by Liptai *et al.* led to higher frequencies.

Armstrong (1969) used a simple elastic energy release model for growing cracks, incorporating the Griffith crack theory. He obtained frequency of emissions of the order of 10^5 Hz, not unreasonable in view of the results of Scholz (1968a) and Mogi (1968). However this model would not seem to apply to the low frequency work of Hardy (1972) and Knill *et al.* (1968). Armstrong does not mention that the Griffith crack theory as applied to the problem appears to set an upper limit on the frequencies, but not a lower limit which is set by whatever features determine the maximum crack length.

Scholz (1968a) has considered the sequence of events leading to acoustic emission in brittle rocks. At low stresses, the emissions are attributed to frictional sliding of preexisting cracks and crushing of pores. At moderate stresses, the material is nearly linearly elastic and few emissions occur. At high stresses, acoustic emissions return caused by propagation of new cracks.

With regard to the propagation aspect, Ang and Williams (1959) studied the elastic radiation from a dislocation propagating with constant velocity. The dislocation treated is a step function discontinuity in the z-component of the displacement field which starts along the line $x = 0$ and propagates with constant velocity in the plane $z = 0$. The elastic medium is initially unstressed. Linear elasticity and plane strain are assumed. The range of velocities from zero to infinity is considered. The problem is formulated in terms of Fourier integral equations and solved in closed form.

Knopoff and Gilbert (1960) have treated rigorously a number of possible emission sources resulting from propagation of a dislocation in strain or displacement along a fault. They also use a unit step function and treat eight fault types. Their results apply to first motion produced by these sources at great distance from the fault. Some of the types of sources considered are: (1) two sides of fault offset with respect to each other and in a direction parallel to propagation direction, (2) dislocation transverse to direction of propagation of fault, (3) sudden expansion or collapse of a lenticular cavity, and (4) sudden discontinuity in shear strain between the two sides of the fault plane. Other types of sources are also included.

Savage and Mansinha (1963) have investigated experimentally the radiation pattern for elastic waves from a tensile fracture in a two-dimensional model. The waves were initiated in the glass sheet by applying a flame to a scratch. The observed patterns from the first motions were compared with the theory of Ang and Williams (1959) and Knopoff and Gilbert (1960). The experimental conditions did not reproduce all of the conditions of either theory but were felt to be close enough to warrant comparison. There was general qualitative agreement, but poor quantitative agreement. It is pointed out that there may be significant problems in applying linear elasticity to fracture processes which certainly must take into account nonlinear effects.

This, together with earlier work, emphasizes the extreme difficulty in learning about the actual acoustic emission source from the received record on the instrumentation. The elastic radiation problem of typical sources is a very significant mathematical problem fraught with difficulties, and then there is the problem of mechanical and electrical filtering in the system itself.

Eshelby (1949) has calculated the energy radiated by a harmonically-vibrating screw dislocation. The result is very simple and could be used as a crude estimate of the acoustic emission energy radiated by a moving single dislocation.

IV. Materials Investigated with Acoustic Emission

A. UNFLAWED METAL SPECIMENS

There has been some work correlating the acoustic emissions with the number of slip events (Schofield, 1961; Fisher and Lally, 1967). The work of Fisher and Lally indicated that slip events occur cooperatively and that dislocation velocities during yielding may be as high as 10^3 cm/sec.

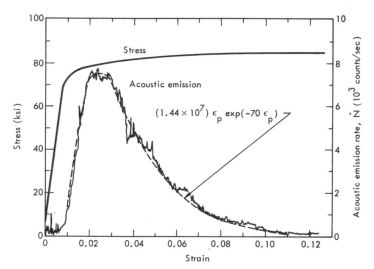

F*IG*. 7. Acoustic emission versus strain in 7075-TS aluminum. Comparison with Gilman's mobile dislocation model.

The results of one of the most beautiful acoustic emission experiments to date is presented in Fig. 7, which shows acoustic emission rate as a function of strain for a 7075-TS aluminum tensile specimen (Dunegan and Harris, 1969). Superimposed on the acoustic emission data is a fit of Gilman's (1966) mobile dislocation model, which gives mobile dislocation density as a function of plastic strain. The equation expressing the mobile density of dislocations is

$$N_m = m\,\varepsilon_p\,e^{-\phi\varepsilon_p}, \tag{1}$$

where N_m is the mobile dislocation density, ε_p the plastic strain, m the dislocation breeding factor, $\phi = H/\sigma$, where H is the hardening coefficient, and σ is the root mean square stress. Using these acoustic emission results, it is very easy to obtain the hardening coefficient, a very important mechanical parameter. The possibility of using acoustic emission measurements to obtain fundamental parameters such as this was dealt a severe blow when it was found that the results for many materials did not fit Gilman's equation well at all. In particular, Dunegan and Harris (1969) and Dunegan and Tatro (1971) describe the lack of fit for an iron–3% silicon tensile specimen. Possible reasons for the lack of fit are presented, e.g. inhomogeneous strain due to Lüder's bands.

Gerberich and Reuter (1969) have also used Gilman's model to determine the values of the work-hardening coefficient of 7075 aluminum with various heat treatments.

The author is not aware of a comprehensive program which may have been undertaken to ascertain how many materials (and under what conditions) will obey Gilman's equation. It would seem to be a very fundamental area of cooperation between engineering and science.

B. Flawed Metal Specimens

Acoustic emission techniques have great potential for detecting cracks and other flaws. These flaws act as stress concentrators and produce localized plastic deformation at macroscopic stress levels well below general yielding. The localized plastic deformation, in general, produces acoustic emissions and the emissions can thus be used to detect the onset of such flaws. Hence acoustic emission, probably better than any other nondestructive test, allows the monitoring of engineering structures for integrity against such flaws. (Structural integrity evaluation will be discussed in a later section.)

Early work in this particular area used rather low sensitivity acoustic emission measurements to determine the "pop-in" stress for cracked (notched) specimens (Romine, 1961; Jones and Brown, 1964). The fracture toughness can be computed from the pop-in stress and the appropriate equations for the stress intensity factor. (Unstable fracture occurs in structural materials when the stress intensity factor of a growing crack, K, is equal to the fracture toughness of the material.)

Recent work with more refined instrumentation has also related acoustic emission to flaws and analyzed the results using fracture mechanics (Dunegan et al., 1968, 1969, 1970; Dunegan and Harris, 1969). There was some work on "pop-in" here but the main thrust was to relate acoustic emission to the stress intensity factor, K, and also to look for crack growth during proof loading. Stresses near the tip of a crack in an elastic solid are completely controlled by the stress intensity factor. (This factor depends on the shape, size, and location of the crack, the specimen geometry, and type of loading.) It is assumed by these workers that plastic deformation is highly localized near the crack tip and hence the plastic zone size will be controlled by K. Thus acoustic emission will depend on K. Dunegan et al. (1968) have proposed a theory for the relation between the acoustic emission from a flawed specimen and the stress intensity factor for the crack. The prediction is that the accumulated acoustic emission count N should be proportional to the fourth power of K. This result applies for through cracks in plates. A similar expression has been obtained by Dunegan et al. (1969) for penny-shaped cracks, but here it is not possible to directly determine K from the acoustic emission without also knowing the flaw size. Thus from theory it would appear that acoustic emission measurements would allow the determination of the flaw condition.

Measurements on single-edge-notch (SEN) fracture toughness specimens have shown the exponent n in the expression

$$N \propto K^n$$

to be more like 4–6 for 7075-T6 aluminum and about 8 for beryllium. Figure 8 shows the results from four specimens which happened to show the K^4 behavior. Work on multiple crack specimens have also shown that the emission is again controlled by K.

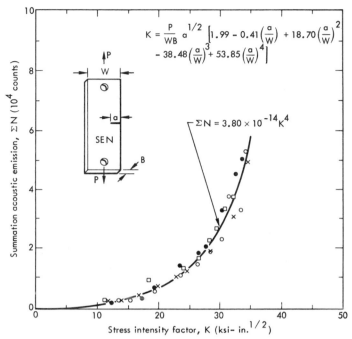

FIG. 8. Accumulated acoustic emission as a function of stress intensity factor. $\times$, $a_0 = 0.550$ in.; $\bigcirc$, $a_0 = 0.500$ in.; $\bullet$, $a_0 = 0.360$ in.; $\square$, $a_0 = 0.325$ in.

Dunegan *et al.* (1969, 1970) have devised a scheme whereby they can say whether cracks are growing at a fixed load or have grown during repeated proof loads. If, while holding a cracked specimen at a proof load, emission occurs, then the crack is propagating. There should be no acoustic emission upon repeated loadings to the proof load. If emissions occur at a load less than the proof load then crack growth must have taken place during the proof loading scheme.

Gerberich and Hartbower (1967) have found some very interesting and potentially useful empirical relations between crack parameters and acoustic emission. It was found that the number and size of the acoustic emissions seemed to bear a unique relationship to the amount of slow crack growth. A semiempirical relationship was developed from elasticity theory. The result was

$$\Delta A \sim \left(\sum g\right)^2 E/K^2,$$

where ΔA is the incremental area swept out by the crack, $\sum g$ is the sum of the stress wave amplitudes associated with the increments of growth, E is the elastic modulus, and K is the applied stress intensity factor. Figure 9 shows the data comparing experiment with the empirical theory for steel, aluminum, and titanium.

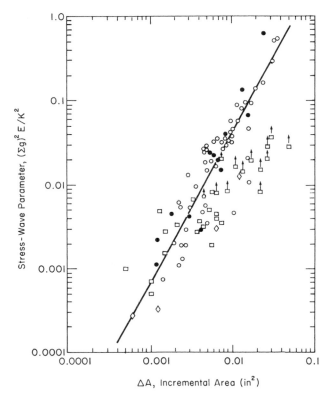

FIG. 9. Stress wave parameter versus incremental crack area for certain materials ○, D6aC steel—high stress intensity fatigue; ●, D6aC steel—environmental cracking under sustained load; □, 7075 T6 aluminum—rising load; ◇, 6Al-4V titanium—rising load.

They also found that a reasonably linear relation exists between $(\sum g)/$ cycle and crack growth increment/cycle. This result is shown in Fig. 10. The slope in this relationship is very much dependent on the material and the condition of the specimen.

Acoustic emission techniques have been used to study slow, subcritical crack growth in D6aC steel under hydrogen embrittling, stress corrosion cracking, and fatigue conditions (Hartbower *et al.*, 1968), under hydrogen embrittling and stress corrosion cracking conditions (Gerberich and Hartbower, 1969), and under stress corrosion cracking conditions (Hartbower *et al.*, 1972). Dunegan and Tetelman (1971a) studied hydrogen embrittlement in 4340 steel using acoustic emission. Katz and Gerberich (1970) studied stress corrosion cracking in titanium alloys and Dunegan and Harris (1974) used acoustic emission on a precracked uranium 0.3% titanium specimen in salt water.

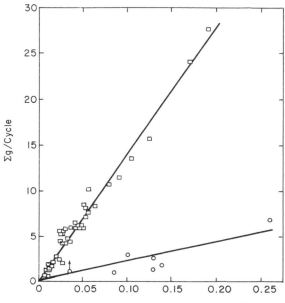

FIG. 10. Sum of stress-wave amplitudes per cycle versus crack growth increment per cycle. □, Specimens 1750-1,2; ○, specimens 1550-1,2,3.

C. Rocks

As mentioned in the Introduction, the area of acoustic emission (often called microseisms) in rocks has been adequately surveyed in two review articles—Knill *et al.* (1968) and Hardy (1972). Hence, an attempt will be made here to summarize the work in rocks very briefly. The very early work was summarized in Section II. Most of this was field work with a very small amount of laboratory work. The bulk of the subsequent knowledge gained in the laboratory came form three workers—Mogi, Scholz, and Hardy.

1. *Laboratory Work*

Mogi (1962a,b, 1968) performed bending tests on various rocks and monitored the acoustic emissions as well as stress and strain. In the 1962 work, a crystal cartridge type pickup was used, so acoustic emissions in the audio range were monitored. The 1968 work employed lead-titanate-zirconate compressional mode disks 3 mm thick to determine the source location; hence emissions used here were in the high kilohertz or low megahertz range. He found some striking similarities between rock emissions (microfracturing) of these laboratory samples and actual earthquakes in regard to their statistical behavior. The buildup of microfracturing (as determined by observing the acoustic emissions) before fracture is very often similar to foreshock behavior in earthquakes. Fracture itself may correspond to the main shock of the earthquake.

Some similarities between the amplitude–frequency behavior in laboratory microfracturing and earthquakes was found. In seismology, this relation is embodied in the Gutenberg–Richter relation (Gutenberg and Richer, 1954)

$$\log N = a + b\,(8 - M),$$

where N is the number of earthquakes, M the instrumental magnitude introduced by Richter, and a and b are constants. Ishimoto and Iida (1939) have a similar relation

$$n(a)da = Ka^{-m}\,da,$$

where a is the maximum trace amplitude of earthquakes, $n(a)da$ the number of earthquakes having a maximum trace amplitude a to $a + da$, and K and m are both constants. There is a relation between b and m in the above formulas, namely

$$b = m - 1.$$

Mogi determined the amplitude–frequency curve for a large number of natural rocks and "artificial rocks" (pine resin, pine resin with mechanical irregularities, granular pumice, or coal) of varying degrees of heterogeniety. If the structure was sufficiently heterogeneous, the amplitude–frequency relation for the laboratory specimens obeyed the above equations. A specimen with a regular structure does not obey them. The value of m varied from 0.3 to 2.7 depending on the structure of the medium and the state of the applied stress. The m-value increased with the degree of heterogeneity. Mogi hopes that by comparing the m-values of lab tests and actual earthquakes, new information can be obtained concerning the heterogeneity and state of stress of the seismic region. The assumption was made that the elastic shocks occur stochastically, with a transition probability which depends on the state of stress. Using this, the frequency curve of elastic shocks under variable stress was calculated, the degree of heterogeniety being an important parameter.

Mogi's 1968 work was involved with an accurate determination of microfracture source location in rocks subjected to bending. Here he needed high frequency emissions for good time resolution. He found that in heterogeneous rocks at low stresses the sources were at random locations. However, before failure, the microfracturing was concentrated in limited regions, one of which initiated fracture. In homogeneous rocks, fracture takes place quite suddenly without the above microfracturing sequence. The spatial distribution of sources has a certain resemblence to that of earthquakes.

Scholz (1968a,b,c) made a detailed study of the microfracturing of brittle rocks using acoustic emission and stress–strain measurements. One of the unique features of Scholz's work (along with that of Mogi, 1968) is his use of the high frequency region (100kHz to 1 MHz); most other work in rocks had been done in the audio range. He also found that microfracturing events in many laboratory samples obeyed the Gutenberg-Richter of the equivalent Ishimoto-Iida relation, and measured b-values for a number of rocks with widely varying properties (Westerly granite, San Marcos gabbro, Colorado rhyolite tuff, marble, Pottsville sandstone, and Rutland quartzite). Results were obtained for uniaxial compression and Westerly granite was also compressed under various confining pressures. Early workers had not used confining pressures. The a-values and b-values were quite different for ductile and brittle rocks. In particular, the value of the coefficient b depends very strongly on the state of stress and to a lesser degree on the rock type properties. Confining pressures, up to 5 kbar, did not produce a significant change in the microfracturing behavior, as indicated by the acoustic emissions. It is important to make such studies at high pressures (and high temperatures) as the properties of rocks change with these parameters; rocks brittle under atmospheric pressure and room temperature become ductile at increased pressure and temperature (Mogi, 1962a). Scholz (1968b) also determined the location of the sources and found (as did Mogi) that the locations of the initial sources were not related to the eventual fault plane. He found that the microfracturing could be directly related to the inelastic part of the stress–strain behavior, i.e. creep in brittle rock at low temperatures is due to time dependent cracking (microfracturing). The behavior could be adequately described by a statistical model in which the rock is treated as a heterogeneous elastic material.

Scholz (1968c) performed an extremely interesting laboratory experiment which is of interest with regard to earthquake aftershocks. He found that microfracturing can still be observed after brittle fracturing of the rock, provided provisions are taken so that the specimen remains intact. If the intact sample is isolated after fracture, microfracturing activity decays hyperbolically in a manner similar to typical earthquake aftershock sequences.

Hardy, at Penn State, has been active in acoustic emission studies for a number of years. Chugh *et al.* (1968) investigated the frequency spectra of the acoustic emissions in Crab Orchard Sandstone, Indiana Limestone, and Barre Granite. The frequency range covered was from about 0.5 to 15 kHz. They concluded that frequency spectra provide only limited information with regard to the actual source of microseismic activity. (Frequency analysis is quite tenuous, for almost every element in the specimen instrumentation system has some frequency dependent properties, which change the nature of the emission as it propagates.) The amplitude–frequency spectra were not sufficiently stress dependent for use in predicting impending failure.

Hardy *et al.* (1970) have made a detailed study of creep and microseismic activity in a number of rocks of the type used by Chugh *et al.* (1968). Previous workers had already indicated that there was a correlation between creep and microseismic activity (Scholz, 1968d; Vinogradov, 1959; Gold, 1960, 1968). It was found that both the axial creep strain and accumulated acoustic emission activity versus time data could be fitted to the Burgers viscoelastic model. For all three rock types, a nearly linear relation existed between emission activity and creep strain. The actual degree of correlation between creep strain and microfracture is not known, but there is certainly strong correlation between the two. A very detailed study would be needed to resolve this question, as rock fracture is a quite complicated process (Bieniawski, 1967a,b, 1969).

Hardy (1969, 1972) was interested in using acoustic emission to assess the stability of underground gas storage reservoirs. Little is known about the fracturing process which goes on below the surface when fluids are injected under pressure into low permeability strata to increase storage capacity. It is hoped to be able to locate unstable regions in the cap rock. Small laboratory size gas storage reservoir models have been developed in order to further understand the process.

Some early work by Goodman (1963) is worth mentioning. Using relatively crude instrumentation, he observed the subaudible noise generated during repeated compressions of rocks. Emissions were observed during unloading as well as loading of the specimens. A load called the point of "accelerated rock noise activity" was found. If the load was not increased beyond this point, then fewer emissions were detected during subsequent loading cycles. If the load was increased beyond this point, the number of emissions may be even larger in subsequent loadings. Some of the acoustic emission activity which disappears with repeated loading cycles, can be recovered if the specimen remains unloaded for a period of time. The major recovery in acoustic emission generating ability occured within the first twelve hours of the rest period. Thus the Kaiser effect, which is so very important in metals, is not generally observed in rocks.

2. *Field Work*

Some work has already been described in Section II. In this section, a very brief comment will be made on some additional field work. The majority of the work has been involved with determining instabilities in underground mines. Cook (1963) used acoustic emission monitoring of the Witwatersrand gold mines of South Africa. He used electromagnetic transducers with a flat response from 15 to 300 Hz, and there were provisions for recording 16 outputs on tape. Sources of the emissions were determined from triangulation studies. Cook concluded that it should be possible to make rockbursts occur at opportune times from information obtained via acoustic emission monitoring.

A good account of relatively recent Soviet research is contained in Vinogradov (1959). It is concerned with acoustic emission monitoring of the Anna lead mine in Czechoslovakia. Work was done at a depth of 1300 m, in a region very prone to rockbursts. Electrodynamic pickups were used and the passband of the system was 800 to 4000 Hz. The emission rate of the pulses did not prove sufficient for rockburst prediction; therefore the data was presented in other ways. The pulse energy (approximated by the square of the amplitudes) was plotted against time; the distribution of the number of pulses as a function of their energy was used, and also other special ways of presenting the data. These latter methods of analyzing the data proved more helpful in predicting rockbursts.

Some of the other recent field works are listed here. Work with regard to coal mines has been carried out by Stas *et al.* (1971), Sasaki and Takata (1970), and Bollinger (1970). Hedley *et al.* (1969) have monitored an iron ore mine and Sasaki and Takata (1970) and Blake and Leighton (1970a,b) have worked in various metal-producing mines.

D. COMPOSITE MATERIALS

There has been great interest lately in the structural use of composite materials. The hope is to develop materials which are lightweight and yet strong. It is extremely important to understand the very complex deformation and fracture processes in such materials. Acoustic emission analysis techniques have been brought to bear on this extremely difficult problem and some encouraging preliminary results have been obtained. The first structural integrity investigations were performed on rocket motor cases made from composite materials (refer to Section III, D).

Liptai *et al.* (1971) and Liptai (1971) give brief treatments on the general nature of composites. Here discussions are given of the three basic types—dispersion-strengthened, particle-reinforced, and fiber-reinforced—and possible sources of acoustic emission in each type. Liptai gives results for acoustic emissions versus repeated loading for rings of Fiberglass with an epoxy resin matrix, and glass-epoxy pressure vessels. He concluded that acoustic emission data are helpful in establishing the mechanisms governing operative fracture modes and assessing structural integrity.

Harris *et al.* (1972) made measurements on Al_3Ni whisker-reinforced aluminum composites. These fall in the category of fiber-reinforced composites. The percentage of broken fibers and acoustic emission were measured as a function of tensile strain. The authors developed a simple model for the number of acoustic emissions as a function of the number of broken whiskers. The experimental data of emissions versus percentage of broken fibers was compared with the theoretical result. The agreement was reasonably good, again indicating that acoustic emission can be used to monitor failures in composites.

Gerberich (1970) used acoustic emission to establish the point at which fiber fractures occured in composites made from ductile stainless steel fibers in an age-hardened aluminum matrix.

Fitz-Randolph *et al.* (1972) investigated boron fiber–epoxy resin composites using acoustic emission and other methods. The indications of this preliminary work were that acoustic emission could be related directly to fracture surface energies.

Hay *et al.* (1972) have worked with stainless steel and tungsten fibers in an aluminum matrix. Various types of loading were used and some preliminary frequency analysis was undertaken by using passband filters of frequencies of 0.1 to 0.3 MHz or 1 to 2 MHz. The stainless steel fiber specimens were noisier than those with tungsten fibers. The emission behavior was very dependent on the type of loading [Liptai, (1971) also found this in his Fiberglass specimens]. This is to be expected because the type of loading determines the types of failure experienced by a composite specimen. They found preliminary indications from their acoustic emission data that in two nominally similar specimens, the one with a considerably lower elastic modulus also had much larger emission at low strains. This could indicate that microcracking was contributing to the lower modulus.

Mehan and Mullen (1971), Mullen and Mehan (1972), and Mullen *et al.* (1971) have been interested in failure modes in epoxy composites and their possible detection and categorization via acoustic emission techniques. The failure mechanisms considered were fiber breakage, matrix cracking, and fiber debonding or pullout. They attempted to identify various failure modes expected to occur in composites and to catalog characteristic acoustic signatures relating to individual failure modes. The best system for this work seemed to be the boron–epoxy system. Some work was done on single filament specimens to isolate the mechanism as much as possible. These authors feel that frequency analysis of the emissions at failure offers promise with regard to ascertaining the reliability of structures and understanding the failure mechanisms. In their work on composite failures, the emissions are handled in such a way that each emission registers only one count, independent of the size of the emission.

E. Concrete

Rüsch (1959) did some early work on acoustic emission in concrete. He used a piezoelectric crystal microphone as a detector for the elastic waves. From the acoustic emission results and measurements of the change in volume and absorption of ultrasonic waves in concrete, he was able to infer that in the typical brief duration tests, considerable disruption of the internal structure occurs at about 75% of the failure load.

Green (1969b) has looked at acoustic emission from laboratory type cylindrical concrete specimens and a scale model prestressed concrete reactor pressure vessel. Acoustic emissions were found to be an indicator of failure processes in concrete, and early warning of total compressive failure was indicated. Resolution of gross cracking, onset of pressure vessel failures and leakage, and prestressing rod failure were all indicated. A preliminary correlation between acoustic emission and material modulus was obtained.

F. Ceramics

Romrell and Bunnell (1970) have used acoustic emission to monitor crack growth in ceramic tubes due to thermal shock. Thorium–yttrium oxide tubes, which are used as an electrolyte to measure oxygen concentration in liquid sodium, were thermally shocked by immersing in molten sodium. The acoustic emission results correlated well with microscopic observation of cracks. A high level of emissions from one tube was due to sodium corrosion of the ceramic. Aluminum oxide tubes were thermally shocked by inserting one end into a furnace and heating to 1250°C. Again, as in the case of the thorium–yttrium tubes, acoustic emissions could be detected well before failure of the tubes.

Gatti *et al.* (1971) observed acoustic emissions in the transparent ceramics Lucalox alumina (G. E. product) and special spinels. Specimens were subjected to 4-point bending. The Lucalox proved to be much "noisier," but there was no clear reason for the difference. Flaws were also introduced by thermal shock (using a flame). The flaws were then extended under load and the acoustic emission monitored. Some work was also done on a thin single crystal sapphire disk. The crack growth was monitored photographically and an attempt was made to correlate crack growth and the acoustic emission results. There appeared to be no simple correlation between acoustic activity and the crack surface generated.

G. Ice

Gold (1960) investigated the creep of ice using acoustic emission techniques. He was interested in the degree to which cracking contributed to the various stages of creep. The cracking activity was determined from the acoustic emission. Later work on ice is found in Gold (1968). Some workers in the acoustic emission field have used Gold's example to prove that cracking will produce acoustic emission, as it is possible to actually observe the cracks forming.

H. Soils

Cadman and Goodman (1967) appear to be the only workers who have used a definite soil material (refer to Section V, C).

Koerner and Lord (1972) investigated soils under controlled stress conditions in the laboratory. The instrumentation was very simple. An accelerometer, of bandwidth from a few hertz to 6000 Hz, picked up the emissions. The output of the accelerometer was fed into a charge amplifier, and the amplifier output went directly into an electronic counter, with a trigger level of 0.10 V. The amplified emissions were also observed on an oscilloscope. The sensitivity of the system was 20 V/g. The soil tested was a slightly organic, medium plasticity, clayey silt. The liquid limit was 55%, the plastic limit 35%, and the shrinkage limit 27%. The soil was remolded and compacted into samples 2.8 in. in diameter and 5 in high. The specimens were tested in unconfined compression. The load was applied in 2.5 lb increments (until failure) with a hydraulically operated hand press. Tests with an aluminum specimen (which does not emit at these low stresses) showed that with the instrumentation sensitivity used, the loading system produced no "artificial emissions."

Four water contents were investigated and the results are shown in Fig. 11. As was expected, the drier, more brittle samples had much higher emission at the same stress level. Experiments with different types of contacts

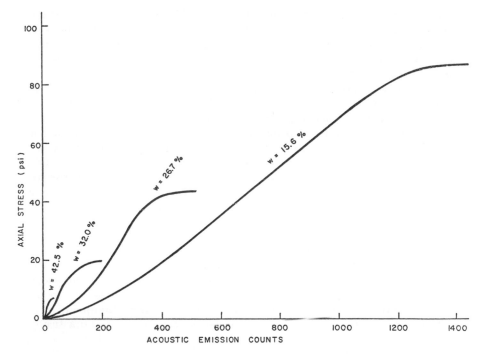

FIG. 11. Acoustic emission versus stress in a medium plasticity, clayey silt at various water contents.

between the specimen and the loading apparatus indicated that the emissions arose from the bulk of the sample and not the contacting surfaces. The stress–acoustic emission curves showed a strong resemblance to the stress–strain curves of the material. Thus the possibility exists of using acoustic emission measurements in the field or the laboratory as an alternative to stress and/or strain measurements.

Further work has been done in dry soils by Lord and Koerner (1974). The instrumentation used was similar to that of Koerner and Lord (1972), the only difference being that an acoustic steel wave guide was used between the sample and the accelerometer. The wave guide was a ¼ in. diam. piece of drill rod about 1 ft long. It was pointed on one end (this end was placed in the soil) and the accelerometer was screwed into the other end. The wave-guide was incorporated into the experimental setup for a very practical reason. A significant use of acoustic emission in soils might be the monitoring of the stability of earth embankments. Emissions might not travel very far in the soil, due to the high attenuation of elastic waves in soils; however, a wave guide (a good conductor of sound) could pick up emissions from any point near its surface and conduct them to the accelerometer. Thus the "range" of each accelerometer is increased many-fold. Wave guides have already been used by other workers for various purposes (Dunegan and Tetelman, 1971b; Vetrano and Jolly, 1972; Speich and Fisher, 1972; Anderson *et al.*, 1972). The soils were tested in unconfined compression using the hand-operated hydraulic press. A dial deflection gage was used to measure strains, Various blends of sand and clayey silt were investigated.

The stress–strain and stress–acoustic emission responses of the five soils tested are presented in Figs. 12 through 16. The curves are typical of the four or five tests made for each type of soil. There is certainly a correlation between cumulative acoustic emissions and strain, but the present results do not give a linear relation between the two as did the work in rocks (Hardy *et al.*, 1970). It would seem that soils produce enough emissions via a wave guide pickup technique to consider seriously the possibility of monitoring earth slopes for stability. There has been much loss of life and many streams have been seriously polluted due to the failure of soil slopes or dikes.

Some preliminary tests on soils using the commercial Dunegan Research Corporation acoustic emission apparatus failed to indicate any emissions whatsoever. From this it was concluded that the emissions in soils are of frequencies much below 100 kHz, and that the frequency response of the present instrumentation was adequate. Viewing the emissions on the oscilloscope indicated that the primary frequencies are in the very low kilohertz region.

It should be mentioned, from a fundamental research standpoint, that soils (along with other materials whose properties can be changed drastically

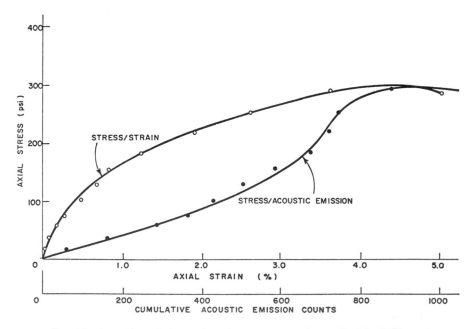

FIG 12. Acoustic emission and strain versus stress in a soil with 100% sand.

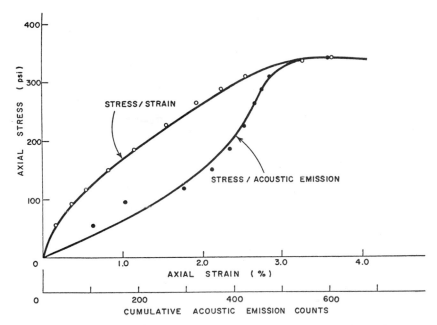

FIG. 13. Acoustic emission and strain versus stress in a soil with 67% sand and 33% clayey silt.

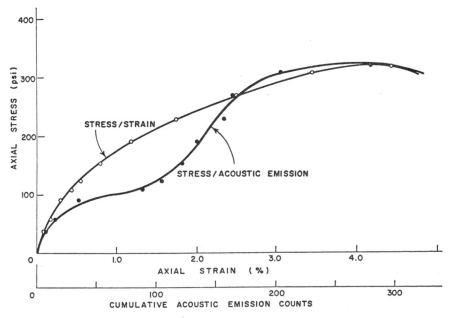

FIG. 14. Acoustic emission and strain versus stress in a soil with 50% sand and 50% clayey silt.

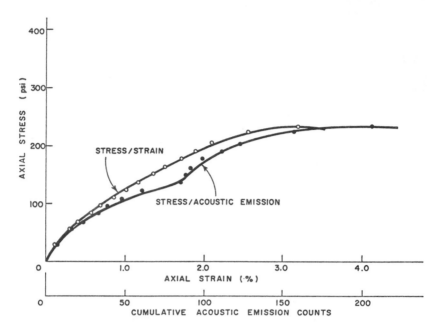

FIG. 15. Acoustic emission and strain versus stress in a soil with 33% sand and 67% clayey silt.

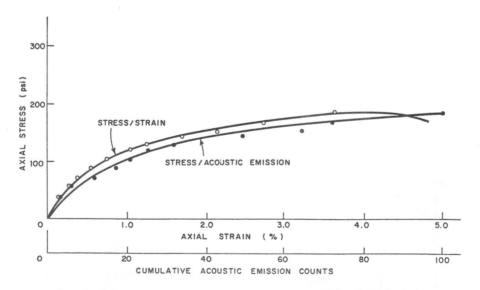

FIG. 16. Acoustic emission and strain versus stress in a soil with 100% clayey silt.

via frabrication) offer a unique opportunity to determine the nature of the source of emissions. For example, Lord and Koerner (1974) found that emissions are greater in the sandy soils than in the fine-grained soils. This suggests that sliding friction is probably more significant than rolling friction in producing acoustic emissions.

I. Wood

There is little work published concerning acoustic emission in wood. In fact, only one article was uncovered in the literature search for this article. Kishinouye (1937) observed the emissions in wooden beams under flexural stress. Unfortunately, at the time of this writing a copy of the article has not been obtained and no details of his work can be given. Both Green *et al.* (1970) and the Jersey Nuclear Co. Report mention observing emissions in wood. The Jersey Report states that emissions from fiber failure in wood are somewhat similar to those from flawed metals, but are larger and lower infrequency.

Meunow of the Law Engineering Testing Co. in Atlanta, Georgia has used acoustic emission to verify the integrity of many civil engineering structures. In performing safety inspections of buildings, the building is water-loaded. Acoustic emission sensors are placed under the load and on a circle with center on the load line. In this manner, areas of soft wood can be found, and in one case an area damaged by an unreported fire was located. (This same technique offers the detection of weld problems in structural steel and areas of deterioration in concrete.) Law also uses acoustic emission to determine the approximate locations of areas of poor bonding in large laminated wooden beams. The suspect areas are then examined by radiography or low frequency ultrasonic techniques, to determine the extent of the nonbond areas. This approach eliminates the need for the much more costly complete examination by the more conventional radiographic and ultrasonic techniques. The complete details of Law's work are not available, due to the private nature of information between client and consultant. It is most interesting that this work was performed in the low kilohertz region, where ambient noise interference would be expected to be very strong.

Because of the lack of published data on wood, the author performed some acoustic emission tests on about 50 small fir specimens. The specimens of size $\frac{1}{2}$ in. thick, $1\frac{1}{2}$ in. wide, and 12 in. long (8 in. between support points) were all cut from two by fours. A center load produced flexural stress in the specimens. The commercially available Dunegan Research Corp. acoustic emission apparatus was used. The sensor was placed about an inch from the center of the specimen. All runs were made with the filter off and an overall gain of 92 dB. Figure 17 shows the accumulated emissions versus the load (expressed as a percentage of failure load) in specimens of different water contents. The specimen of 46% water had a curve almost identical with that of the as-received specimens. These latter specimens failed at an average load of 230 lb. The orientation between the direction of loading and the grain

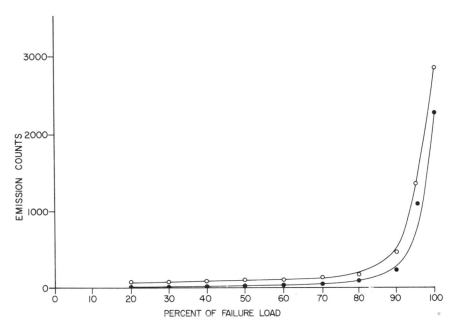

FIG. 17. Acoustic emission versus load in clear fir as a function of water content.
○, Dry wood 0% H_2O (failure at average of 435 lb); ●, as-received 11% H_2O (failure at average of 352 lb).

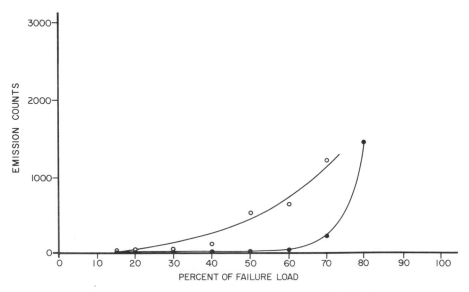

FIG. 18. Acoustic emission versus load in clear dry fir during first and second load application. ○, Trial 1; ●, trial 2; failure at 485 lb.

varied from 40° to 10°. Four specimens were run at each water content. A typical total count before failure was about 5000. The time duration between initial load application and failure was about 20 min.

Some preliminary data was gathered with regard to irreversibility in emission versus load curves. Figure 18 shows the results for a clear beam, while Fig. 19 depicts the case for a beam with a sizeable knot in it. Both experiments were for completely dry wood. Although there is not a strict "Kaiser effect," there is strong irreversibility which could be of significant

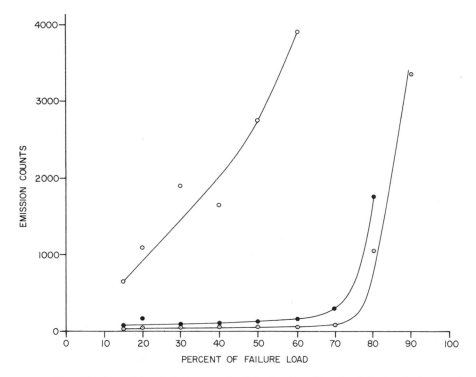

FIG. 19. Acoustic emission versus load in dry fir with a knot during first, second and third load application. ○, Trial 1; ●, trial 2; ◉, trial 3. Failure at 435 lb.

use in nondestructive testing. The knot seemed to be the source of significant emission at quite low stresses.

These results are strictly of a preliminary nature and a great deal more work must be done to get a reasonably clear picture of the acoustic emission behavior of woods. For example, time effects may be important.

Undoubtedly many more materials (unknown to the author) have been investigated and will be investigated in the future. For example, plastics must have received some attention by now.

V. Processes Studied with Acoustic Emission

A. WELDING

Notvest (1966) was the first worker to apply acoustic emission techniques to the welding process. Crack sensitivity (in D6 Ac steel) during the welding cycle was evaluated by a restrained joint cracking test. The cracking was associated with transformation reactions, and was detected by acoustic emission measurements. Test welds were considered to have no cracks when no acoustic emissions were recorded. He was able to develop certain thermal weld cycles that produced no emissions and applied this knowledge to the weld fabrication of Titan III rock motor cases. Fabrication costs were reduced due to the reduction of rejects and shorter process cycles.

Day (1969) and Jolly (1969) at Battelle-Northwest have investigated the welding process in 304 and 316 stainless steel using acoustic emission. Butt welds on plates were used. Transducers were attached near the weld line, and then the emissions monitored as the weld cooled. With his instrumentation, Day (1969) was able to detect longitudinal, shear, and Rayleigh waves. Intentional defects were introduced into the welds by adding small amounts of titanium or other noncompatible filler. Other "tricks" were also used to vary weld quality. The defects were analyzed by conventional radiography and metallography. Acoustic emission was found to occur some 20 to 45 sec after the defect region began to solidify. Typically, the acoustic emission rate reached a maximum within one minute following welding of the defect. (The acoustic emission–time profile will of course depend on the size of the piece being welded.) The acoustic emission data was successfully correlated with the defects as indicated radiographically and metallographically. Cracks and gross porosity can be detected via acoustic emission. A "good" weld produced something like 200 counts whereas a weld with significant cracks might result in over 10,000 counts. Acoustic emission results predicted cracks in some welds which were not seen in radiographs. However, subsequent metallographic analysis did show cracks, thus validating the acoustic emission method. There are some problems to be overcome in using this method to test welds, e.g. scale flaking off the surface of some welds produces a number of emissions which are not indicative of weld integrity. It seems as though the difficulties could be overcome, and the method offers great promise for detecting weld defects under actual working conditions. In some applications very long welds must be made, and it is very costly to have to rip out the entire length of the weld if a defect is subsequently discovered at the beginning of the pass. Acoustic emission should allow a defect to be detected before the weld has increased significantly in length. Triangulation might be used to pinpoint the defect location.

Hartbower et al. (1972) monitored the welds in high strength structural steel during, immediately after welding, and for periods of up to 18 days after welding. In this type material, fabricators are often required to wait several days after weld completion to assure that cold (delayed) cracking is completed before nondestructive inspection. If the duration of cold cracking could be determined, then the production time between fabrication and testing could be saved. Acoustic emission offers the possibility of monitoring the cracking as it actually occurs. Cruciform specimens were used to induce weld cracking and actual shipyard welding practice was followed. Some welds were found to produce acoustic emission bursts for over 400 hr after welding. The acoustic emission was correlated with the amount of actual cracking as determined by metallography. (In this work it was found that cracking slag on the weld gave significant emission.)

B. Martensitic Transformations

Martensitic transformations are diffusionless, shear-type reactions that take place through the cooperative movement of very many atoms. The transformation takes place very rapidly and is autocatalytic, i.e. the formation of each plate or crystallite triggers the nucleation of other plates in adjoining regions. Thus the martensitic transformation would appear to be a well-defined source of acoustic emissions.

Liptai *et al.* (1969) determined the acoustic emission associated with the martensitic transformations in gold–47.5 at.% cadmium, indium–22 at.% thallium, cobalt, and plutonium. The specimens were tested in single crystal or large grain form and were $\frac{1}{4}$ in. by 2.5 in. long. The gold–cadmium specimen was investigated in most detail. The martensitic cubic→orthorhombic transformation occured at about 71°C on heating and 60°C on cooling. About one million counts were observed both on heating and cooling. There were more emissions recorded on heating than cooling and they were over one order of magnitude larger. The indium–thallium system has a martensitic face-centered-tetragonal (fct) → face-centered-cubic (fcc) structure at about 27°C. The energy of the emissions were equivalent to those of the gold–cadmium system. Cobalt has a martensitic hexagonal-close-packed → fcc structure change at about 417°C. Exploratory experiments showed that emissions were generated on both heating and cooling. Plutonium shows the $\alpha \to \beta$ transformation at 112°C which is consistent with a martensitic transformation. Again emissions were observed on heating and cooling. These workers also studied the eutectoid decomposition at 133°C in tin–19 at.% cadmium. This is a nucleation-and-growth type transformation. No acoustic emissions were observed even though the gain used was two orders of magnitude larger than that used for the martensitic transformation runs. This result was not entirely unexpected since nucleation-and-growth transformations proceed at low, diffusion-controlled growth rates, where there is probably sufficient time for stresses to be relaxed. It would appear that acoustic emission observations would indicate, among other things, what type of transformation was occurring.

Beattie (1971) studied acoustic emission in the indium–23 at.% thallium alloy. He found that emissions occured at the fcc → fct phase transition only in well annealed samples. Much larger amounts of emission were observed at temperatures below 210°K.

Brown and Liptai (1972) used the martensitic transformation in the gold–47 at.% cadmium alloy as a possible standard source of acoustic emissions. The prepackaged sample was sent to various laboratories active in acoustic emission research. "The results indicated that although many counting and display techniques were used to characterize the acoustic emissions from the martensitic phase transformation during heating and cooling, all of the methods produced equivalent information. No significant differences were noted in characterizing the transformation." It was found that the system could not be used as a standard source because the acoustic emission activity changed with the repeated thermal cycling.

Speich and Fisher (1972) have performed a detailed study of the martensitic transformation in a Fe–28% Ni–0.11% C alloy. Acoustic emission, electrical resistivity, and quantitative metallographic techniques were used. Simultaneous measurement of the electrical resistance and the acoustic emission permitted the determination of the volume of martensite formed per acoustic emission. The metallography was used to determine the number of plates involved in each acoustic emission. The results for this alloy indicated that about fifteen martensite plates are involved in each acoustic emission. Also the volume of martensite formed per acoustic emission decreases with increasing volume fraction of martensite. It appears that acoustic emission studies will be valuable in studying the kinetics of the martensite transformation because they provide, unlike other techniques, essentially a plate-by-plate record of the transition.

C. Slope Stability

Acoustic emission techniques have been used to determine the stability of slopes. Beard (1962) used the "Seismitron" developed by Crandell of Liberty Mutual Insurance. This is a quartz crystal geophone. The number of microseisms per minute was used as a test for stability. Several slopes in California suspected of being unstable were monitored. When the microseismic rate was sufficiently high, the slope subsequently failed.

McCauley (1965) of the California Division of Highways monitored highway construction slopes around the San Luis Reservoir in the Central Coast Ranges of California. There were significant problems with background noise, but some difinite conclusions could be drawn. The acoustic emission rates were high during construction and decreased after construction. The rates were higher during the rainy season than during the dry period. No major slides occured during the monitoring, but small local failures seemed to be indicated on the emission records.

Goodman and Blake (1966) at the University of California, with the support of the California Department of Highways, addressed themselves to landslide and slope stability problems. They constructed and field tested their own rock noise detector instrumentation. They found that rock noises from landslides and slopes are usually in the 100 to 1000 Hz range, and that creeping landslides definitely generate emissions. Triangulation studies encountered difficulty in determining the source location due to extreme velocity variations, and high attenuation in the softer materials. The probes had a range of about 100 ft in the materials studied. There were definite ambient noise problems, but rock noise monitoring did give warning of sudden movements in landslides and rockslides.

Cadman and Goodman (1967) built laboratory model "slope" to perform careful triangulation studies. A box of $60 \times 60 \times 60 \times 124$ cm was filled with moist sand. Four Rochelle salt, bender, crystal transducers (geophones) were placed at known locations in the sand. The sand "slope" was then tilted slowly to failure and the output of the geophones recorded on magnetic tape. The tape was then played through a high frequency oscillograph and the arrivals of the various pulses timed. From these arrival times, the foci of the emissions were determined. It was found (as in the studies in rock) that the source locations close to failure were much nearer the real failure surface than were those of earlier noises. This laboratory work gives insight in to where to place transducers in the actual monitoring of slopes in the field.

Paulsen *et al.* (1967) instituted a slope stability program at the U.S. Borax and Chemical Corporation open pit mine at Boron, California. It was a joint venture between U. S. Borax and the U.S. Bureau of Mines. The equipment was essentially of the Obert (1941) type with an overall gain of about 1000. Recording was by a galvanometer type pen recorder. The number of emissions occurring in a 24 hr period was used as a measure of slope stability. When microseismic activity became high (and also a slight bed movement was indicated on an extensometer), corrective measures were taken. The effectiveness of the corrective measures can be directly determined via the rock noise records after the measures were taken. The authors feel that the emission monitoring is a valuable aid in the slope stability problem.

Wisecarver *et al.* (1969) of the U.S. Bureau of Mines in, cooperation with the U.S. Borax and Chemical Corporation and the Kennecott Copper Corporation, have used acoustic emission monitoring to ascertain the stability of large, open-pit slope walls. The early work used Obert (1941) Rochelle salt type transducers and vacuum tube electronics. Later work used lead zirconate titanate transducers and solid state electronics. This latter instrumentation had a gain of about 10,000 and a flat frequency response from 50 to 5000 Hz. Because of mining operations noise, recordings were usually taken during hours of minimum mining noise. Slopes were monitored at the Boron mine near Boron, California, and the Kimberly, Liberty, and Tripp-Veteran open pit mines near Ely, Nevada. At Kimberly, monitoring was started when the pit was steepened from 40° to about 60°. The emission records indicated that the pit wall adjusted to the steeper slope. The measurements at the Liberty pit showed increased emission after nearby earthquakes. The rock noise rate in the pit wall rose significantly after each quake and then returned to normal within 24 to 48 hr. At the Tripp-Veteran pit monitoring with 10 geophones was undertaken in an area which was expected to fail and in which tension cracks had formed. During the monitoring, the rock mass moved down the slope of the pit wall 5 ft but did not break away. (Most of the displacement occurred on one day.) The most active geophones produced an average rate of 750 noises/hr on the day of maximum movement. The high was 2500/hr. Within 8 hr of the maximum movement the rate had dropped to 200/hr and 10 days later the rate was 3/hr. Microseismic rates measured at locations greater than 300 ft from the failure could not be used for failure detection. The authors concluded that the technique, although still in the research stage, offers a satisfactory means for monitoring slope stability and determining the effect of corrective measures.

D. MAGNETIC EFFECTS

It has been pointed out in the acoustic emission literature (Spanner, 1970) that there might be emissions associated with the magnetization process in ferromagnetic materials. The phenomenon would be the acoustic effect accompanying the well-known Barkhausen magnetic effect. The Barkhausen effect (Bozorth, 1929, 1951; Bozorth and Dillinger, 1930) is concerned with the small, discontinuous changes in overall magnetization of a ferromagnetic sample caused by the sudden change in spin direction of a relatively small number of atoms residing in a volume of some 10^{-5} to 10^{-6} cm^3. There should be some acoustic emissions generated during the Barkhausen discontinuities due to magnetoelastic coupling (Lord, 1967). Search of the literature failed to reveal any such study.

The work presented below is meant to be a very limited first study of this "acoustic Barkhausen effect." The Barkhausen effect literature was searched in order to set up a meaningful experiment. Pieces of nickel 200 rod*, $\frac{1}{8}$ in. in diameter and 1 ft long were used. They were used in either the as-received condition (actual condition unknown—probably cold worked) or annealed by holding at 1100°C in a hydrogen atmosphere for a few hours and then furnace-cooled. The solenoid used to magnetize the sample was one inch in diameter and ten inches long. (Refer to Fig. 20.) The windings of the solenoid were such that the axial field at its center is calculated to be

$$H(\text{Oe}) = 30 \, i \, (\text{A}),$$

where H is the magnetic intensity and i is the current. The coil could accommodate 5 A without overheating. An applied magnetic field of about 3 Oe is needed to overcome the maximum demagnetizing field of this long rod sample. In the figures to follow the magnetizing force will be indicated in coil current, not magnetic field. Only relative values are of interest here so coil current is a good way to describe the magnetizing force. A Dunegan Research Corporation acoustic emission apparatus was used. The amplifica-

---

* The nickel rod was purchased from A. D. MacKay, Inc., 198 Broadway, New York, N.Y.

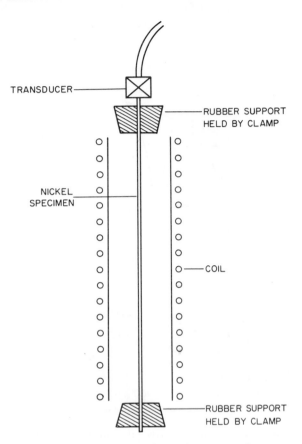

FIG. 20. Schematic drawing of experimental setup to observe acoustic emission during magnetization.

tion was 92 dB and no filters were used. All measurements were made at room temperature.

Prior to making measurements on the samples, the transducer was placed about $\frac{1}{2}$ in. from the top of the specimen and the magnetic field was varied from zero to maximum in both directions many times. Not a single emission was detected. Thus the transducer itself was totally insensitive to magnetic field changes.

The hysteresis (B–H) curves were determined using the standard ballistic galvanometer approach (Bozorth, 1951; Wehr *et al.*, 1969; Bates, 1961) with the coil of Fig. 20. Before each acoustic emission test the sample was demagnetized by passing a relatively large 60 Hz current through the solenoid and then slowly reducing the current to zero. This procedure was repeated many times to insure a fully demagnetized specimen.

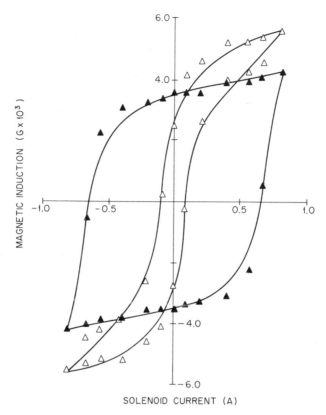

FIG. 21. Hysteresis loops for as-received and annealed nickel rods. ▲, As-received; △, annealed.

Figure 21 shows the hysteresis curves of the as-received and annealed specimens. The as-received nickel rod was quite active acoustically. The first application of the magnetic field after demagnetizing produced a very small acoustic emission count (from 3 to 15). No additional counts were observed when the field was reduced to zero. Upon application of the reverse field, many emissions were produced starting between −0.60 and −0.65 A in the coil. Again no emissions were observed when the current was reduced to zero. Another reversal produced many emissions occurring between +0.60 and +0.65 A. These same results were obtained through as many current reversals as desired.* The number of emissions depended very strongly on the rate of current application. The rate dependence is shown in Fig. 22. Figure 22 applies for both directions of current applications. The emissions versus coil current, for various rates of current application, are given in Fig. 23. Emis-

* Twenty reversals were as many as were tried.

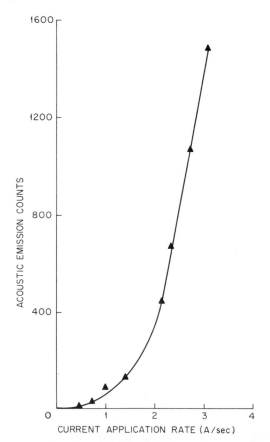

Fig. 22. Acoustic emission counts versus rate of solenoid current application in as-received nickel rod.

sions always start at a current magnitude of between 0.60 and 0.65 A and end between 0.75 and 0.87 A, depending on the rate of current application. Reference to Fig. 21 shows that emissions only occur at the steep parts of the hysteresis loop.

Preliminary observations indicate that the acoustic emissions are of the continuous type (Fig. 24) at the onset of acoustic activity, and become discrete bursts (Fig. 25) near the end of acoustic emission generation. Crude frequency analysis from oscilloscope pictures indicate that many of the emissions are in the 100 to 200 kHz range. (The statements in this paragraph should be considered of a very preliminary nature.)

The material after anneal shows almost no acoustic emission at any position on its hysteresis loop, at any rate of current application. The most that could be obtained was about 20 counts.

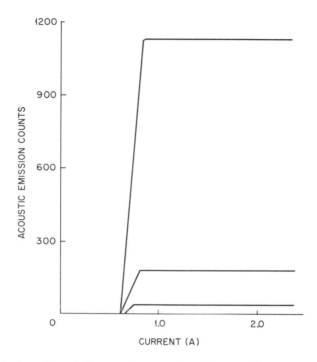

Fɪɢ. 23. Acoustic emission count versus solenoid current in as-received nickel rod.

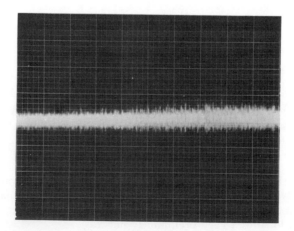

Fɪɢ. 24. Oscilloscope picture of the onset of acoustic emissions in as-received nickel rod. Vertical scale: 0.5 V/cm; horizontal scale: 1 msec/cm.

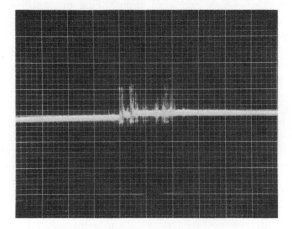

FIG. 25. Oscilloscope picture of acoustic emissions near the end of acoustic generation. Vertical scale: 1 V/cm; horizontal scale: 1 msec/cm.

The data is generally consistent with the Barkhausen effect. Larger Barkhausen effects are seen in cold-worked than annealed material (Forster and Wetzel, 1941).

The exact condition of the as-received material is unknown, but the hysteresis loops of Fig. 21 certainly show that the as-received material is much "harder" magnetically than the "soft" annealed material. The annealed rod was quite flexible and could be bent very easily, whereas the as-received material was much more rigid.

The maximum Barkhausen noise in nickel occurs at the steep portion of the hysteresis loop (Bozorth, 1929). The acoustic emissions also occur at the steep part of the loop. The more rapid application of the magnetic field should cause the spins to change direction more rapidly, thus giving a larger inertial effect for generation of the acoustic emissions.

It is difficult to say at this point what fundamental* or practical† importance can be attached to a subsequent detailed analysis of this "acoustical Barkhausen effect." Fortunately the effect is rather large compared to background noise, thus making the measurements themselves rather easy. It is worthwhile to point out that there is an analogous Barkhausen effect in ferroelectrics (Chynoweth, 1958), and possible acoustic emissions could be

---

* It would seem that the simplest domain configurations [e.g. a picture frame specimen Lord (1967)] would have to be used here to study the fundamentals of this particular magneto-elastic generation of elastic waves.

† It might be possible to make a "magnetic-hardness-tester" from this effect. For example, a simple acoustic emission rig could be used to tell if the material is "hard" (lots of emissions) or "soft" (few emissions). However, the effect is probably quite size-dependent, as is the Barkhausen effect.

observed there also. [*Note added in proof:* This magnetic work has been published in *Lett. Appl. Eng. Sci.* **2**, 1 (1974). Buckman (1972) has detected acoustic emissions in ferroelectric crystals.]

VI. Structural Integrity

Almost every acoustic emission measurement is involved in essence, in determining the integrity of some structure, e.g. a tensile specimen. However, in this section, the use of acoustic emission in actual engineering structures will be considered. Most structural failures result from the growth of a crack; hence the fundamental work on crack growth, described in Section IV B, is of the utmost importance here. In order to use the techniques for structural integrity analysis, it is mandatory to know the relation between crack behavior and acoustic emission in all materials of interest. The location of the flaws can be determined by triangulation techniques (such as are used in seismology) with the use of four transducers. [Good introductions to triangulation studies in the simplest geometry are found in Cadman and Goodman (1967), Obert and Duvall (1961), and Hardy and Chugh (1969).] The safety factor of the structure can be ascertained by comparing the acoustic emission behavior with that of a like structure which has been loaded to failure.

The largest effort so far has gone into developing and using the techniques to evaluate pressure vessels, usually employing the proof test techniques. The Aerojet-General Corporation group was the first to use this technique (Section III, D) and they used it on a number of systems (Green *et al.*, 1964, 1966; Srawley, 1966; Green, 1966; Wildermuth, 1967; Reuter *et al.*, 1968; Hartbower and Crimmins, 1968; Hartbower *et al.*, 1969). Srawley (1966), from his recordings, was able to locate the failure source in the motor case by triangulation after the test was over. Hartbower and Crimmins (1968) discuss nineteen failures which they feel could have been prevented with the use of acoustic emission techniques. (These failures are described in Liptai *et al.*, 1971.)

In 1966, Hutton and Parry started to develop acoustic emission techniques with the idea of applying the technology to the nondestructive testing of nuclear reactor pressure vessels and other pressure components. The ability to monitor the integrity of a nuclear reactor is of the utmost importance for it is very undesirable to shut down a reactor, but yet public safety must be of first importance. A good review of their work on structural integrity (and the work of others) is found in Hutton and Parry (1971) and Parry (1971). They used the shear wave component of the acoustic emission due to its sharp, distinct leading edge, thus making source location determinations more precise. The frequency range used was normally between 90 and 300 kHz. This range eliminates a good share of the low frequency ambient noise, but is still low enough so that the emissions are not attenuated too strongly as they travel in the material. Figure 26 shows a block diagram of the electronic situation. After amplification, the signals from the various

FOUR CHANNEL BLOCK DIAGRAM

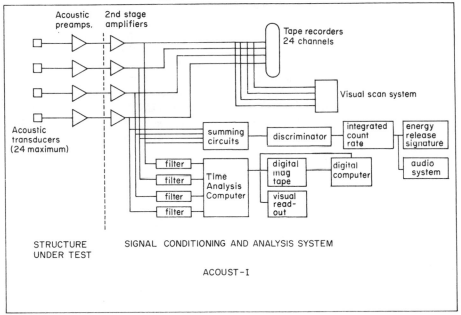

FIG. 26. A complete setup for determining the integrity of a structure (Jersey Nuclear group's instrumentation).

transducers are fed into a special time analysis digital computer. This computer determines the emission source locations by hyperbolic triangulation techniques, which have been developed for many common structural shapes. In addition to source location, the circuitry also gives the acoustic energy released by the flaws, and this allows for a determination of whether a flaw is significant to the integrity of the structure. The results are available on a real time basis and all data is recorded for future analysis. Jersey Nuclear Co. workers (1971) used the technique on a representative commercial power reactor vessel manufactured by Rolls Royce. The vessel was clad with stainless steel. Acoustic emission was monitored during a preservice hydrotest. Ten of thirteen suspect areas revealed by acoustic emission were immediately confirmed by ultrasonic examination. One area agreed with construction records. The remaining two areas were inaccessible to ultrasonic examination. These results were very encouraging.

Parry (1971) has shown that leaks in pressure systems can be located using acoustic emission techniques. A brief account is given of the measurements on long runs of steam, gas, and chemical transmission pipes. This work also showed that corrosion areas could be detected on a gas transmission pipe.

Reactor noise presents a problem to the use of acoustic emission under actual operating conditions. Hutton (1969) has simulated reactor hydraulic noise and showed that the emissions and noise could be separated if frequencies above 1.5 MHz are used. He looked at the noise from the turbulent flow with and without cavitation. Vetrano and Jolly (1972) have reviewed the problems of in-service measurements on reactor pressure vessels.

A unique application of acoustic emission was applied to the study of the elevated temperature burst tests on reactor piping materials (Hutton, 1968). Acoustic emission was monitored as a function of pressure so that burst could be predicted. This gave ample advanced notice, so that a high speed camera could be turned on to record the final stages of crack growth. This procedure gave considerably more advance notice than strain gage monitoring.

The General Dynamic (Fort Worth) group has been active in acoustic emission since 1968. The object of the work is the development of a system which will be able to detect cracks in aircraft structures during fatigue cycling. This is an extremely difficult task, but one of the highest priority, since most noncombat aircraft crashes can be traced to crack growth of one kind or another. The big problem is that of friction noise from rivets and bolts, which can be much larger than the acoustic emissions from growing fatigue cracks. The results of their work can be found in Nakamura (1969, 1971) and Nakamura *et al.* (1971a,b,c, 1972). In an attempt to eliminate background structural noise, one or more master transducers are located in the area to be monitored and the area is surrounded by a number of slave transducers. The slave transducers intercept noise from outside the region and inactivate the masters for a short time. The rejection of more than 30,000 outside noise signals to detect one real emission is not uncommon. There are certainly problems with this system. The masters may be inactivated when a real acoustic emission occurs. The system cannot distinguish between friction noise and real emissions within the area being monitored. Frequency filtering can be used to some advantage to help eliminate the background noise. The amplitude distribution (the number of events at a given amplitude) appear to be quite different for rubbing noise and emissions from a growing crack. This fact may be of significant help in separating the noise from the true emissions.

Harris and Dunegan (1971) have written a short review and present some results on the structural integrity testing of pressure vessels. They emphasize (as have others) that the monitoring can be performed in two ways. The structure can be continuously monitored to detect emissions generated from crack growth, but a faster method is to periodically over pressurize (proof test) the structure and observe the acoustic emissions. If no crack growth has taken place since the last pressurization, no emissions will be observed until the previous maximum pressure is exceeded. If emissions are observed at a lower pressure, it is an indication that crack growth is proceeding. This is nicely summed up in Fig. 27. The figure represents the acoustic emission behavior versus pressure in a welded 4130 steel pressure

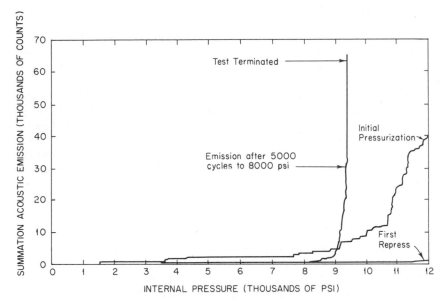

Fig. 27. Acoustic emission results obtained during proof testing and fatigue of a pressure vessel.

vessel of dimensions 6 in. long, 4 in. o.d., and $\frac{1}{4}$ in. wall thickness. End caps were welded on. An intentional defect was machined into the vessel. The vessel was initially pressurized to 12,000 psi, which turned out to be a convenient proof pressure. The first repressurization produced little emission. The vessel was next cycled 5000 times to 8000 psi and proofed. The emissions increased rapidly once the 8000 psi value was exceeded. The rapid buildup in emissions is a clear indication that failure will soon occur due to growing cracks. The proof and cycling pressures were then reduced and failure finally occurered at 7500 psi after many thousands of fatigue cycles. Thus acoustic emission gives a new nondestructive method for determining the fatigue life of a vessel.

VII. Potpourri of Topics (Brief Descriptions)

In this section a number of topics will be merely listed, so that the interested reader will at least know that such work has been performed. The relegation of topics to this section in no way indicates lesser importance of the work. The subjects discussed earlier in the article were those with which the author was most familiar.

A. Kaiser Effect

Work was performed by Kerawalla (1965) in carbon steel, from which he concluded that the Kaiser effect: (a) is not universal, (b) is associated only with continuous type emission, and (c) is not evident for loads below the upper yield point.

Spanner (1970) indicated that normal annealing procedures will effectively return most of the "first stress" emission characteristics to essentially all metals. Spanner also indicated that some metals and alloys do not exhibit any measurable Kaiser effect.

Dunegan and Tatro (1967) have used the Kaiser effect in aluminum to make a tell-tale pressure transducer. The passive transducer determines the maximum pressure to which it was subjected.

B. Emissioms During Unloading

Sankar (1969) found that emissions occurring during unloading were related to Bauschinger effect (an immediate reduction in flow stress after the metal was lightly deformed in the opposite direction). Some metals exhibit much more emission upon unloading then loading. Kerawalla (1965) found that emissions generated during unloading were all of the burst type.

C. Effects of Rate of Loading

Dunegan and Green (1972) and Fisher and Lally (1967) have examined the effect of rate of loading on the acoustic emission behavior and found it to be quite rate sensitive.

D. Application of Acoustic Emission to Civil Engineering Structures

Meunow of the Law Engineering Testing Company in Atlanta, Georgia has used acoustic emission to investigate the structural integrity of many civil engineering type structures. [A brief resume of his work is contained in Hutton (1972) and Liptai *et al.* (1971).] He has inspected large cranes, major bridges, wooden structures, and the prestressing effects of rods in concrete.

E. Effects of Nuclear Reactor Irradiation

Michaels and Fraser (1967) determined the effect of nuclear reactor irradiation on the acoustic emission of zircaloy-2 tube specimens. The irradiated material was found to be less ductile than the as-received tubing, and had a greater acoustic emission activity. The emission rate at failure for irradiated tubing was some four times that of as-received material.

F. Measurement of Surface Coating Thickness from Acoustic Emission

Dunegan and Tetelman (1971b) found that the acoustic emission of aluminum alloys depended on the thickness of the anodized coating. They then used acoustic emission as a means of determining coating thickness.

G. Detection of Boiling

The detection of the onset of boiling in liquid sodium in reactors has been attempted by acoustic means. Details can be found in Woodward and Stephens (1971) and in Anderson et al. (1972).

H. Investigation of Honeycomb Material with Acoustic Emission

Green et al. (1970) determined the structural integrity of honeycomb material from acoustic emission measurements.

I. Determination of Bond Quality from Acoustic Emission Measurements

Acoustic emission behavior has been found to be a reliable indicator of bond quality (Schmitz and Frank, 1965; Beal, 1967; Pollack, 1968).

J. Creep Effects and Acoustic Emission

Tatro and Liptai (1962) observed continuous acoustic emission when solder wire was subjected to a constant load. Dunegan and Harris (1969) have observed acoustic emissions in beryllium during creep. Adams et al. (1969) have also studied the problem of creep in beryllium. Adams has continued his creep–acoustic emission studies at Tulane University, and some of his results are reported in Liptai et al. (1971) and Engle and Dunegan (1969).

K. Fatigue Effects and Acoustic Emission

Many of the investigations concerning structural integrity were concerned with fatigue effects. Some particular fatigue studies will be emphasized here. Hutton (1970) found in stainless steel that some acoustic emission effects began at 25% of the fatigue cycles required to form a visible surface crack. Acoustic emission measurements showed microcrack formation at 10% of fatigue life in tension–tension fatigue tests in aluminum. Hutton suggested that this fact could be used to significantly reduce fatigue testing time. Hartbower et al. (1968) have investigated low cycle fatigue effects in cracked fracture toughness specimens of D6AC steel specimens. Dunegan et al. (1970) have also dealt with fatigue of cracked fracture toughness specimens. Dau (1971) also talks of the fatigue problem, and Anderson et al. (1972) indicated

that acoustic emission measurement give an indication of microcrack generation. Mitchell (1965) also dealt with fatigue effects.

L. Effect of Temperature on Acoustic Emission

Anderson *et al.* (1972) have worked on the development of sensors
which can be used at high temperatures (and also used in a radioactive,
sodium-immersed environment). They have also performed some acoustic
emission measurements at elevated temperatures in a stainless steel fatigue
test specimen.

Dunegan and Tetelman (1971b) have employed a "wave guide" to test
a Rene 41 tensile specimen containing a transverse weld, at 1400°F in a
gleeble machine. They were able to detect crack initiation from the acoustic
emission results.

Baker (1968) has investigated the effect of temperature ($-100°$ to
$+250°F$) on acoustic emission response of maraging steels and D6 AC steel
in various heat-treated conditions.

Green (1969c) studied crack growth in various metals at 165°F via
acoustic emission.

Green *et al.* (1970) observed that during loading 2014-T6 (T1G welded
with 4043 filler wire) produced a significantly larger number of emissions at
$-320°$ and $-423°F$ than comparable specimens tested at room temperature.

Crussard *et al.* (1958) observed that brittle failure of steel, at $-193°C$,
was preceded by bursts of acoustic energy. They observed acoustic emissions
when Lüders bands formed (corresponding to the Portevin–Le Chatelier
effect). It might just be possible to determine the critical temperature at
which body-centered-cubic metals become brittle by means of a nondestructive acoustic emission method.

Tetelman and Chow (1972) have measured the acoustic emission as a
function of stress at $-321°F$ in a low carbon steel.

VIII. Conclusions and Suggestions for Further Work

A. Fundamental Area

In a fundamental sense, acoustic emission studies are extremely interesting and have great potential in the understanding of mechanical properties
of solids. Unfortunately, not much new information has been obtained from
recent acoustic emission studies due to the extreme difficulties in obtaining
details of the actual source mechanism. For example, there does not appear
to be definitive information as to whether longitudinal or transverse wave
generation predominates at the source. Some work has been done by Kroll
and Tatro (1964), Egle (1965), Egle and Tatro (1967), Hutton (1968, 1969)
and the Jersey Nuclear Co., but advances in this fundamental area will
undoubtedly have to wait for a breakthrough of some type in under-

standing the source mechanism. It would seem that an extension of the work of Savage and Mansinha (1963) should be undertaken.

One area in which acoustic emission may yield some new information is in regard to the study of the kinetics of certain processes in fine detail. For example, a close look could be taken of the martensite transformation, creep effects, the curing of materials (e.g. concrete drying), etc.

The effect of low and high temperatures on the acoustic emission response of most materials is not well known. However, many materials in service are exposed to extreme temperatures. Of special interest might be an investigation of the ductile to brittle transition in body-centered-cubic metals, and the temper brittleness problem in steels.

The fatigue problem is a very large one (both in a fundamental and practical vein). Work has shown that acoustic emission techniques can give very early indication of fatigue failure, and hence a nondestructive indication of fatigue life. This work should be followed up in great detail for, if successful, it could save enormous amounts of time in conventional fatigue tests. It might also contribute information to the difficult fundamental area of fracture mechanics.

No references were found, in the acoustic emission studies investigated, of the liquid metal embrittlement problem. This would seem to be a fruitful area.

It would be interesting to determine in detail the magnetoelastic effects accompanying the magnetic Barkhausen effect.

In the area of rocks, it would seem to be of significant fundamental interest to determine the relative importance of emissions in the various frequency regimes. Also of interest here is the effect of pressure and temperature on the acoustic emission behavior of rocks, as the ductility changes significantly with these factors.

B. APPLIED AREA

It is in the area of practical applications that acoustic emission really shines. Structural integrity studies have proved very successful, and it is almost certain that new types of structures will be monitored with the acoustic emission technique. This work has a decided advantage over conventional techniques, such as radiography and ultrasonics, in that a large area can be monitored and the defects located. Acoustic emission will undoubtedly find great use in the area of welding. The weld defects can be determined very quickly and hence eliminate long costly ripouts of the subsequent weld line. Acoustic emission will be very useful in determining the time scale of delayed cracking in welds. Acoustic emission detection of leaks in piping should also prove to be very popular. The extremely difficult work of Nakamura in determining acoustic emissions in the presence of large amounts of fastener and other background noise is ultimately a critical one. It should be pursued with great vigor in spite of its very difficult nature.

Bozorth, R. M. (1929). *Phys. Rev.* **34**, 772.

Bozorth, R. M. (1951). "Ferromagnetism," p.524 and 843. Van Nostrand-Reinhold, Princeton, New Jersey.

Bozorth, R. M., and Dillinger, J. F. (1930). *Phys. Rev.* **35**, 773.

Brown, A. E., and Liptai, R. G. (1972). *Amer. Soc. Test. Mater., Spec. Tech. Publi.* **505**.

Buchman, P. (1972). *Solid State Electron.* **15**, 142.

Cadman, J. D., and Goodman, R. E. (1967). *Science* **15**, 1182.

Chugh, Y. P., Hardy, H. R., Jr., and Stefanko, R. (1968). *Proc. Symp. Rock Mech., 10th, 1968, Austin, Texas.*

Chynoweth, A. G. (1958). *Phys. Rev.* **110**, 1316.

Cook, N. G. W. (1963). *Proc. Symp. Rock Mech., 5th, 1962* p. 493.

Crussard, D., Lean, J. B., Plateau, J., and Bachet, C. (1958). *C. R. Acad. Sci.* **246**, 2845.

Dau, G. J. (1971). *Ind. Res.* **70**, 40.

Day, C. K. (1969). "An Investigation of Acoustic Emission from Defect Formation in Stainless Steel Weld Coupons," Rep. BNWL-902. Batelle-Northwest, Richland, Washington.

Drouillard, T. F. (1974). "Acoustic Emission: A bibliography of 1970–1971–1972." ASTM Spec. Tech. Publ. Amer. Soc. Test. Mater., Philadelphia, Pennsylvania.

Dunegan, H. L., and Green, A. T. (1972). *Amer. Soc. Test. Mater., Spec. Tech. Publ.* **505**.

Dunegan, H. L., and Harris, D. O. (1969). *Ultrasonics* **7**, 160.

Dunegan, H. L., and Harris, D. O. (1974). *In* "Experimental Techniques in Fracture Mechanics," (A. S. Kobayashi, ed.). Soc. Exp. Stress Anal. (to be published).

Dunegan, H. L., and Tatro, C. A. (1967). *Rev. Sci. Instrum.* **38**, 1145.

Dunegan, H. L., and Tatro, C. A. (1971). *In* "Techniques of Metal's Research" (R. F. Bunshah, ed.), Vol. V, Part 2, pp. 273–312. Wiley (Interscience), New York.

Dunegan, H. L., and Tetelman, A. S. (1971a). *Eng. Fract. Mech.* **2**, 387.

Dunegan, H. L., and Tetelman, A. S. (1971b). *Res./Develop.* **22**, 20.

Dunegan, H. L., Tatro, C. A., and Harris, D. O. (1964). "Acoustic Emission Research," Rep. UCID-4868, Rev. 1. Lawrence Radiation Laboratory, Livermore, California.

Dunegan, H. L., Harris D. O., and Tatro, C. A. (1968). *Eng. Fract. Mech.* **1**, 105.

Dunegan, H. L., Harris, D. O., and Telelman, A. S. (1969). "Prediction of Fatigue Lifetime by Combined Fracture Mechanics and Acoustic Emission Techniques," Rep. AFFDL TR 70-144, p. 459. Air Force Flight Dynamics Lab.

Dunegan, H. L., Harris, D. O., and Tetelman, A. S. (1970). *Mater. Eval.* **28**, 221.

Egle, D. M. (1965). Ph.D. Thesis, Tulane University, New Orleans, Luisiana.

Egle, D. M., and Tatro, C. A. (1967). *J. Acoust. Soc. Amer.* **4**, 321.

Engle, R. B. (1966). Ph.D. Thesis, Michigan State University, Ann Arbor (University Microfilms, P. O. Box 1346, Ann Arbor, Mich. 48106- 6707535).

Engle, R. B., and Dunegan, H. L. (1969). *Int. J. Nondestruct. Test.* **1**, 109.

Eshelby, J. D. (1949). *Proc. Roy. Soc., Ser. A* **197**, 396.

Fisher, R. M., and Lally, J. S. (1967). *Can. J. Phys.* **45**, 1147.

Fitz-Randolph, J., Phillips, D. C., Beaumont, P. W. R., and Tetelman, A. S. (1972). *J. Mater. Sci.* **7**, 289.

Forster, F., and Wetzel, H. (1941). *Z. Metallk.* **33**, 115.

Frederick, J. R., and Felbeck, H. J. (1972). *Amer. Soc. Test. Mater., Spec. Tech. Publ.* **505**.

Gatti, A., Mehan, R. L., and Noone, M. J. (1971). "Development of a Process for Producing Transparent Spinal Bodies," Final Rep., Contract N00019 71 C 0216. Naval Air Systems Command, General Electric Laboratories, King of Prussia, Pennsylvania.

Gerberich, W. W. (1970). *J. Mater. Sci.* **5**, 283.

Gerberich, W. W., and Hartbower, C. E. (1967). *Int. J. Fract. Mech.* **3**, 185.

Gerberich, W. W., and Hartbower, C. E. (1969). "Monitoring Crack Growth of Hydrogen Embrittlement and Stress Corrosion Cracking by Acoustic Emission," Proc. Conf. Fund. Aspects of Stress Corrosion Cracking, Ohio State Univ., Columbus, Ohio. Nat. Ass. Corrosion Eng., Houston, Texas.

Gerberich, W. W., and Reuter, W. G. (1969). "Theoretical Model of Ductile Fracture Instability Based on Stress-Wave Emission," ONR Contract Rep. N00014 66 00340 Aerojet-General Corp., Sacramento, California.

Gillis, P. P. (1971). *Mater. Res. Stand.* **11**, 11.

Gillis, P. P. (1972). *Amer. Soc. Test. Mater., Spec. Tech. Publ.* **505**.

Gilman, J. J. (1966). *Proc. U. S. Nat. Congr. Appl. Mech., 5th, 1966* pp. 385–403.

Gold, L. W. (1960). *Can. J. Phys.* **38**, 1137.

Gold, L. W. (1968). *Res. Nat. Res. Counc. Can., Div. Bldg. Res., Pap.* **368**.

Goodman, R. E. (1963). *Geol. Soc. Amer. Bull.* **74**, 487.

Goodman, R. E., and Blake, W. (1966). *Highw. Res. Rec.* **119**.

Green, A. T. (1966). *NASA Contract. Rep.* **NASA CR-61161**.

Green, A. T. (1969a). *Nucl. Safety* **10**, 4.

Green, A. T. (1969b). "Stress Wave Emission and Fracture of Prestressed Concrete Reactor Vessel Materials," USAEC Contract No. W-7405-eng.-26 Eng. Rep. Aerojet-General Corp. Sacramento, California.

Green, A. T. (1969c). "Development of a Nondestructive Testing Technique to Determine Flaw Criticality," ARPA Contract F33615-68 C 1705. Aerojet-General Corp., Sacramento, California.

Green, A. T., Lockman, C. S., and Steele, R. K. (1964). *Mod. Plast.* **41**, 137.

Green, A. T., Lockman, C. S., Brown, S. J., and Steele, R. K. (1966). *NASA Contract Rep.* **NASA CR-55472**.

Green, A. T., Dunegan, H. L., and Tetelman, A. S. (1970). "Nondestructive Inspection of Aircraft Structures and Materials Via Acoustic Emission," Tech. Rep. DRC-107. Dunegan Res. Corp., Livermore, California.

Gutenberg, B., and Richter, C. F. (1954). "Seismicity of the Earth." Princeton Univ. Press, Princeton, New Jersey.

Hardy, H. R., Jr. (1969). "Microseismic Activity and Its Application to the Natural Gas Industry," Transmission Conf. Proc., Cat. No. X 59969, pp. T147 to T156. Amer. Gas Ass., New York.

Hardy, H. R., Jr., and Chugh, Y. P. (1969). "Application of Microseismic Techniques in Mining," Paper presented at Mine Mechanization Symposium, Varanasi, India, March 1969.

Hardy, H. R., Jr., Kim, R. Y., Stefanko, R., and Wang, Y. J. (1970). *Proc. Symp. Rock Mech., 11th, 1969* p. 377.

Hardy, H. R., Jr. (1972). *Amer. Soc. Test. Mater., Spec. Tech. Publ.* **505**.

Harris, D. O., and Dunegan, H. L. (1971). "Verification of Structural Integrity of Pressure Vessels by Acoustic Emission and Periodic Proof Testing," Tech. Rep. DRC 71. Dunegan Res. Corp., Livermore, California.

Harris, D. O., Tetelman, A. S., and Darwish, F. A. (1972). *Amer. Soc. Test. Mater., Spec. Tech. Publ.* **505**.

Hartbower, C. E., and Crimmins, P. P. (1968). "Fracture of Structural Metals as Related to pressure Vessel Integrity and In-Service Monitoring," Rep. Aerojet-General Corp., Sacramento, California.

Hartbower, C. E., Gerberich, W. W., and Crimmins, P. P. (1968). *Weld Res. (New York)* p. 1.

Hartbower, C. E., Climent, F. J., Morais, Co., and Crimmins, P. P. (1969). "Stress Wave Analysis Technique Study on Thick-Walled Type A 302 B Steel Pressure Welds," Rep. NAS 9-7759. Aerojet-General Corp., Sacramento, California.

Hartbower, C. E., Reuter, W. G., Morais, C. F., and Crimmins, P. P. (1972). *Amer. Soc. Test. Mater.*, *Spec. Tech. Publ.* **505.**

Hay, D. R., Ruzauskas, E. J., and Pollack, N. (1972). "Acoustic Emission in Metal-Matrix Composites," Presented at Refractory Composites Working Group, Houston, Texas, Feb. 1 and 2.

Hedley, D. G. F., *et al.* (1969). *Proc. Can. Rock Mech. Symp.*, *5th. 1968.* p. 105.

Hodgson, E. A. (1942). *Bull. Seismol. Soc. Amer.* **32,** 249.

Hodgson, E. A. (1943). *Trans. Can. Inst. Mining Met.* **46,** 313.

Hodgson, J. H. (1947). *Bull. Seismol. Soc. Amer.* **37,** 5.

Hodgson, E. A. (1958). "Dominion Observatory Rockburst Research—1938–1945." Canadian Dept. of Mines and Technical Surveys, Ottawa.

Hutton, P. H. (1968). *Mater. Eval.* **26,** 125.

Hutton, P. H. (1969). *Nondestruct. Test.* **2,** 111.

Hutton, P. H. (1970). "Acoustic Emission—What it is and its Application to Evaluate Structural Soundness of Solids," Rep. BNWL-SA-2983. Battelle-Northwest, Richland, Washington.

Hutton, P. H. (1972). *Amer. Soc. Test. Mater.*, *Spec. Tech. Publ.* **505.**

Hutton, P. H., and Parry, D. L. (1971). *Mater. Res. Stand.* **11,** 25.

Ishimoto, M., and Iida, K. (1939). *Bull. Earthquake Res. Inst.* **17,** 433.

Jersey Nuclear Co. "Acoustic Analysis Technology Applied to the Nondestructive Test Evaluation of Commercial Structures and Structural Materials." Jersey Nuclear Co., Richland, Washington.

Jersey Nuclear Co. (1971). "An Approach to Pre-Service and In-Service Inspection of Nuclear Reactor Coolant Systems Using Acoustic Emission Analysis." Jersey Nuclear Co., Richland, Washington.

Jolly, W. D. (1969). *Weld. J.* (*New York*) **48,** 21.

Jones, M. H., and Brown, W. F., Jr. (1964). *Mater. Res. Stand.* **4,** 120.

Kaiser, J. (1950). Ph.D. Thesis, Technische Hochschule, Munich.

Kaiser, J. (1953). *Ark. Eisenhuettenwesen* **25,** 43.

Katz, Y., and Gerberich, W. W. (1970). *J. Fract. Mech.* **6,** 219.

Kerawalla, J. N. (1965). Ph.D. Thesis, University of Michigan, Ann Arbor.

Kishinouye, F. (1937). "Frequency-Distribution of the Ito Earthquake Swarm—An Experiment in Fracture," Bull. Earthquake Res. Inst., Tokyo University.

Knill, J. L., Franklin, J. A., and Malone, A. W. (1968). *Int. J. Rock Mech. Mining Sci.* **5,** 87.

Knopoff, L., and Gilbert, F. (1960). *Bull. Seismol. Soc. Amer.* **50,** 117.

Koerner, R. M., and Lord, A. E., Jr. (1972). *J. Soil Mech. Found.*, *Div. ASCE* **98,**161.

Kroll, R. J., and Tatro, C. A. (1964). *Exp. Mech.* **4,** 129.

Liptai, R. G. (1963). Ph.D. Thesis, Michigan State University, East Lansing.

Liptai, R. G. (1971). "Acoustic Emission from Composite Materials," Rep. URCL-72657. Lawrence Radiation Lab., Livermore, California.

Liptai, R. G. (1972). *Amer. Soc. Test. Mater.*, *Spec. Tech. Publ.* **505.**

Liptai, R. G., and Harris, D. O. (1971). *Mater. Res. Stand.* **11,** 8.

Liptai, R. G., Dunegan, H. L., and Tatro, C. A. (1969). *Int. J. Nondestruct. Test.* **1,** 213.

Liptai, R. G., Harris, D. O. Engle, R. B., and Tatro, C. A. (1971). *Int. J. Nondestruct. Test.* **3,** 215.

Lord, A. E. (1967). *Acustica* **18,** 187.

Lord, A. E., Jr., and Koerner, R. M. (1974). *J. Test. Eval.*

Lynnworth, L. C., and Bradshaw, J. E. (1971). *Mater. Res. Stand.* **11**, 33.

McCauley, M. L. (1965). *Bull. Ass. Eng. Geol.* **2**, 1.

Malone, A. W. (1965). M.Sc. Thesis, Department of Geology, Imperial College, London.

Mason, W. P. (1950). "In Piezoelectric Crystals and their Application to Ultrasonics." Van Nostrand-Reinhold, Princeton, New Jersey.

Mason, W. P., McSkimin, H. J., and Shockley, W. (1948). *Phys. Rev.* **73**, 1213.

Mehan, R. L., and Mullen, J. V. (1971). *J. Compos. Mater.* **5**, 266.

Michaels, T. E., and Fraser, M. C. (1967). "Acoustic Emission Behavior of Zircaloy-2 Pressure Tubing Under Applied Stress," BNWL-545. Battelle-Northwest, Richland, Washington.

Mitchell, L. D. (1965). Ph.D. Thesis, University of Michigan, Ann Arbor. (University Microfilm, P. O. Box 1346, Ann Abor, Mich. 48016, #6606657).

Mogi, K. (1962a). *Bull. Earthquake Res. Inst.* **40**, 125.

Mogi, K. (1962b). *Bull. Earthquake Res. Inst.* **40**, 831.

Mogi, K. (1968). *Bull. Earthquake Res. Inst.* **46**, 1103.

Mullen, J. V., and Mehan, R. L. (1972). *J. Mater.* (to be published).

Mullen, J. V., Mazzio, V. F., and Mehan, R. L. (1971). "Basic Failure Mechanisms in Advanced Composites," NASA Rep. 2420-NO3. General Electric Laboratories, King of Prussia, Pennsylvania.

Nakamura, Y. (1969). "Development of an Acoustic Emission Monitoring System," Rep. ERR FW 901. General Dynamics, Fort Worth, Texas.

Nakamura, Y. (1971). *Mater. Eval.* **29**, 8.

Nakamura, Y., Veach, C. L., and McCauley, B. O. (1971a). "Amplitude Distribution of Acoustic Emission Signals," Rep. ERR FW 1176. General Dynamics, Fort Worth, Texas.

Nakamura, Y., Veach, C. L., and McCauley, B. O. (1971b). *Amer. Soc. Test. Mater., Spec. Tech. Publ.* **505**.

Nakamura, Y., Hagenmeyer, J. W., and Veach, C. L. (1971c). "Acoustic Emission Observations of Delayed Cracking in Steel Specimens Containing Interference—Fit Fasteners—A Perliminary Study," Rep. ERR FW 1185. General Dynamics, Fort Worth, Texas.

Nakamura, Y., McCauley, B. O., and Veach, C. L. (1972). "Study of Acoustic Emission During Mechanical Tests of Large Flight Weight Tank Structure," Rep. MSC-04800, FZK-390. General Dynamics, Fort Worth, Texas.

Notvest, K. (1966). *Weld. J.* (*New York*) **45**, 173-S.

Obert, L. (1941). *U.S., Bur. Mines, Rep. Invest.* **R1-3555**.

Obert, L., and Duvall, W. (1942). *U.S., Bur. Mines, Rep. Invest.* **R3654**.

Obert, L., and Duvall, W. (1945a). *U.S., Bur. Mines, Rep. Invest.* **R1-3797**.

Obert, L., and Duvall, W. (1945b). *U.S., Bur. Mines, Rep. Invest.* **R1-3803**.

Obert, L., and Duvall, W. (1957). *U S., Bur. Mines, Bull.* **575**.

Obert, L., and Duvall, W. (1961). *U.S., Bur. Mines, Rep. Invest.* **R1-5882**.

Parry, D. L. (1971). "Acoustic Emission Integrity Analysis of Pressure Vessels and Piping." Jersey Nuclear Co., Richland, Washington.

Paulsen, J. C., Kistler, R. B., and Thomas, L. L. (1967). *Mining Cong. J.* **53**, 28.

Pollack, A. A. (1968). *Ultrasonics*, **6**, 88.

Reuter, W. G., Green, A. T., Hartbower, C. E., and Crimmins, P. P. (1968). "Monitoring of Crack Growth in Ti 6A1 4V. Alloy by the Stress-Wave Analysis Technique," Rep. NAS 9-7759. Aerojet-General Corp., Sacramento, California.

Romine, H. E. (1961). "Determination of the Driving Force for Crack Initiation from Acoustic Records of G_c Tests in High Strength Materials for Rocket Motor Cases," NWL Rep. No. 1779, U.S. Naval Weapons Lab.

Romrell, D. M., and Bunnell, L. R. (1970). *Mater. Eval.* **28**, 267.

Rüsch, H. (1959). *Zem.-Kalk-Gips* **12**, 1.

Sankar, N. G. (1969). Ph.D. Thesis, University of Michigan, Ann Arbor.

Sasaki, K., and Takata, A. (1970). *In* "Rock Mechanics in Japan," Vol. 1, p. 167. Jap. Soc. Civil Eng., Tokyo.

Savage, J. C., and Mansinha, L. (1963). *J. Geophys. Res.* **68**, 6345.

Schmitz, J., and Frank, L. (1965). "Nondestructive Testing for Evaluation of Strength of Bonded Material," Rep. NASA CR 67983. General American Transportation Corp., Niles, Illinois.

Schofield, B. H. (1961). "Acoustic Emission Under Applied Stress," ARL 150 (AD 274 484). Aeronautical Research Laboratory, Office of Aerospace Research, U.S. Air Force (can be obtained from National Technical Information Service, 5285 Port Royal Road, Springfield, Va. 22151.)

Schofield, B. H. (1963a). "Acoustic Emission Under Applied Stress," Tech. Doc. Rep. No. ASD TDR 63 509 Part I (AD 419 177). Air Force Materials Laboratory, Aeronautical Systems Division, Air Force Systems Command Wright-Patterson Air Force Base, Ohio.

Schofield, B. H. (1963b). *In* "Proceedings of the Symposium on Physics and Nondestructive Testing," pp. 63–91. Southwest Research Institute, San Antonio, Texas.

Schofield, B. H. (1964). "Acoustic Emission Under Applied Stress," Tech. Doc. Rep. No. ASD TDR 63 509 Part II (AD 600 233). Air Force Materials Laboratory Research and Technology Division, Air Force Systems Command, Wright-Patterson Air Force Base, Ohio.

Scholz, Ç. H. (1968a). *J. Geophys. Res.* **73**, 1447.

Scholz, C. H. (1968b). *Bull. Seismol. Soc. Amer.* **58**, 399.

Scholz, C. H. (1968c). *J. Geophys. Res.* **73**, 3295.

Scholz, C. H. (1968d). *Bull. Seismol. Soc. Amer.* **58**, 1117.

Sedgwick, R. T. (1968). *J. Appl. Phys.* **39**, 1728.

Spanner, J. C. (1970). M. S. Thesis, Dept. of Metallurgy, Washington State University, Pullman.

Speich, G. R., and Fisher, R. M. (1972). *Amer. Soc. Test. Mater., Spec. Tech. Publ.* **505**.

Srawley, J. E. (1966). *NASA Tech. Memo.* **NASA TMX-1194**.

Stas, B., *et al.* (1971). *Proc. Symp. Rock Mech., 12th, 1970* p. 109

Tatro, C. A. (1959). "Sonic Techniques in the Detection of Crystal Slip in Metals", Status rep. Division of Engineering Research, College of Engineering Michigan State University, East Lansing.

Tatro, C. A. (1971). *Mater. Res. Stand.* **11**, 17.

Tatro, C. A. (1972). *Amer. Soc. Test. Mater., Spec. Tech. Publ.* **505**.

Tatro, C. A., and Liptai, R. G. (1962). "Acoustic Emission from Crystalline Substances," *In Proc. Symp. Phys. Nondestructive Testing*, pp. 145–174. Southwest Research Inst., San Antonio, Texas.

Tatro, C. A., and Liptai, R. G. (1963). *In* "Proceedings of the Fourth Annual Symposium on Nondestructive Testing of Aircraft and Missile Components," pp. 287–346. Southwest Research Institute, San Antonio, Texas.

Tetelman, A. S., and Chow, R. (1972). *Amer. Soc. Test. Mater., Spec. Tech. Publ.* **505**.

Vetrano, J. B., and Jolly, W. D. (1972). *Mater. Eval.* **30**, 9.

Vinogradov, S. D. (1959). *Bull. Acad. Sci. USSR, Geophys. Ser.* p. 313.

Wehr, M. R., Davis, F. K., Chen, H. S. C., Estilow, U.S., and Smith, F. J. (1969). "Experiments in Physics," p. 101. Drexel Inst. Tech. Press, Philadelphia, Pennsylvania.

Wildermuth, D. (1967). "Pressure Testing of AFRM 017 Service Propulsion Systems Fuel Tank Utilizing Aerojet-General Corp. Stress-Wave Analysis Technique," Rep. NAS 9-6766. Aerojet-General Corp., Sacramento, California.

Wisecarver, D. W., Merrill, R., and Stateham, R. M. (1969). *Trans. AIME* **244**, 378.

Woodward, B., and Stephens, R. W. B. (1971). *Ultrasonics* **9**, 21.

BIBLIOGRAPHY*

Agarwal, A. B. L. (1968). "An investigation of the behavior of the acoustic emission from metals and a proposed mechanism for its generation," Ph. D. Thesis. University of Michigan, Ann Arbor.

Agarwal, A. B. L., Frederick, J. R., and Felbeck, D. K. (1970). Detection of plastic microstrain in aluminum by acoustic emission. *Metal. Trans.* **1**, 1069.

Anderson, T. T., and Gate, T. A. (1968). Acoustic boiling detection in reactor vessels. Presented at 15th Nuclear Science Symposium, Montreal, Canada.

Armstrong, B. H. (1968). "Exploratory Study of Acoustic Emission Prior to Earthquakes." Rep. 320-3249. IBM Scientific Center, Palo Alto, California.

Balderston, H. L. (1972). The broad range detection of incipient failure using the acoustic emission phenomena. *Amer. Soc. Test. Mater., Spec. Tech. Publ.* **505**.

Beard, F. D. (1961). "Predicting Slides in Cut Slopes," p. 72. Western Construct.

Bill, R. C. (1970). "An Acoustic Emission Study of the Deformation Mechanisms of Polycrystalline Aluminum and Copper," Ph.D. Thesis. University of Michigan, Ann Arbor.

Blake, W. (1972). Rock burst mechanics. *Quart. Colo. Sch. Mines* **67**, 1.

Blake, W., and Duvall, W. (1969). Some fundamental properties of rock noises. *Trans. AIME* **244**, 288.

Bolles, M. M., Bass, B. R., Thompson, H. A., and Adams, K. H. (1966). "Final Report on Acoustic Emission at Tulane Univ.," Rep. URCL-13231. Lawrence Radiation Laboratory, Livermore, California.

Borchers, H., and Kaiser, J. (1958). Acoustic effect in the phase transformation in the lead-tin system. *Z. Metallk.* **49**, 95 (in German).

Borchers, H., and Tensi, H. M. (1960). A better piezoelectric method for the investigation of processes in metals during mechanical deformation and phase changes. *Metallk.* **51**, 212 (in German).

Borchers, H., and Tensi, H. M. (1962). Piezoelectric impulse measurements during the mechanical deformation of AlMg 3 and Al 99. *Z. Metallk.* **53**, 692.

Broadbent, C. D., *et al.* (1968). "Design and Application of Microseismic Devices," Paper presented at Fifth Canadian Symp. on Rock Mech., University of Toronto.

Brown, J. W. (1965). "An Investigation of Microseismic Activity in Rock Under Tension," M.S. Thesis. Mining Engineering, Penn. State University, University Park, Pennsylvania.

Brown, J. W., and Singh, M. M. (1966). An investigation of microseismic activity in rock under tension. *Trans. AIME* **233**, 255.

Buchheim, W. (1958). "Geophysical Methods for the Study of Rock Pressure in Coal and Potash-Salt Mining," Int. Strat. Control Congress, Leipzig, pp. 222–223.

Cadman, J. D. (1967). "Subaudible Noise in Small Scale Landslides," Thesis. University of California, Berkeley.

* A great deal of the Russian work has not been given here. Much of this work is given in Antsyferov (1966).

Cadman, J. D., Goodman, R. E., and Van Alstine, C. (1967). "Research on Subaudible Noise in Landslides," Rep. Eng. Dept., University of California, Berkeley.

Chambers, R. H., and Hoenig, S. H. (1969). " New Techniques in Nondestructive Testing by Acoustic and Exo-electron Emission," Semi-Annu. Progr. Rep. AD-691-230. Eng. Exp. Sta., University of Arizona, Tuscon.

Chugh, Y. P. (1968). "An Investigation of Frequency Spectra of Microseisms Emitted From Rock Under Tension in the Range 300-15,000 cps," M.S. Thesis. Mining Engineering, Penn. State University, University Park, Pennsylvania.

Cook, N. G. W. (1965). The failure of rock. *Int. J. Rock Mech. Mining Sci.* **2**, 389.

Cross, N. O. (1970). "Acoustic Emission Technique for Insuring Safe Hydrostatic Tests of Pressure Vessels," Rep. 70-PET-31. Amer. Soc. Mech. Eng.

Cross, N. O., Louschin, L. L., and Thompson, J. L. (1972). Acoustic emission testing of pressure vessels for petroleum refineries and chemical plants. *Amer. Soc. Test. Mater., Spec. Tech. Publ.* **505**.

Darwish, F. A. I. (1969). "The effect of Temperature and Fiber Orientation on the Strength and Deformation Characteristics of Fiber Composites," Ph.D. Dissertation. Dept. of Materials Science, Stanford University, Stanford, California.

Day, C. K. (1969). "An Investigation of the Plastic Bursts of Microstrain in Zinc as Sources for Acoustic Emission," M.S. Thesis, Washington State University, Pullman.

Day, C. K. (1969). "An Investigation of Acoustic Emission from Defect Formation in Stainless Steel Weld Coupons," BNWL-902. Battelle-Northwest, Richland, Washington.

Dunegan, H. L., and Green, A. T. (1971). Factors effecting acoustic emission response from materials. *Mater. Res. Stand.* **11**, 21.

Dunegan, H. L., and Green, A. T. (1972). Factors affecting acoustic emission response from materials. *Amer. Soc. Test. Mater., Spec. Tech. Publ.* **505**.

Dunegan, H. L., and Harris, D. O. (1968). "Acoustic Emission-A New Nondestructive Testing Tool," Rep. URCL-70750. Lawrence Radiation Laboratory, Livermore, California.

Dunegan, H. L., Harris, D. O., and Tatro, C. A. (1964). "Acoustic Emission Research Status Report," USAEC Report UCLL-4868 (Rev. 1). University of California, Lawrence Radiation Laboratory, Livermore, California.

Dunegan, H. L., Brown, A. E., and Krauss, P. L. (1968). " Piezoelectric Transducers for Acoustic Emission Measurement," Rep. URCL-50553. Lawrence Radiation Laboratory, Livermore, California.

Duvall, W., and Stephenson, D. E. (1965). Seismic energy available from rock bursts and underground explosions. *Trans. AIME* **232**, 235.

Fitch, C. E., Jr. (1969). "Acoustic Emission Signal Analysis in Flat Plates," BNWL-1008. Battelle-Northwest, Richland, Washington.

Fitz-Randolph, J. M. (1971). "Acoustic Emission Characterization of the Fracture Process in a Boton-Epoxy Composite," M.S. Thesis. School of Eng., University of California, Los Angeles.

Fowler, K. A. (1971). Acoustic emission simulation test set. *Mater. Rec. Stand.* **11**, 35.

Fowler, K. A., and Papadakis, E. P. (1972). Observation and analysis of simulated ultrsonic acoustic emission waves in plates and complex structures. *Amer. Soc. Test. Mater., Spec. Tech. Publ.* **505**.

Franklin, J. A. (1962). " Some Effects of Loading Technique on the Acoustic Emissions from Stressed Rock," M.S. Thesis. Geology Dept., Imperial College, London University, London.

Frederick, J. R. (1969). " Use of Acoustic Emission in Nondestructive Testing," Semi-Annu. Rep. 01971-1P. University of Michigan, Ann Arbor (also 01971-2-7).

Frederick, J. R. (1970). Acoustic emission, emission as a technique for nondestructive testing. *Mater. Eval.* **28**, 43.

Gerberich, W. W., and Hartbower, C. E. (1966). "Feasibility Study for Measuring Fatigue-Crack Growth Rate in Welded Hy-80 Steel Using Stress-Wave Emission," Final Rep. Contract M600 (167)-69434(X), FBM. Aerojet-General Corp., Sacramento California.

Gerberich, W. W., Zackay, V. F., Parker, E. R., and Porter, D. (1970). The role of grain boundaries on crack growth. "Ultrafine-Grain Metals," p. 259. Univ. of Syracuse Press, Syracuse.

Gillis, P. O. (1972). Dislocation motions and acoustic emissions. *Amer. Soc. Test. Mater., Spec. Tech. Publ.* **505**.

Goodman, R. E., and Blake, W. (1964). "Microseismic Detection of Potential Earth Slumps and Rock Slides," Rep. MT-64-6. Inst. Eng. Res., University of California, Berkeley.

Green, A. T. (1967). "Testing of Glass Hemispheres Using Aerojet-General Stress Wave Analysis Technique," AD 825542. Aerojet-General Corp., Sacramento, California.

Green, A. T. (1969). "Stress-Wave Emission Generated During the Hydrostatic Compression Testing of Glass Spheres," Rep. N00014-69-C-0333 (Naval Ship Research and Development Center, Washington, D.C.). Aerojet-General Corp., Sacramento, California.

Green, A. T. (1969). "Detection of Incipient Failures in Pressure Vessels by Stress-Wave Emissions," Tech. Rep. DRC-108. Dunegan Corp., Livermore, California.

Green, A. T. (1971). "Stress-Wave Emission and Fracture of Prestressed Concrete Reactor Vessel Materials," Tech. Rep. DRC 71-3. Dunegan Research Corp., Livermore, California.

Green, A. T., and Hartbower, C. E. (1970). "Development of a Nondestructive Testing Technique to Determine Flaw Criticallity," ARPA Order No. 1244, Program Code 8010, Contract K33615-68-C-1705. Aerojet-General Corp., Sacramento, California.

Green, A. T., Lockman, C. S., and Haines, H. K., (1963) "Acoustical Analysis of Filament-Wound Polaris Chambers," Rep. 0672-01F. Aerojet-General Corp., Sacramento, California.

Green, A. T., Hartbower, C. E., and Lockman, C. S. (1965). "Feasibility Study of Acoustic Depressurization System," Rep. NAS 7-310. Aerojet-General Corp., Sacramento, California.

Green, A. T., Lockman, C. S., Brown, S. J., and Steele, R. K. (1966). "Feasibility Study of Acoustic Depressurization System," NASA CR-55472. Aerojet-General Corp., Sacramento, California.

Green, A. T., Dunegan, H. L., Tetelman, A. S. (1970). "Nondestructive Inspection of Aircraft Structures and Materials Via Acoustic Emissions." Dunegan Research Co., Livermore, California.

Hall, A. (1959). "Microseismic Measurements," M.H. No. 30. Akad. Eng. Wissensch. FKO, Stockholm.

Harding, S. T. (1970). "A Least Squares Seismic Location Technique and Error Analysis," Int. Rep. — IR/71-8. Dept. Mineral Eng., Penn State University.

Harding, S. T. (1970). "A Critical Evaluation of Microseismic Studies," Int. Rep. RML-IR/70-24. Dept. Mineral Eng., Penn State University, University Park, Pennsylvania.

Harding, S. T., and Hardy, H. R., Jr. (1970). "Acoustic Emission During Transient and Secondary Creep, and During Creep Recovery," Presented at the Joint Annual Meeting of the Seismological Society and Geological Society of America, Milwaukee, Wisconsin.

Harding, S. T. *et al.* (1971). "Investigation of Crack Initiation in Rock Discs Loaded in Diametric Compression Using Acoustic Emission Techniques," Int. Rep. RML-IR/71-14. Dept. Mineral Eng., Penn State University, University Park, Pennsylvania.

Hardy, H. R., Jr. (1969). "Applications of Acoustic Emissions in Rock Mechanics," Presented at the Acoustic Emission Working Group Meeting, Philadelphia, Pennsylvania.

Hardy, H. R., Jr. (1970). "Stability Studies on Gas Storage Resevoir Models," Paper presented at Underground Storage Session, American Gas Association Transmission Conf., Denver, Colorado (Conf. Proc. AGA Cat. No. X59970, Fall 1970, pp. T132–T139).

Hardy, H. R. Jr., and Chugh, Y. P. (1968). "Consideration of Similitude Requirements for Gas Storage Resevoir Model Studies," Res. Prog. Rep. AGA 68-3. Dept. Mining, Penn State University, University Park, Pennsylvania.

Hardy, H. R. *et al.* (1970). Creep and microseismic activity in geologic materials. *Proc. Symp. Rock Mech., 11th, 1969.*

Hartbower, C. E. (1969). "Application of SWAT to the Nondestructive Inspection of Welds," Tech. Note. Aerojet-General Corp., Sacramento, California.

Hartbower, C. E., Gerberich, W. W., and Crimmins, P. P. (1966). "Characterization of Fatigue Crack Growth by Stress Wave Emission," Contract NAS 1-4902, Final Rep. CR-66303. NASA Langley Research Center; also in *Int. J. Fract. Mech.* **3** 185 (1967).

Hartbower, C. E., Gerberich, W. W., and Crimmins, P. P. (1966). "Monitoring Subcritical Crack Growth by an Accoustic Technique," Presented at Lockheed Symp. on Weld Imperfections.

Hartbower, C. E., Gerberich, W. W., and Crimmins, P. O. (1967). "Mechanisms of Slow Crack Growth in High Strength Steels," AFML-TR-67-26, Vol. 1. Aerojet-General Corp., Sacramento, California.

Hartbower, C. E., Gerberich, W. W., Reuter, W. G., and Crimmins, P. P. (1967). "Stress-Wave Characteristics of Fracture Instability in Constructional Alloys," Rep. USN-C-0340, U.S. Navy; also AD-674-881, Aerojet-General Corp., Sacramento, California.

Hartbower, C. E., Gerberich, W. W., and Liebowitz, H. (1968). Investigation of crack-growth stress-wave relationships. *Eng. Fract. Mech.* **1**, 291.

Hartbower, C. E., Gerberich, W. W., and Crimmins, P. O. (1969). "Mechanisms of Slow Crack Growth in High Strength Steels," Vol. 2, Contract AF 33 (615)-2788. Aerojet-General Corp., Sacramento, California.

Hartbower, C. E., Green, A. T., and Crimmins, P. O. (1969). "Correlation of Stress-Wave Emission Characteristics With Fracture in Aluminum Alloys," Rep. 1246 Q-5 NASA Contract NAS 8-21405. Aerojet-General Corp., Sacramento, California.

Helfrich, H. K. (1965). Automatically recorded microseismic observations. *Felsmechanik* und *Ingenieurgeol.* **4** (in German).

Herron, T. J. (1956). "The Detection and Delineation of Subsurface Subsidence by Seismic Methods," Thesis. Geophysics Dept., Michigan College of Mines and Technology.

Hodgson, K., and Jough, N. C. (1967). The relationship between energy release rate, damage and seismicity in deep mines. In "Failure and Breakage of Rock" (C. Fairhurst, ed.), p. 194. AIME, New York.

Hutton, P. H. (1967). "Acoustic Emission in Metals as a NDT Tool," Presented at the 27th Nat. Conf. of SNT, Cleveland, Ohio.

Hutton, P. H. (1968). "Nuclear Reactor System Noise Analysis, Dresden-1 Reactor, Commonwealth Edison Co.," BNWL-867 and BNWL-933. Battelle-Northwest, Richland, Washington.

Hutton, P. H. (1969). "Integrity Surveillance of Pressure Systems by Means of Acoustic Emission," BNWL-SA-2194. ASME,

Hutton, P. H. (1969). "Detection of Incipient Failure in Nuclear Reactor Pressure Systems Using Acoustic Emission," Rep. BNWL-997. Battelle-Northwest, Richland, Washington.

Hutton, P. H. (1970). "Acoustic Emission Applied to Determination of Structural Integrity," BNWL-SA-3147 (presented at the 11th Open Meeting of Mechanical Failures Prevention Group at Williamsburg, Virginia). Battelle-Northwest, Richland, Washington.

Hutton, P. H. (1967–1970). "Crack Detection in Pressure Piping by Acoustic Emission," Nuclear Safety Quarterly Progress Reports BNWL 537, BNWL 1266, BNWL 1315-1, and others. Battelle-Northwest, Richland, Washington.

Hutton, P. H. (1971). "Identification of Worn Bullet Forming Dies by Acoustic Signature in Process," BNWL-SA-3813 (presented at the 8th Symposium on Nondestructive Evaluation in Aerospace, Weapons Systems and Nuclear Applications). Battelle-Northwest, Richland, Washington.

Hutton, P. H. (1971). "Nuclear Reactor Background Noise versus Flaw Detection by Acoustic Emission," BNWL-SA-3820 (presented at the 8th Symposium on Nondestructive Evaluation, San Antonio, Texas). Battelle-Northwest, Richland, Washington.

Hutton, P. H., and Ord, R. N. (1970). Acoustic emission. "Research Techniques in Nondestructive Testing," p. 1. Academic Press, New York.

Hutton, P. H., and Pederson, H. N. (1969). "Crack Detection in Pressure Piping by Acoustic Emission," Nuclear Safety Quarterly Progress Report May–July 1969, BNWL-1187. Battelle-Northwest, Richland, Washington.

Jersey Nuclear Co. "Acoustic Analysis Technology Applied to the Nondestructive Test Evaluation of Commercial Structures and Structural Materials." Jersey Nuclear Co., Richland, Washington.

Jersey Nuclear Co. "An Approach to Pre-Service and In-Service Inspection of Nuclear Reactor Coolant Systems Using Acoustic Emission Analysis." Jersey Nuclear Co., Richland, Washington.

Jersey Nuclear Co. "NDT-Acoustic Test of Hydrogeneration Sphere," Rep. Jersey Nuclear Co., Richland, Washington.

Joffe, A. (1928). "The Physics of Crystals." McGraw-Hill, New York. One of the first to indicate sounds from deformation.

Jolly, W. D. (1968). "Acoustic Emission from Weld Defects," NDT-Memo-68-2. Pacific Northwest Laboratory, Richland, Washington.

Jolly, W. D. (1969). "The Application of Acoustic Emission to in-Process Weld Inspection," BNWL-SA-2212. Battelle-Northwest, Richland, Washington.

Jolly, W. D. (1969). "The Use of Acoustic Emission as a Weld Quality Monitor," Rep. BNWL-SA 2727. Battelle-Northwest, Richland, Washington.

Jolly, W. D. (1970). *Mater. Eval.* **28**, 135.

Jolly, W. D. "An In-Site Weld Defect Detector-Acoustic Emission," BNWL-817. Pacific Northwest Laboratory, Richland, Washington.

Kaiser, J. (1957). On noise generation during melting and solidification of metals. *Fortschr. Ingenieur wiss.* **23**, 38 (in German).

Kerawalla, J. N. (1965). "An Investigation of the Acoustic Emission from Commercial

Ferrous Materials Subjected to Cyclic Tensile Loading," Ph.D. Thesis. University of Michigan, Ann Arbor (University Microfilms, P. O. Box 1346, Ann Arbor, Michigan, 48016- #6606632).

Konstantinova, A. G. (1959). The shape of elastic pulses accompanying rock breaking. *Bull. Acad. Sci. USSR, Geophys. Ser.* pp. 421–426.

Konstantinova, A. G. (1960). Time distribution of elastic pulse energy during destruction of rocks. *Bull. Acad. Sci. USSR, Geophys. Ser.* pp. 1056–1061.

Konstantinova, A. G. (1962). The connection between the energy of elastic pulses generated in the destruction of solids and the stress and dimensions of the ruptures. *Bull. Acad. Sci. USSR, Geophys. Ser.* pp. 135–137.

Konstantinova, A. G., Mysyna, L. G., and Ivanov, U. S. (1965). Analysis of seismoacoustical processes accompanying strong sudden bursts of coal and gas. *Bull. Phys., Solid Earth Ser.* 11, 767–770 (translation Amer. Geophys. Un.).

Kroll, R. J. (1962). "Stress Waves in Test Specimens due to Simulated Acoustic Emission," Ph.D. Thesis. Michigan State University, East Lansing.

Leighton, F., and Duvall, W. I. (1972). A least squares method for improving the source location of rock noise. *U.S., Bur. Mines, Rep. Invest* **R1-7626**.

Liptai, R. G. (1963). "Acoustic Emission-A Surface Phenomenon," pp. 287–341. Southwest Research Inst., San Antonio, Texas.

Liptai, R. G., Harris, D. O., and Tatro, C. A. (1972). An introduction to acoustic emission. *Amer. Soc. Test. Mater., Spec. Tech. Publ.* **505**, 3.

McCabe, W. M. (1972). "An Acoustic Study of Failure in Rock," Informal Rep., Drexel University.

McCauley, D. D. (1972). "Development and Testing of a Wide Temperature Acoustic Emission Sensor," Rep. ERR-FW-1308. General Dynamics, Fort Worth, Texas.

McCullogh, B. B. (1965). "Acoustic Emission—An Experimental Method," M.S. Thesis. Tulane University, New Orleans, La.

Merrill, R. (1958). Method of determining strength of mine roof. *U.S., Bur. Mines, Rep. Invest.* **R1-5406**.

Merrill, R. H. (1968). Bureau contribution to slope angle research at the Kimberly Pit, Ely, Nevada. *Trans. AIME* **241**, 513.

Modes, M. C. (1973). "Acoustic Emission and Weld Defects," Senior Thesis. Dept of Metallurgy, Drexel University.

Mogi, K. (1962). The fracture of a semi-infinite body caused by an inner stress origin and its relation to earthquake phenomena. Papers No. 1 and 2. *Bull. Earthquake Res. Inst* **40**, 815; **41**, 595

Mogi, K. (1963). "Some Discussions of Earthquake Phenomena from the Standpoint of Fracture Theory," Geophysical Papers Dedicated to Prof. Kenso Sassa. Kyoto University, Kyoto, Japan.

Mogi, K. (1963). Some discussions on aftershocks, foreshocks and earthquake swarms— the fracture of a semi-infinite body caused by an inner stress origin and its relation to the earthquake phenomena. *Bull. Earthquake Res. Inst.* **41**, 615.

Nagumo, S., and Hoshino, K. (1967). Occurrence of micro-fracturing shocks during rock deformation with a special reference to activity of earthquake swarms. *Bull. Earthquake Res. Inst.* **45**, 1295.

Nakamura, Y. (1969). "Instruction Manual for the ARL-6 Acoustic Emission Monitoring System," Rep. FZM-5420. General Dynamics, Fort Worth, Texas.

Nakamura, Y., McCauley, B. O., Gardener, A. H., Redonond, J. C., Hagemeyer, J. W., and Burton, G. M. (1969). "Development of an Acoustic Emission Monitoring System," Rep. ERR-FW-1021. General Dynamics, Fort Worth, Texas.

Nakamura, Y., Veach, C. L., and McCauley, B. O. (1972). Amplitude distribution of acoustic emission. *Amer. Soc. Test. Mater., Spec. Tech. Publ.* **505.**

Obert, L., and Duvall, W. I. (1967). "Rock Mechanics and the Design of Structures in Rock." Wiley, New York.

Oliver, J., Ryall, A., Brune, J. N., and Slemmons, D. B. (1966). Microearthquake activity recorded by portable seismographs of high sensitivity. *Bull. Seismol. Soc. Amer.* **56,** 899.

Parry, D. L. (1967). "Nondestructive Flaw Detection by use of Acoustic Emission," USAEC Rep. IDO-17230. Phillips Petroleum Co., Idaho Falls, Idaho

Parry, D L (1967). Nondestructive flaw detection in nuclear power installations. *Trans. Amer. Nucl. Soc.* **10,** 330.

Parry, D. L. (1968). Nondestructive flaw detection in nuclear power installations. *In* "Incipient Failure Diagnosis for Assuring Safety and Availability of Nuclear Power Plants," Conf. 671011, pp. 107–126.

Parry, D. L. (1972). "Acoustic Integrity Analysis in the Oil and Chemical Industry." Paper presented at a Session on Corrosion Inspection Methods. Div. Refining, American Petroleum Institute, New York.

Parry, D. L., and Robinson, D. L. (1970). "Incipient Failure Detection by Acoustic Emission," Development and Status Rep. IN-1398. Idaho Nuclear Corp., Idaho Falls, Idaho.

Parry, D. L., and Robinson, D. L. (1970). "Elk River Acoustic Leak Detection Test," IN-1371. Idaho Nuclear Corp., Idaho Falls, Idaho.

Persson, T., and Hall, B. (1958). "Micro-Seismic Measurements for Predicting the Risk of Rock Failure and the Need for Reinforcement in Underground Cavities." Roy. Inst. Tech., Stockholm.

Pollack, A. A. (1967). "Stress-Wave Emission During Stress Corrosion Cracking of Titanium Alloys," Boeing Doc. D1-82-0658. Boeing Sci. Res. Lab., Seattle, Washington.

Pollack, A. A. (1970). "Acoustic Emission from Solids Undergoing Deformation," Ph.D. Thesis. University of London.

Pollack, A. A., and Radon, L. C. (1970). "Acoustic Emissions in the Fracture Toughness Test of a Mild Steel," Rep. FG-26. Dept. Mech. Eng., Imperial College, London.

Press, F. (1958). Elastic wave radiation from faults in ultrasonic models. *Publ. Dominion Obser., Ottawa* **20** (2), 271.

Radon, L. C., and Pollack, A. A. (1970). "Acoustic Emissions and Energy Transfer during Crack Propagation," Rep. FRR-31. Mech. Eng. Dept., Imperial College, London.

Rathbun, D. K. Beattie, A. G., and Hiles, L. A. (1971). "Filament Wound Materials Evaluation with Acoustic Emission," Rep. SCL-DC-70-260. Sandia Lab., Livermore, California.

Redstreake, W. (1963). Metals sound off on fatigue. *Iron Age* **192,** 97.

Reuter, W. G., Crimmins, P. P., and Hartbower, C. E. (1968). "SWAT Measurements of Stress Relief Cracking in 2 1/4 Cr-1 Mo Weldments," Rep. Aerojet-General Corp., Sacramento, California.

Riznichenko, Y. V. (1958). The study of seismic conditions. *Bull. Acad. Sci. USSR, Geophys. Ser.* pp. 615–622.

Romine, H. E. (1959). "Acoustic of Straining and Fracture in Notched Sheet Specimens," Materials Advisory Board Department, MAB 156-M. Div. Eng. Ind. Res., Nat. Acad. Sci.—Nat. Res. Counc., Washington, D.C.

Romrell, D M., and Bunnell, R. A. (1970). "Acoustic Emission Monitors Crack Growth in Ceramics," BNWL-SA-3064. Battelle-Northwest, Richland, Washington.

Rummel, F., and Angenheister, G. (1964). "Investigation on the Occurence of Noise Effects in a Landslide at Kaunertal." Inst. Appl. Geophys., University of Munich.

Schofield, B. H. (1960) "Acoustic Emission Under Applied Stress," Nos. 7–10, PB158743–158745. Lessells & Associates, Waltham, Massachusetts (available from National Technical Information Service, 5285 Port Royal Road, Springfield, Virginia 22151).

Schofield, B. H. (1961). "Acoustic Emission Under Applied Stress," Progr. Rep. No. 11, AD2255144. Lessells & Associates, Waltham, Massachusetts (available from National Technical Information Service, 5285 Port Royal Road, Springfield, Va. 22151).

Schofield, B. H. (1963). "Acoustic Emission from Metals—It's Detection, Characteristics and Source," Proc. Symp. Phys. Nondestruct. Test., San Antonio, Texas.

Schofield, B. H. (1965). "Investigation of Applicability of Acoustic Emission," AFW-TR-65-106. Lessells & Associates, Waltham, Massachusetts.

Schofield, B. H. (1966). "A Study of the Applicability of Acoustic Emission Pressure Vessel Testing," Lessells & Associates, Waltham, Massachusetts.

Schofield, B. H. (1972). Research on the sources and characteristics of acoustic emission. *Amer. Soc. Test. Mater., Spec. Tech. Publ.* **505**

Schofield, B. H., Bareiss, B. A., and Kyrala, A. (1958). "Acoustic Emission under Applied Stress," Rep. WADA-TR-58-194; AD 155676. Lessells & Associates, Waltham, Massachusetts (available from Library of Congress PB 151215).

Scholz, C. H. (1967). Frequency-magnitude relationship of microfracturing events during the triaxial compression of rock. *Trans. Amer. Geophys. Union* **48**, 205.

Scholz, C. H. (1968). "Microfracturing of Rock in Compression," Ph.D. Thesis. Massachusetts Institute of Technology, Cambridge, Massachusetts.

Shamina, O. G. (1956). Elastic impulses during the destruction of rocks. *Bull. Acad. Sci. USSR, Geophys. Ser.* pp. 513–518.

Shoemaker, P. S. (1961). " Acoustic Emission, and Experimental Method," M.S. Thesis. Michigan State University, East Lansing.

Sibek, V., Simane, J., and Buben, J. (1964). "Methods of Research into Rockbursts in the Czechoslovak Socialist Republic," pp. 1–12.

Simane, J. (1963). "The Use of the Seismoacoustic Method in Investigations of Rock Bursts," 15th Mining and Metallurgy Day at Berkakademie, Freiberg.

Stateham, R. M., and Vanderpool, J. S. (1971). Microseismic and displacement investigations in an unstable slope. *U.S., Bur. Mines, Reep. Invest.* **R1-7470**.

Steele, R. K., Green, A. T., and Lockman, C. S. (1965). "Structural Seismology Techniques for Prevention of Failure of Rocket Chambers." CPIA-Interagency Chemical Rocket Propulsion Group.

Steele, R. K., Green, A. T., and Lockman, C. S. (1968). Acoustic monitoring of hydrotests. *In* "Weld Imperfections" (A. R. Pfuger and R. E. Lewis, eds.), p. 361. Addison-Wesley, Reading Massachusetts.

Stephens, R. W. B., and Pollack, A. A. (1971). Waveforms and frequency spectra of acoustic emissions. *J. Acoust. Soc. Amer.* **50**, 904.

Susuki, K., Sasaki, Z., Siohara, Z., and Hirota, T. (1965). "A New Approach to the Prediction of Failure by Rock Noise," Proc. Int. Conf. on Strata Control and Rock Mechanics, p. 1. Columbia University, New York.

Tatro, C. A. (1957). "Sonic Techniques in the Detection of Crystal Slip in Metals," *Eng. Res.*, Vol. 1, Prog. Rep., Engineering Experiment Station, College of Engineering, Michigan State University, East Lansing.

Tatro, C. A. (1960). "Acoustic Emission from Crystalline Materials Subjected to External Loads," Prog. Rep., Div. Eng. Res., Michigan, State University East Lansing.

Tetelman, A. S. (1971). Acoustic emission testing and microfracture processes. *Mater. Res. Stand.* **11**, 13.

Tetelman, A. S., and Chow, R. (1972). Acoustic emission testing and microcracking processes. *Amer. Soc. Test. Mater., Spec. Tech. Publ.* **505**.

Veach, C. L. (1970). "Development of Mk II (Rise Time) Acoustic Emission Monitoring Unit," Rep. ERR-FW-1021. General Dynamics, Fort Worth, Texas.

Veach, C. L. (1971). "Development of Coincidence Acoustic Emission Monitoring Unit," Rep. ERR-FW-1175. General Dynamics, Fort Worth, Texas.

Veach, C. L., and Nakamura, Y. (1971). "Development of Acoustic Emission Monitoring Techniques," Research Summary ARR-15. General Dynamics, Fort Worth, Texas.

Vinogradov, S. D. (1957). Acoustic observations in collieries of the Kizelsk Coal Basin. *Bull. Acad. Sci. USSR, Geophys. Ser.* No. 6.

Vinogradov, S. D. (1959). On the distribution of the number of fractures in dependence on the energy liberated by the destruction of rocks. *Bull. Acad. Sci. USSR, Geophys. Ser.* **12**, 1850 (Engl. Trans. AGO 1292).

Vinogradov, S. D. (1962). Experimental study of the distribution of the number of fractures in respect to the energy liberated in the destruction of rocks. *Bull. Acad. Sci. USSR, Geophys. Ser.* p. 171 (Engl. Transl. AGU 119).

Vinogradov, S. D. (1964). "The Use of Acoustics in Observing Processes of Rock Disintegration." Moscow (in Russian).

von Rotter, D., and Thiele, B. (1963). Underground seismoacoustic studies in iron ore mines. *Z. Berg.—Huetten Wiss.* **1**, 14 (in German).

Waite, E. V. (1970). "Acoust-S-A Digital Program for Acoustic Triangulation of Spherical Vessels," IN-1369. Idaho Nuclear Co., Idaho Falls, Idaho.

Waite, E. V., and Moore, K. V. (1968). "Acoust-A Digital Program for Acoustic Triangulation of Nuclear Vessels," IDO-17280. Idaho Nuclear Co., Idaho Falls, Idaho.

Watanabe, M. (1963). "The Occurrence of Elastic Shocks During Destruction of Rocks and Its Relation to the Sequence of Earthquakes," Geophysical papers dedicated to Professor Kenzo Sassa. Kyoto University, Kyoto, Japan.

Subject Index

A

Acoustic emission, 289–339
  applied area in, 337–338
  Barkhausen effect and, 329
  bond quality from, 335
  burst vs. continuous type in, 294–295,
    299
  in ceramics, 311–312
  in civil engineering structures, 334
  commercial sources of, 338
  in composite materials, 310–311
  in concrete, 311
  crack growth in, 305
  creep effects in, 335
  defined, 290–291
  early work and background in, 294–301
  in earthquakes, 306–307
  fatigue effects and, 335–336
  field work in, 309–310
  flawed metal specimens in, 303–306
  fundamental area of, 336–337
  geophone in, 291
  historical work in, 291–293
  honeycomb material in, 335
  in ice, 312
  Kaiser effect in, 294–295, 334
  magnetic effects in, 324–330
  martensitic transformation in, 321–322
  materials investigated in, 301–319
  materials used in, 291
  metal specimens in, 301–306
  microseisms in, 291–292
  models of emissive sources for, 299–300
  modern instrumentation in, 297–299
  in nuclear reactor irradiation, 334
  in nuclear vessel tests, 330–333
  piezoelectric transducer in, 296–298
  "pop-in" stress for, 303
  processes studied with, 320–330
  rate of loading in, 334
  reactor boiling detection in, 335
  in rocks, 291–292, 306–310
  slope stability and, 322–324
  in soils, 312–317
  structural integrity work in, 296–297,
    330–333, 337
  surface coating thickness and, 335
  temperature effects in, 336
  triangulation studies in, 293
  types and models of, 299–301
  in underground gas storage reservoirs,
    309
  unflawed metal specimens in, 301–303
  during unloading, 334
  in welding, 320–321
  in wood, 317–319
Acoustic Emission Working Group, 338–
  339
Acoustics
  Laplace and Webster regions in, 118–119
  matched asymptotic expansion in, 110–
    124, 145–146
  nonlinear, 125–143
  piston problem in, 135–143
  shock reflection from wall in, 132–135
  slowly varying guide in short wave-
    length limit in, 121–125
  Webster Horn equation in, 75, 111–114
  Webster variables in, 113–116
Acoustic waveguides, see Waveguides
Aerojet-General Corp., 330
Amplitude frequency effect, in quartz
  crystal resonators, 272–278
Anisotropic solids, ultrasonic diffraction
  in, 155–158
Aslamazov-Larkin theory, for fluctuations
  in thin helium films, 62
Asymptotic matching principle, 82–83, 144
  see also Matched asymptotic expansion

Attenuation correction, in ultrasonic
    diffraction, 173, 177
Average superfluid density, in third sound
    studies, 57–59

B

Barker code correlator
    errors inherent in, 228–229
    frequency and velocity misalignments
        in, 228–230
    in phase coded surface waves, 218–220,
        223–230
Barkhausen effect, in acoustic emission,
    329
Beam spreading, in diffraction, 153
Boundary layer, in matched asymptotic
    expansion, 80
Broadband pulses, in ultrasonic diffrac-
    tion, 191–195
Buffer rods, in ultrasonic diffraction,
    175–186
Bureau of Mines, U.S., 323
Burgers equation
    in acoustic wave formation, 130–133
    in matched asymptotic expansion, 146
    in time-harmonic piston problem, 136
Burgers region, defined, 130
Burst emissions, in acoustic emissions,
    293–295, 299

C

Ceramics, acoustic emission in, 311–312
Civil engineering studies, acoustic emission
    in, 334
Coded pulses, 215–223
Coded signals, elastic surface wave filters
    for, 213–214
Composite materials, acoustic emission in,
    310–311
Concrete, acoustic emission in, 311
Crack growth, in acoustic emissions, 305
Creep
    acoustic emission and, 335
    microseismic activity and, 308

D

D'Alembert's paradox, 72
Dc electric field, influence of on quartz
    crystals, 283–287
Dielectric coefficients, for quartz crystals,
    261–265
Diffraction
    see also Scattering
    beam spreading in, 153
    Fresnel region in, 152
    from large apertures, 152
    by thick plate, 106–109
    two-step solution in, 153
    ultrasonic, see Ultrasonic diffraction
Diffraction corrections, in ultrasonic
    diffraction, 173–186
Dispersion, in ultrasonic diffraction,
    174–175
Distinguished limits, in scattering prob-
    lems, 96
Dot product, in ultrasonic diffraction,
    157–158

E

Earthquakes, rock behavior in, 306–307
Echo amplitude correction, in ultrasonic
    diffraction, 176
Elastic coefficients, for piezoelectric
    quartz crystals, 252–255
Elastic surface waves
    Barker code correlator for, 218–230
    fixed sequence generator/correlator for,
        233
    programmable sequence generator for,
        231–235
    pulse compression filters for, 236–241
    pulse expansion filters for, 239–240
    velocity dispersion for, 236–237
Elastic surface wave filters, 213–215,
    236–241

F

Fatigue effects, in acoustic emission,
    335–336
Fay solution, in time-harmonic piston
    problem, 141

Ferroelectrics, Barkhausen effect in, 329–330
Fixed code sequence generator, for elastic surface waves, 233
Flat films, third sound in, 2–27
FM pulse compression filters, 236–241
Forcing function, in matched asymptotic expansion, 80
Fourth sound, in helium films, 9
Fraunhoffer region, in ultrasonic diffraction, 152
Fresnel region, in ultrasonic diffraction, 152
Fubini solution, in time-harmonic piston problem, 138

G

General Dynamics Corp., 332
Geophone, in acoustic emissions, 291
GPJM theory, in superfluid helium films, 61–62

H

Hard strip, scattering by, 94–98
Helium
  equations of motion for, 16–22
  liquid phase of, 1–2
  superfluid, *see* Helium II
  thermal and viscous penetration depth of, 21
Helium II, flow of, 1–2
Helium films
  critical velocity of, 2
  and equations of motion for gas, 16–22
  and equations of motion for substrate, 15–16
  fourth sound in, 9
  He^3-He^4 combinations in, 49–51
  normal fluid motion and attenuation in, 55–57
  saturated and nonsaturated, 32–33
  substrate roughness in, 38–40
  superfluid, 1–64
  surface waves in, 2
Helium III films, third sound in, 49–51
Helium IV films, third sound in, 49–51

Hopf-Cole transformation, 131
Hydrodynamics, third sound in, 5–6
"Hysteresis" effect, in quartz crystals, 259–261

I

Ice, acoustic emission in, 312
I.D., *see* Interdigital transducers
Impedance concept, in matched asymptotic expansion, 89–91
Interdigital transducers
  grid structure of, 217–218, 222–223
  phase code surface elastic wave generation by, 215–216

J

Jersey Nuclear Company, 331

K

Kaiser effect, in acoustic emissions, 294–295, 334
Kapitza resistance, 15–16
Kennecott Copper Corp., 323

L

Laplace region, in acoustics, 118
Lead diodes, in programmable sequence generator, 233
Long wavelength approximation, in acoustic waveguides, 111

M

Magnetic effects, in acoustic emission, 324–330
Matched asymptotic expansions (MAE), 69–147
  in acoustic streaming, 143
  acoustic waveguides in, 110–125
  asymptotic matching principle in, 82–83

boundary layer in, 80
defined, 70
dimensional reasoning in, 73
and eigenfunctions in closed cavity, 121
exact closed form expressions in, 107
forcing function in, 80
in future of acoustics, 145–146
Hopf-Cole transformation in, 131
impedance concept in, 89–91
"matching on overlap" in, 144
Navier-Stokes equation in, 75
in nonlinear acoustics, 125–143
origin and development of, 71–72
physical reasoning in, 129
Reynolds number in, 71–73
scattering and diffraction problems
in, 93–110
scattering matrices and impedances
in, 74–75
second-order model equation in, 91–93
techniques of through one-dimensional
expansions, 76–88
three-dimensional problems in, 109–110
time-harmonic piston problem in,
135–143
Metals, martensitic transformation in, 321
Minimum singularity principle, 96
in scattering by soft strip, 103
Microseisms
in acoustic emissions, 291–292
field work in, 309–310
laboratory work in, 306–309
source of, 293
MOS transistors, limitations of, 231–232

N

Navier-Stokes equations, 7–9
for compressible perfect gas, 126
in matched asymptotic expansion, 75,
135–136, 146
in nonlinear acoustics, 126
in shock reflection from wall, 132–135
NDT transducers, in ultrasonic diffraction,
182–183
Nickel rod, acoustic emission studies of,
326–328
Nonlinear acoustics
matched asymptotic expansion in,
125–143

weak shock waves in, 126–132
Nuclear reactor irradiation, effects of on
acoustic emission of specimens, 334

O

One-dimensional expansions, matched
asymptotic expansion in, 76–88

P

Perturbation methods, matched asymptotic expansions as, 70
Phase-coded elastic surface waves,
decoding in, 217
Phase-coded signals, 215–223
Physical acoustics, 69–147
linear, 75
Piezoelectric effects, for quartz crystals,
255–261
Piezoelectric energy, equation for, 255
Piezoelectric quartz crystals, 255–262
see also Quartz crystals
nonlinear effects in, 245–287
Piezoelectric third order coefficients, for
quartz crystals, 286
Piezoelectric transducer, in acoustic
emission, 296–298
Piston problem
simplified treatment of, 127
time-harmonic, 135–143
Poincaré expansion, in matched
asymptotic expansion, 79
Poisson equation, in scattering and
diffraction problems, 100
Programmable sequence generator, for
elastic surface waves, 231–235
Pulse compression, codes used in, 215–223
Pulse compression filters, for elastic
surface waves, 236–241
Pulse compression radar, 213
Pulse-echo experiments, in ultrasonic
diffraction, 165

Q

Quartz crystal resonators
amplitude frequency effect in, 272–278
electric conductivity influence in, 279–
281

equivalent electrical circuits for, 278–283

linear equivalent electric circuit for, 278–279

nonlinear effects in, 266–278

nonlinear equivalent circuit for, 281–283

vibrating, 266–267

Quartz crystals

AT- and BT-cuts in, 247, 264–265

characteristic coefficients for, 252–266

conductibility in, 265–266

damping and conductivity coefficients for, 265–266

dielectric coefficients for, 261–265

diffusion phenomena in, 283–284

elastic coefficients for, 252–255

equation of equilibrium in, 248–249

frequency applied field relation for, 284–287

fundamental equations of, 247–252

"hysteresis" effect in, 259–261

influence of applied dc electric field in, 283–287

long-term drift in, 246

nonlinear phenomena in, 246–247

piezoelectric coefficients and, 255–261, 286

second-order coefficients for, 262

signal-to-noise ratio in, 246

stabilities for, 246

strain definition in, 247–248

strains in, 247–249

stress-strain relation in, 250

X-, Y-, and Z-cuts in, 263–264

Quartz crystal units, evolution of, 245–246

Quartz watches, 246

R

Radar systems

Barker correlator or coder in, 224

pulse compression, 213

Reactor vessels, acoustic emission studies of, 330–333, 337

Resonators

quartz, *see* Quartz crystal resonators

third sound, 41–49

Reynolds number

in matched asymptotic expansion, 71–72

in time-harmonic piston problem, 137

Rock bursts, prediction of, 293

Rock noises, 291, 306–310

see also Acoustic emission

in slope stability studies, 322–323

Rocks

creep and microseismic activity in, 308

curves for amplitude-frequency relationships in, 307

Rolls-Royce, Ltd., 331

S

Saturated films, third sound in, 32–33

Scattering

see also Diffraction

distinguished limit in, 96

by hard strip, 94–98

minimum singularity principle in, 96, 103

by soft strip, 103–106

Scattering and diffraction problems

higher order approximations in , 99–103

in matched asymptotic expansion, 93–110

three-dimensional, 109–110

Seismoacoustics, 293

Sequence generator, programmable, 231–235

Shock reflection, from wall, 132–135

Short wavelength limit, slowly varying guide in, 121–125

Single apertures, ultrasonic diffraction from, 151–208

Slope stability, acoustic emission studies of, 322–324

Soft strip, scattering by, 103–106

Soils, acoustic emission in, 312-317

SOS (silicon on sapphire) diodes

characteristics of, 234

limitations of, 232

Spectrum analysis, in ultrasonic diffraction, 180–186

Square generator/correlator, for elastic surface waves, 234–235

Structural integrity, acoustic emission studies of, 330–333, 337

Substrate

equation of motion for, 15–16

roughness of in third sound experiments, 38–40

Superfluid density, average, 57–59
Superfluid helium film
    see also Helium films
    linearized equations of motion for, 7–9
    third sound in, 1–64
Superfluidity, onset of in third sound,
        62–63
Surface waves
    elastic, 213–215
    phase coded, 215–223
    in ultrasonic diffraction, 206–208

T

Thick plate, diffraction by, 106–109
Thin films
    Aslamazov-Larkin theory of, 62
    third sound in, 27–30
Third sound
    attenuation of, 62–63
    averaged hydrodynamic equations in,
        9–15
    average film temperature in, 11
    average superfluid density in, 57–62
    chemical potential in, 13
    combined equations of, 22–23
    detailed theory of, 4–7
    elementary theory of, 3–4
    energy in, 52–55
    experiments in, 32–38
    general results in, 23–27
    hydrodynamics of, 5–7
    microscopic theories of, 57–63
    in mixed helium III and IV films,
        49–55
    normal fluid motion and attenuation in,
        55–57
    onset of superfluidity in, 62–63
    resonators for, 41–49
    in saturated films, 32–33
    substrate roughness in, 38–40
    in superfluid helium films, 1–64
    temperature dependence in, 34
    temperature variations and evaporation
        in, 6
    theory of in flat films, 3–23
    in thick films, 30–32
    in thin films, 27–30
    in unsaturated films, 33

velocity of, 36
    wavefronts in, 5
Time-harmonic piston problem, 135–143
    Fay solution in, 141
    Fubini solution in, 138
    initial shock region in, 140
    old age solution in, 142
Transducer
    in Barker correlator, 226
    echo from in ultrasonic diffraction, 202
    pressure profiles for in ultrasonic
        diffraction, 200–201
Transducer loss, in ultrasonic diffraction,
        186–187, 191
Transistors, as transmission gates, 231–232

U

Ultrasonic diffraction
    amplitude and loss in, 165–171
    for anisotropic solids, 155–158
    attenuation corrections in, 173, 177
    broadband pulses in, 191–205
    buffer rods in, 175–186
    computations in, 160–165
    diffraction corrections in, 173–186
    diffraction loss and phase in, 189,
        195–205
    dispersion introduced by, 174–175
    echo amplitude correction in, 176
    echo amplitude peaks in, 171
    experiments in, 165–172
    finite-width specimens in, 205–206
    formulation for pressure and phase
        in, 153–157
    formulation for spatial phase dot
        product in, 157–158
    Fraunhoffer region in, 152
    Fresnel region in, 152
    input amplitude profile in, 186–191
    loss peaks in, 171
    NDT transducers in, 182–183
    phase and time delay experiments in,
        177–178
    phase profiles in, 199
    physical limits in, 159–160
    pressure profiles in, 197, 200
    pulse-echo experiments in, 165
    radiation field in, 192–195
    for single apertures, 151–208

spectrum analysis in, 180–186
theory of, 153–160
transducer loss in, 186–187, 191
two areas of, 152
velocity corrections in, 174
Ultrasonic surface waves, 206–208
Unsaturated films, third sound in, 33
U.S. Borax and Chemical Corp., 323

V

Velocity dispersions, for elastic surface
waves, 236–237
Vibrating plates
boundary conditions for, 270–271
fundamental equations for, 266–268
in quartz crystal resonators, 266–278
wave propagation equation for, 268–270

W

Waveguides
acoustic, 110–125

long wavelength approximation in, 111
slowly varying guide in short wave-
length limit, 121
Webster expansions in, 119
Webster Horn equation in, 111
Weak shock waves, in nonlinear acoustics,
126–132
Webster Horn equation, in acoustic wave-
guides, 75, 111–112, 114
Webster variables, in acoustics, 113–115
Welding, acoustic emission and, 320–321
Westinghouse Aerospace and Electronic
Systems Division, 223
WKB method, in matched asymptotic
expansion, 75
Wood, acoustic emission in, 316–319

Z

Zero thickness problem, in diffraction by
thick plate, 108

Contents of Previous Volumes

Volume I, Part A—Methods and Devices

Wave Propagation in Fluids and Normal Solids—*R. N. Thurston*

Guided Wave Propagation in Elongated Cylinders and Plates—*T. R. Meeker and A. H. Meitzler*

Piezoelectric and Piezomagnetic Materials and Their Function in Transducers—*Don A. Berlincourt, Daniel R. Curran, and Hans Jaffe*

Ultrasonic Methods for Measuring the Mechanical Properties of Liquids and Solids—*H. J. McSkimin*

Use of Piezoelectric Crystals and Mechanical Resonators in Filters and Oscillators—*Warren P. Mason*

Guided Wave Ultrasonic Delay Lines—*John E. May, Jr.*

Multiple Reflection Ultrasonic Delay Lines—*Warren P. Mason*

Volume I, Part B—Methods and Devices

The Use of High- and Low-Amplitude Ultrasonic Waves for Inspection and Processing—*Benson Carlin*

Physics of Acoustic Cavitation in Liquids—*H. G. Flynn*

Semiconductor Transducers—General Considerations—*Warren P. Mason*

Use of Semiconductor Transducers in Measuring Strains, Accelerations, and Displacements—*R. N. Thurston*

Use of $p-n$ Junction Semiconductor Transducers in Pressure and Strain Measurements—*M. E. Sikorski*

The Depletion Layer and Other High-Frequency Transducers Using Fundamental Modes—*D. L. White*

The Design of Resonant Vibrators—*Edward Eisner*

Volume II, Part A—
Properties of Gases, Liquids and Solutions

Transmission of Sound Waves in Gases at Very Low Pressure—*Martin Greenspan*

Phenomenological Theory of the Relaxation Phenomena in Gases—*H. J. Bauer*

Relaxation Processes in Gases—*H. O. Kneser*

Thermal Relaxation in Liquids—*John Lamb*

Structural and Shear Relaxation in Liquids—*T. A. Litovitz and C. M. Davis*

The Propagation of Ultrasonic Waves in Electrolytic Solutions—*John Stuehr and Ernest Yeager*

Volume II, Part B—Properties
of Polymers and Nonlinear Acoustics

Relaxations in Polymer Solutions, Liquids and Gels—*W. Philoppoff*

Relaxation Spectra and Relaxation Processes in Solid Polymers and Glasses—*I. L. Hopkins and C. R. Kurkjian*

Volume Relaxations in Amorphous Polymers—*Robert S. Marvin and John E. McKinney*

Nonlinear Acoustics—*Robert T. Beyer*

Acoustic Streaming—*Wesley Le Mars Nyborg*

Use of Light Diffraction in Measuring the Parameter of Nonlinearity of Liquids and the Photoelastic Constants of Solids—*L. E. Hargrove and K. Achyuthan*

Volume III, Part A— Effect of Imperfections

Anelasticity and Internal Friction Due to Point Defects in Crystals—*B. S. Berry and A. S. Nowick*

Determination of the Diffusion Coefficient of Impurities by Anelastic Methods—*Charles Wert*

Bordoni Peak in Face-Centered Cubic Metals—*D. H. Niblett*

Dislocation Relaxations in Face-Centered Cubic Transition Metals—*R. H. Chambers*

Ultrasonic Methods in the Study of Plastic Deformation—*Rohn Truell, Charles Elbaum, and Akira Hikata*

Internal Friction and Basic Fatigue Mechanisms in Body-Centered Cubic Metals, Mainly Iron and Carbon Steels —*W. J. Bratina*

Use of Anelasticity in Investigating Radiation Damage and the Diffusion of Point Defects—*Donald O. Thompson and Victor K. Paré*

Kinks in Dislocation Lines and Their Effects on the Internal Friction in Crystals —*Alfred Seeger and Peter Schiller*

Volume III, Part B—Lattice Dynamics

Use of Sound Velocity Measurements in Determining the Debye Temperature of Solids—*George A. Alers*

Determination and Some Uses of Isotropic Elastic Constants of Polycrystalline Aggregates Using Single-Crystal Data— *O. L. Anderson*

The Effect of Light on Alkali Halide Crystals—*Robert B. Gordon*

Magnetoelastic Interactions in Ferromagnetic Insulators—*R. C. LeCraw and R. L. Comstock*

Effect of Thermal and Phonon Processes on Ultrasonic Attenuation—*P. G. Klemens*

Effects of Impurities and Phonon Processes on the Ultrasonic Attenuation of Germanium, Crystal Quartz, and Silicon—*Warren P. Mason*

Attenuation of Elastic Waves in the Earth —*L. Knopoff*

Volume IV, Part A— Applications to Quantum and Solid State Physics

Transmission and Amplification of Acoustic Waves in Piezoelectric Semiconductors—*J. H. McFee*

Paramagnetic Spin–Phonon Interaction in Crystals—*Edmund B. Tucker*

Interaction of Acoustic Waves with Nuclear Spins in Solids—*D. I. Bolef*

Resonance Absorption—*Leonard N. Liebermann*

Fabrication of Vapor-Deposited Thin Film Piezoelectric Transducers for the Study of Phonon Behavior in Dielectric Materials at Microwave Frequencies— *J. de Klerk*

The Vibrating String Model of Dislocation Damping—*A. V. Granato and K. Lücke*

The Measurement of Very Small Sound Velocity Changes and Their Use in the Study of Solids—*G. A. Alers*

Acoustic Wave and Dislocation Damping in Normal and Superconducting Metals and in Doped Semiconductors—*Warren P. Mason*

Ultrasonics and the Fermi Surfaces of the Monovalent Metals—*J. Roger Peverley*

Volume IV, Part B—Applications to Quantum and Solid State Physics

Oscillatory Magnetoacoustic Phenomena in Metals—*B. W. Roberts*

Transmission of Sound in Molten Metals— *G. M. B. Webber and R. W. B. Stephens*

Acoustic and Plasma Waves in Ionized Gases—*G. M. Sessler*

Relaxation and Resonance of Markovian Systems—*Roger Cerf*

Magnetoelastic Properties of Yttrium-Iron Garnet—*Walter Strauss*

Ultrasonic Attenuation Caused by Scattering in Polycrystalline Media—*Emmanuel P. Papadakis*

Sound Velocities in Rocks and Minerals: Experimental Methods, Extrapolations to Very High Pressures, and Results— *Orson L. Anderson and Robert C. Liebermann*

Volume V

Acoustic Wave Propagation in High Magnetic Fields—*Y. Shapira*

Impurities and Anelasticity in Crystalline Quartz—*David B. Fraser*

Observation of Resonant Vibrations and Defect Structure in Single Crystals by X-ray Diffraction Topography—*W. J. Spencer*

Wave Packet Propagation and Frequency-Dependent Internal Friction—*M. Elices and F. Garcia-Moliner*

Coherent Elastic Wave Propagation in Quartz at Ultramicrowave Frequencies —*John Ilukor and E. H. Jacobsen*

Heat Pulse Transmission—*R. J. von Gutfeld*

Volume VI

Light Scattering as a Probe of Phonons and Other Excitations—*Paul A. Fleury*

Acoustic Properties of Materials of the Perovskite Structure—*Harrison H. Barrett*

Properties of Elastic Surface Waves— *G. W. Farnell*

Dynamic Shear Properties of Solvents and Polystyrene Solutions from 20 to 300 MHz—*R. S. Moore and J. H. McSkimin*

The Propagation of Sound in Condensed Helium—*S. G. Eckstein, Y. Eckstein, J. B. Ketterson, and J. H. Vignos*

Volume VII

Ultrasonic Attenuation in Superconductors: Magnetic Field Effects—*M. Gottlieb, M. Garbuny, and C. K. Jones*

Ultrasonic Investigation of Phase Transitions and Critical Points—*Carl W. Garland*

Ultrasonic Attenuation in Normal Metals and Superconductors: Fermi-Surface Effects—*J. A. Rayne and C. K. Jones*

Excitation, Detection, and Attenuation of High-Frequency Elastic Surface Waves —*K. Dransfeld and E. Salzmann*

Interaction of Light with Ultrasound: Phenomena and Applications Spin-Phonon Spectrometer—*R. W. Damon, W. T. Maloney, and D. H. McMahon*

Volume VIII

Spin–Phonon Spectrometer—*Charles H. Anderson and Edward S. Sabisky*

Landau Quantum Oscillations of the Velocity of Sound and the Strain Dependence of the Fermi Surface— *L. R. Testardi and J. H. Condon*

High-Frequency Continuous Wave Ultrasonics—*D. I. Bolef and J. G. Miller*

Ultrasonic Measurements at Very High Pressures—*P. Heydemann*

Third-Order Elastic Constants and Thermal Equilibrium Properties of Solids— *J. Holder and A. V. Granato*

Interaction of Sound Waves with Thermal Phonons in Dielectric Crystals—*Humphrey J. Maris*

Internal Friction at Low Frequencies Due to Dislocations: Applications to Metals and Rock Mechanics—*Warren P. Mason*

Volume IX

Difference in Electron Drag Stresses on Dislocation Motion in the Normal and the Superconducting States for Type I and Type II Superconductors—*M. Suenaga and J. M. Galligan*

Elastic Wave Propagation in Thin Layers —*G. W. Farnell and E. L. Adler*

Solid State Control Elements Operating on Piezoelectric Principles—*F. L. N-Nagy and G. C. Joyce*

Monolithic Crystal Filters—*W. J. Spencer*

Design and Technology of Piezoelectric Transducers for Frequencies Above 100 MHz—*E. K. Sittig*

Volume X

Surface Waves in Acoustics—*H. Uberall*

Observation of Acoustic Radiation from Plane and Curved Surfaces—*Werner G. Neubauer*

Electromagnetic Generation of Ultrasonic Waves—*E. Roland Dobbs*

Elastic Behavior and Structural Instability of High-Temperature A-15 Structure Superconductors—*Louis R. Testardi*

Acoustic Holography—*Winston E. Kock*

A	5
B	6
C	7
D	8
E	9
F	0
G	1
H	2
I	3
J	4